Empowering Wheat Cultivation with GIS, Digital Approaches and Artificial Intelligence

Nusret Zencirci • Faheem Shehzad Baloch
Jin-Ying Gou • Velimir Mladenov
Sotirios Fragkostefanakis • Marta da Silva Lopes
Ephrem Habyarimana • Hakan Ulukan
Asuman Kaplan Evlice
Editors

Empowering Wheat Cultivation with GIS, Digital Approaches and Artificial Intelligence

Springer

Editors
Nusret Zencirci
Department of Biology
Bolu Abant Izzet Baysal University
Bolu, Türkiye

Faheem Shehzad Baloch
Department of Biotechnology, Faculty Science
Mersin University
Mersin, Türkiye

Jin-Ying Gou
Frontiers Science Center for Molecular Design Breeding
China Agricultural University
Beijing, China

Velimir Mladenov
Faculty of Agriculture
University of Novi Sad
Novi Sad, Serbia

Sotirios Fragkostefanakis
Institute of Molecular Biosciences
Goethe University Frankfurt
Frankfurt, Germany

Marta da Silva Lopes
Sustainable Field Crops Programme
Institute of Agrifood Research and Technology
Barcelona, Spain

Ephrem Habyarimana
ICRISAT
Hyderabad, Telangana, India

Hakan Ulukan
Department of Field Crops
Ankara University
Dışkapı, Ankara, Türkiye

Asuman Kaplan Evlice
Sivas Science and Technology University
Sivas, Türkiye

ISBN 978-3-031-99953-6 ISBN 978-3-031-99954-3 (eBook)
https://doi.org/10.1007/978-3-031-99954-3

This Springer imprint is published by the registered company Springer Nature Switzerland AG
The registered company address is: Gewerbestrasse 11, 6330 Cham, Switzerland

Dedication goes to:
both
farmers who feed the world,
agricultural scientists who contributed
agriculture since the beginning of the history
most likely started in Göbeklitepe, Türkiye

Preface

Since Green Evolution, world agricultural study and breeding programs have witnessed excellent progress for elite cultivars with higher yields, stronger disease resistance, better resilience, and higher quality. There are still around one billion people in the world hunger pandemic, and another one billion will increase by 2050, a huge potential threat to world food security. Climate change and fast-evolution diseases pose substantial challenges to the sustainable production of wheat, a staple of around 40% of the world's population. It is imperative to employ faster, more open, more collaborative, and even disruptive breeding strategies to accelerate wheat breeding to feed the world in the future.

Creating and studying artificial variations through biotechnology (BT) and natural variations in wild resources are the basis for wheat genetics/breeding and significantly increase wheat research and breeding depth and breadth. The significant achievements and further developments in genomic biotechnologies (GT) decode the wheat genome (pangenome) and generate massive genetic information for wheat research and breeding. The phenotypes (PT) associated with yield, resilience, metabolism, and human health-promoting effects can provide new directions for fundamental wheat research and even synthetic-biotechnology-driven breeding, forming positive feedback circles. The emergence and fast development of artificial intelligence (AI)-aided technologies (AT) offer faster, high-throughput, new strategies and paradigms for dissecting genetic effects, exploring functional genes, and predicting regulatory elements, functional domains, and phenotypes. It is time to integrate AT, BT, GT, and PT for artificial intelligence-aided wheat breeding.

The objective of this book, *Empowering Wheat Cultivation with GIS, Digital Approaches and Artificial Intelligence*, edited by Nusret Zencirci, Faheem Shehzad Baloch, Jin-Ying Gou, Velimir Mladenov, Sotirios (Akis) Fragkostefanakis, Marta da Silva Lopes, Ephrem Habyarimana, Hakan Ulukan, and Asuman Kaplan Evlice, is to address the challenges mentioned above by presenting a comprehensive concept, resource, examples, and new trends on how to combine GIS, digital approaches, and artificial intelligence to accelerate wheat breeding.

This book encompasses different tools of wheat breeding technologies and their implications, especially for the reader to follow. The book does not deal with descriptions and recommendations of different genes/markers or with the various wheat cultivars because they are outside the scope of the present objective. The information for markers/genes is available in updated research and reviews on corresponding research fields. Furthermore, the resources are fast-growing, and revolutionary algorithms are still emerging; therefore, the concept, protocol, platform, and practice of intelligent precision design breeding will become more mature as time goes on. Most researchers, companies, and breeders focus on genotype to phenotype (G to P). We believe that accumulating information and developing digital approaches and AI algorithms will lead to the successful prediction of genotype by phenotype (P to G) and, eventually, phenotype by phenotype (P to P).

This book will be of utmost importance to students, young scientists, extension workers, and progressive farmers dealing with wheat breeding. Professionals working on wheat breeding and basic research should find the book informative and reader-friendly. There is no doubt that, given the current world food crisis, this book is a timely and much-needed publication.

Beijing, China Jin-Ying Gou

Contents

About the Editors

Nusret Zencirci is a graduate of Çukurova University, Faculty of Agriculture, Department of Field Crops. He worked as a wheat breeder in Central Research Institute for Field Crops, Ankara, and served as National and International Coordinator to Wheat programs including International Winter Wheat Improvement Program carried by Türkiye, International Maize and Wheat Improvement Center (CIMMYT) and International Center for Agricultural Research in the Dry Areas (ICARDA). He is a molecular biology professor now at Bolu Abant İzzet Baysal University, Science and Art Faculty, Biology Department, Bolu, Türkiye. He is a member of various editorial boards in various international journals. He is, with his team, breeder of 22 bread and durum wheat cultivars. His research interests are Cereals, Biotic and Abiotic Stresses, Plant Genetic Resources, and Wheat Breeding.

Faheem Shehzad Baloch is working as Professor of plant genetics and genomics at Department of Biotechnology, Faculty of Science, Mersin University, Mersin, Turkiye and currently working as visiting professor at Jeju National University, South Korea through Brain Pool program. He received his PhD with a dissertation on "QTL mapping in wheat" from University of Çukurova, Adana, Türkiye, in 2012. He has more than 13 years of teaching and research experience in the plant phenomics, genetics and genomics, and biotechnology for plant improvement program. He led and participated in many projects funded by national and international organizations. He supervised and co-supervised MSc, PhD, and postdoc candidates from various countries. He has over 170 publications in the web of sciences (WOS) database. He has co-edited 13 books and written over 14 book chapters on important aspects of molecular genetics in relation to plant species. He has an extensive array of citations with over 6060 times as per Google Scholar with an h-Index of 40. He serves as an Editorial Board member of several impacted journals in WOS.

Jin-Ying Gou is a Professor at China Agricultural University, College of Agriculture and Biotechnology. He has more than 10 years of teaching and research experience in wheat genetics, metabolomics, and biotechnology. He led many projects funded by national organizations. He supervised over 30 MSc, PhD, and postdoc candidates from various countries. He is serving as an Editorial Board member of *BMC Plant Biology*, *Molecular Breeding*, and *Wheat Omics*. He has more than 40 SCI publications in WOS with an extensive array of citations with over 3000 times as per Google Scholar.

Velimir Mladenov is a distinguished scientist working at the Faculty of Agriculture, University of Novi Sad, Serbia. He devoted his career to Plant Breeding and Seed Science. Besides university career, simultaneously he developed his path in seed business companies as a consultant. He currently leads Plant Breeding laboratory at the University and serves as a President of the Faculty of Agriculture, University of Novi Sad (UNSFA), council. He has been active in developing global collaborations to tap into expertise of plant scientists worldwide – such as COST actions, where he serves in two (ReCrop: Reproductive Enhancement of CROP resilience to extreme climate (CA22157) and EPI CATCH – EPIgenetic mechanisms of Crop Adaptation To Climate cHange (CA19125)) as a Science Communication Coordinator. He is extremely active in Horizon project calls and at this moment as a participant at two big consortiums: (i) Developing Intercropping for Agrifood Value Chains and Ecosystem Services Delivery in Europe and Southern Countries—IntercropValuES and (ii) Bioeconomy Excellence Alliance for Stimulating Innovative and Inclusive Green Technology-Beaming. Besides that, he stands as a foreign consultant in Kazakhstan national agronomy projects. Under his supervision, seven students are working on their PhDs in Serbia and abroad. He has published widely in the areas of crop genetics and plant breeding in wheat and corn mostly.

Sotirios Fragkostefanakis is a researcher specializing in plant abiotic stress physiology and molecular biology. He conducted his PhD studies in the Department of Horticulture Genetics and Biotechnology of the Mediterranean Agronomic Institute of Chania, Greece, and obtained his doctoral degree from the School of Agriculture, Aristotle University of Thessaloniki, Greece. He is currently leading the laboratory of "Plant Heat Stress Response and Thermotolerance," which is part the group of Cell and Molecular Biology of Plants at the Institute of Molecular Biosciences, Goethe University of Frankfurt, Germany. His research interest lies in the identification of the molecular mechanisms that enable plant survival under heat stress conditions. His team focuses on the characterization of transcription factors and RNA splicing regulators that mediate gene expression under stress conditions. The team's research activities extend to exploring the core mechanisms that are responsible for the maintenance of cellular and protein homeostasis when plants are exposed to high temperatures. The results of Dr. Fragkostefanakis and his team provide valuable insights into the fundamental biology of stress responses and offer potential tools for enhancing crop resilience in the face of global challenges posed by climate change. He has authored several manuscripts in the fields of stress molecular biology, and is member of several international consortia. He is currently Chair of the RECROP COST Action (CA22157) focusing on crop reproductive resilience under extreme climate.

Marta da Silva Lopes is a senior researcher and head of the Sustainable Field Crops Program at the Institute of Agrifood Research and Technology (IRTA) in Lleida, Spain. With a PhD in Plant Biology from the University of Barcelona and a master's in Plant Breeding from IAMZ-CIHEAM, she brings over 20 years of expertise in crop physiology and genetic improvement, including 10 years at CIMMYT. Her research focuses on breeding resilient crops for drought- and heat-prone environments and developing sustainable agricultural systems. She has led efforts to automate and digitize breeding pipelines and phenotyping protocols across multiple experimental stations, significantly improving the efficiency and cost-effectiveness of large-scale research programs. She has authored over 50 SCI-indexed publications, contributed to book chapters, and participates in numerous national and EU-funded projects such as CROPDIVA, ROOT2RES, and LEGENDARY. As the lead of the DROUGHT HUB research group, she coordinates interdisciplinary work on climate adaptation and biodiversity. Her commitment to innovation is also reflected in patented crop varieties and close collaborations with industry. A passionate mentor, she currently supervises seven PhD students and a postdoctoral.

Ephrem Habyarimana leads ICRISAT's research on sorghum breeding to improve productivity and income generation for the smallholders. He has more than 25 years of experience in crop science, genetics, and plant breeding of sorghum, integrating diverse technologies and skills, i.e., agroecological and genomic modeling, big data analytics, e.g., phenomics, genomics for the sustainable, resilient, cost-effective, environment and biodiversity friendly crop improvement, and for optimizing the breeding program to deliver higher rates of genetic gains. He trained in North and Latin America, Africa, Asia, and Europe, and authored several specialized scientific articles as documented under Scopus and elsewhere, e.g., https://scholar.google.com/citations?user=2GfX1AUAAAAJ&hl=en and https://www.researchgate.net/profile/Ephrem-Habyarimana. He is recognized as inventor by European Commission in the fields of Exploratory (https://www.innoradar.eu/innovator/999531360) and market ready (https://www.innoradar.eu/innovation/36956) DeepTech innovations, and he was finalist for the 2019 European DataSci & Artificial Intelligence Awards.

Hakan Ulukan graduated from The University of Ankara, Faculty of Agriculture, Department of Field Crops, Dışkapı, Ankara, Türkiye, in 1987. After graduation he began MSc, and upon the completion he was sent to Italy for 7 months on "Cereal Agriculture." After returning, he started his PhD thesis in 1990, and completed it in 1997 on "wheat rust diseases." Now, he is an academic staff in the same department. His study areas are cereals (major and minor), wheat rusts diseases, wheat landraces, wheat growing and breeding, plant genetic resources, biotic/abiotic stresses, global warming effect(s) on field crop(s)' production, agronomy, ecology, etc.

Asuman Kaplan Evlice is an Associate Professor at Sivas Science and Technology University, Department of Field Crops, Türkiye. She previously worked as a research assistant at Sütçü İmam University (Kahramanmaraş) and researcher at Transitional Zone Agricultural Research Institute (Eskişehir) and Central Research Institute for Field Crops (Ankara). She has documented skills and expertise in agronomy and quality, particularly wheat quality, with more than two decades of research works in cereal and legume breeding with a particular interest in wheat. She has, with her team, developed 5 bread wheat, 3 durum wheat, 11 barley, and 3 lentil cultivars so far. She has numerous articles and book chapters and works as a leader and researcher in many projects.

Chapter 1
Securing Wheat Cultivation for Global Food Safety

Asuman Kaplan Evlice and Nusret Zencirci

Abstract Wheat is a cornerstone of global food security with about 799 million tonnes/year of production, serving as a staple food for 40% of the world's population. However, with the global population projected to reach 9.8 billion by 2050, agricultural production must increase by 70–100% to meet rising food demands. Wheat cultivation faces challenges from climate change, population growth, and the need for sustainable practices, necessitating innovative solutions to enhance yield, resilience, and quality.

Artificial intelligence (AI has emerged as a transformative tool in addressing these challenges. By leveraging machine learning, deep learning, and other AI-driven technologies, agriculture is witnessing advancements in yield prediction, pest and disease management, and climate adaptation. AI models, such as artificial neural networks and random forest regression, have demonstrated superior accuracy in forecasting wheat yield. Additionally, AI facilitates precision agriculture through real-time monitoring, optimizing resource use, and improving crop health. In breeding, AI accelerates the development of stress-tolerant wheat varieties by integrating genomic, phenotypic, and environmental data, significantly reducing the time required for traditional breeding methods.

Despite its potential, the application of AI in agriculture faces challenges, including the need for extensive datasets, computational resources, and skilled personnel. Nonetheless, AI-driven innovations hold immense promise for enhancing wheat productivity, resilience, and sustainability, ensuring food security in the face of climate change and a growing global population. This chapter explores the role of AI in wheat cultivation, focusing on its applications in yield prediction, pest and disease management, and climate adaptation, while addressing challenges for sustainable agriculture.

A. K. Evlice (✉)
Faculty of Agricultural Sciences and Technology, Department of Field Crops, Sivas Science and Technology University, Sivas, Türkiye
e-mail: asuman.kaplanevlice@sivas.edu.tr

N. Zencirci
Department of Biology, Bolu Abant İzzet Baysal University, Bolu, Türkiye

N. Zencirci et al. (eds.), *Empowering Wheat Cultivation with GIS, Digital Approaches and Artificial Intelligence*, https://doi.org/10.1007/978-3-031-99954-3_1

Keywords Wheat · Food safety · Artificial intelligence (AI)

Abbreviations

AI	artificial intelligence
ANNs	artificial neural networks
CNN	convolutional neural network
DL	deep learning
EVI	enhanced vegetation index
FAO	Food and Agriculture Organization
FAOSTAT	FAO Statistical Database
GBLUP	genomic best linear unbiased prediction
GIS	geographic information system
GPS	global positioning system
GS	genomic selection
GWAS	genome-wide association studies
IoT	internet of things
LSTM	long short-term memory
MAE	mean absolute error
ML	machine learning
MLR	multiple linear regression
RFR	random forest regression
RMSE	root mean squared error

1.1 Introduction

Wheat (*Triticum* spp.) is a crop that is fundamental to human civilization. The homeland of wheat is the region called the Fertile Crescent, which includes parts of Türkiye, Iran, Iraq, Syria, Lebanon, Palestine, and Israel (Özberk et al., 2016). Archaeo-botanical studies show that wheat was first cultivated in Şanlıurfa Göbekli Tepe, located in southern Türkiye, around 10,000–8000 BC (Dietrich et al., 2012). Wheat subsequently expanded into regions south of Russia, including Afghanistan, Turkmenistan, Uzbekistan, Kazakhstan, and Azerbaijan, as well as parts of Ethiopia, eastern North Africa (encompassing Egypt and Sudan), and areas around the northern Mediterranean coast (Köksel & Cetiner, 2015).

Wheat, which remains important today, is one of the most extensively cultivated cereals globally (more than 220 million hectares), with production of about 799 million tonnes annually. Major producers include China (17%), India (14%), Russia (11%), the United States (6%), Australia (5%), France (5%), Canada (4%), Pakistan (4%), and Türkiye (3%), which collectively account for a significant portion of global wheat output (FAOSTAT, 2025).

Wheat is used to produce a variety of food products, including bread, pasta, noodles, and pastries, catering to different cultural preferences (Ciudad-Mulero et al., 2020). As a key source of carbohydrates, protein, and micronutrients, wheat significantly contributes to food security, particularly in regions heavily reliant on cereal-based diets (An et al., 2023). It accounts for approximately 20% of the total caloric intake globally, serving as a dietary staple for 40% of the world's population (Acevedo et al., 2018). It is estimated that the global population increases by approximately 100 million people each year and will reach 9.8 billion by 2050 and 11.2 billion by 2100 (United Nations, 2025). According to projections by the Food and Agriculture Organization of the United Nations (FAO), agricultural production needs to grow by 2–3% each year (Ray et al., 2013) and by roughly 70–100% overall (Godfray et al., 2010) to meet the food demands of the projected global population by 2050. In addition, it is predicted by FAO that cereals, which contribute almost half of the daily calorie and protein intake in low- and middle-income countries, will continue to play a critical role in food security until 2050 (Grote et al., 2021), and to keep pace with this increase, annual cereal production, which is 2.1 billion tonnes today, needs to increase to approximately three billion tonnes (Umesha et al., 2018). Wheat, as a staple crop, will play a pivotal role in fulfilling this demand. However, wheat cultivation faces unprecedented challenges that threaten its productivity and sustainability. Challenges arising from climate change, along with associated abiotic and biotic stresses, necessitate significant efforts to enhance resistance or tolerance to these stresses, while also improving grain yield and quality. These advancements are crucial for ensuring food security in a rapidly growing global population (Acevedo et al., 2018).

Artificial intelligence (AI) has become a revolutionary tool for tackling these challenges. This technology mimics human cognitive abilities, enabling it to acquire knowledge, perform tasks, and make decisions. In agriculture, AI is utilized across multiple stages of the food system, including crop and livestock production, postharvest handling, food processing, and waste management (Gardezi et al., 2023). AI-driven innovations play a vital role in increasing wheat yield, strengthening resistance to pests and diseases, and enhancing overall adaptability to climate variability, ultimately ensuring food security amid global climate change (Mushtaq et al., 2024).

This chapter explores the role of AI in securing wheat cultivation, focusing on its applications in yield prediction, pest and disease management, and climate adaptation, while addressing challenges and future directions.

1.2 Wheat Production and Trade

Wheat is currently the most extensively cultivated cereal globally, with over 220 million hectares planted each year across diverse climatic conditions and numerous geographic regions (FAOSTAT, 2025).

Table 1.1 World wheat yield, production and export quantity data

Growing seasons	Yield (kg/ha)	Production (million tonnes)	Export quantity (million tonnes)
2013–2014	3319.5	728.8	174.0
2014–2015	3327.2	741.8	170.9
2015–2016	3418.3	748.4	190.1
2016–2017	3557.3	772.6	196.6
2017–2018	3444.5	732.6	190.9
2018–2019	3575.0	763.3	186.3
2019–2020	3499.6	760.1	198.6
2020–2021	3522.9	774.6	201.1
2021–2022	3704.4	811.4	191.0
2022–2023	3625.0	799.0	199.1

Computed by authors based on FAOSTAT online database, 05 February 2025

The data in Table 1.1 present global wheat yield, production, and export quantities from the 2013–2014 to 2022–2023 growing seasons. Over this period, there has been a noticeable upward trend in both yield and production. Wheat yield increased from 3319.5 kg/ha in 2013–2014 to a peak of 3704.4 kg/ha in 2021–2022, followed by a slight decline to 3625.0 kg/ha in 2022–2023. Production showed a similar trend, rising from 728.8 million tonnes to a peak of 811.4 million tonnes in 2021–2022, with a minor decrease to 799.0 million tonnes in 2022–2023. Export quantities fluctuated more, with the highest volume of 201.1 million tonnes recorded in 2020–2021, followed by a small decline to 199.1 million tonnes in 2022–2023.

Wheat is one of the most important crops for global food security. In many developing countries, it serves as a primary food source, particularly in regions where it constitutes over 50% of daily caloric intake (Shiferaw et al., 2013). The shares of countries in wheat production and trade are given in Table 1.2 and Fig. 1.1. In the 2022–2023 growing season, China led global wheat production with an impressive 136.6 million tonnes (17.1% of the total production), followed by India at 110.6 million tonnes (13.8%). The Russian Federation ranked third with 91.5 million tonnes (11.5%), reinforcing its dominance in both production and export markets. The United States (49.31 million tonnes, 6.2%) and Australia (41.2 million tonnes, 5.1%) rounded out the top five producers. Other significant producers included France (36.0 million tonnes), Canada (32.0 million tonnes), Pakistan (28.2 million tonnes), and Türkiye (22.0 million tonnes), reflecting strong agricultural outputs. Ukraine (21.6 million tonnes) also maintained substantial production level despite regional challenges. In Europe, nations such as Germany (21.5 million tonnes), Poland (12.9 million tonnes), and Romania (9.6 million tonnes) contributed significantly. Meanwhile, countries like Brazil (7.7 million tonnes) and Argentina (12.6 million tonnes) showcased notable wheat cultivation in South America. Several countries, including Kuwait (25 tonnes) and New Caledonia (12 tonnes), had minimal production, highlighting their reliance on wheat imports. Similarly, nations

Table 1.2 World wheat yield, production and export quantity data

Countries	Production (million tonnes)	% of Global production	Countries	Export (million tonnes)	% of Global export
China	136.6	17.1	Russia	31.6	15.8
India	110.6	13.8	Australia	29.3	14.7
Russia	91.5	11.5	Canada	25.6	12.8
USA	49.3	6.2	USA	17.9	9.0
Australia	41.2	5.2	Ukraine	16.2	8.1
France	36.0	4.5	France	13.5	6.8
Canada	32.0	4.0	Romania	8.2	4.1
Pakistan	28.2	3.5	Kazakhstan	7.2	3.6
Türkiye	22.0	2.8	Poland	7.0	3.5
Ukraine	21.6	2.7	Germany	6.8	3.4
Germany	21.5	2.7	Bulgaria	6.4	3.2

Computed by authors based on FAOSTAT online database, 05 February 2025

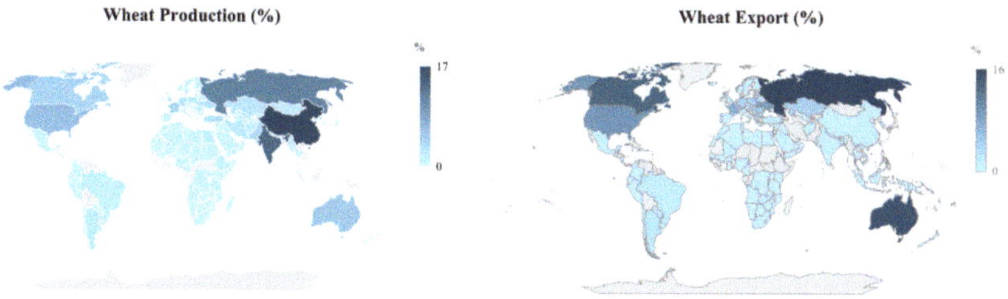

Fig. 1.1 Shares of countries in wheat production and exports (%)

across Africa and Southeast Asia, such as Nigeria (52,757 tonnes) and Thailand (1321 tonnes), produced relatively small quantities, often due to climatic and agricultural limitations.

The Russian Federation was the world's leading wheat exporter, with approximately 31.6 million tonnes, accounting for 15.8% of global exports. Australia followed closely with 29.3 million tonnes (14.7%), and Canada ranked third with 25.6 million tonnes (12.8%), showcasing their strong agricultural export sectors. Other major exporters included the United States (17.9 million tonnes, 9.0%) and Ukraine (16.2 million tonnes, 8.1%), both of which also maintained significant shares, reinforcing their roles in the global wheat market. European countries like France, Romania, Poland, and Germany also made significant contributions, each exporting between 6.7 and 13.5 million tonnes. Smaller exporters, such as Luxembourg, Norway, and Ireland, exported less than 30,000 tonnes annually. At the lower end of the spectrum, countries like Bangladesh, Namibia, Congo, and Afghanistan exported minimal quantities, with some exporting less than 1 tonne. This data highlights the dominance of Russia, Australia, and Canada in the global wheat export market, with a clear disparity between major and minor exporters worldwide in 2022–2023 growing season.

1.3 AI in Wheat to Enhance Food Security

AI, a field within computer science, is recognized as one of the three most advanced technologies of the twenty-first century. This is largely due to the continuous enhancement of computing power in recent decades, coupled with the development of sophisticated algorithms and their accessibility through open-source platforms, which have enabled AI to progress rapidly and find successful applications across various scientific disciplines (Ayres et al., 2021). Over time, AI has evolved into a distinct discipline with its own theoretical and practical foundations. In reality, AI encompasses a broad spectrum of fields, including computer vision, machine learning (ML), and natural language processing, among others (Abioye et al., 2021; An et al., 2023).

Various algorithms can be utilized in AI technology, including fuzzy logic, artificial neural networks (ANNs), swarm intelligence, expert systems, cognitive science, and logic programming (Kumar & Thakur, 2012; Taneja et al., 2023), reinforcement learning (Pantazi et al., 2016). Each of these approaches offers distinct advantages and limitations, making the selection of an algorithm dependent on the specific problem or task being addressed (Taneja et al., 2023). Overall, a key objective of AI research is to enable machines to execute complex tasks that typically require human intelligence (Chen et al., 2020).

AI has the potential to significantly impact modern agriculture by enhancing food production and reducing food losses (Khan et al., 2025). Currently, it is utilized in various agricultural applications, including crop monitoring, irrigation management, disease detection, fertilization, task automation, image and data processing, yield prediction, decision support systems, weed control, and resource optimization. Nevertheless, it is important to recognize the limitations of AI's effectiveness in these areas (Ahmad et al., 2024).

1.3.1 AI in Wheat Breeding and Genetic Improvement

Wheat, a fundamental pillar of global food security, is increasingly challenged by population growth, climate change, and the demand for sustainable farming methods. Plant breeding technologies offer a promising partial solution to address these emerging challenges (Chao et al., 2025). Traditional breeding methods, while effective, are often time-consuming, labor-intensive, and limited in their capacity to handle large numbers of individuals (Crossa et al., 2024). AI has emerged as a transformative tool in wheat breeding, offering innovative solutions to accelerate genetic improvement and enhance crop resilience (Mushtaq et al., 2024). Recent developments in AI have facilitated nonlinear modeling techniques that can capture complex interactions in genomic selection (GS) and genome-wide association studies (GWAS), thereby enhancing genomics-assisted breeding (Robles-Zazueta et al., 2024).

AI, particularly deep learning (DL) models, has significantly enhanced the accuracy and efficiency of GS. For instance, multimodal DL models that integrate genomic, phenotypic, and environmental data have demonstrated superior performance in predicting complex traits such as yield, disease resistance, and drought tolerance (Crossa et al., 2024). A study comparing DL models with traditional genomic best linear unbiased prediction (GBLUP) models found that DL outperformed GBLUP in predicting yield and other agronomic traits, achieving higher accuracy in multi-environment trials (Montesinos-López et al., 2024). Machine learning models that combine genotype and phenotype data with information from unmanned aerial systems have also shown great promise in predicting yield under various environmental conditions. This approach accelerates the development of improved wheat varieties with better performance and greater resilience to climate challenges (Sharma et al., 2023). Moreover, AI techniques have enabled the integration of speed breeding, significantly shortening the time needed for conventional breeding while ensuring greater accuracy and efficiency in developing stress-tolerant wheat varieties (Bhat et al., 2023). These AI-driven strategies have the potential to tackle food security issues by creating climate-smart wheat varieties with improved productivity and resilience (Mushtaq et al., 2024).

1.3.2 AI in Yield Prediction and Crop Management

Crop production is essential for maintaining food security and supporting economies worldwide (Bansal et al., 2024). AI has emerged as a transformative tool in agriculture, offering innovative solutions for yield prediction and crop management (Goel & Pandey, 2024).

AI can be employed in several ways to anticipate crop productivity. A frequently employed approach involves the utilization of remote sensing and satellite imagery (Yadav & Yadav, 2019). AI-driven predictive models have revolutionized yield forecasting by integrating diverse datasets, including genetic, environmental, and agronomic data. Subsequently, AI programs can scrutinize this data and discern patterns that can be utilized to forecast agricultural yields (Ramesh et al., 2023). AI techniques, particularly ML and DL, have revolutionized yield prediction by leveraging large datasets and complex algorithms. Convolutional neural networks (CNNs), long short-term memory (LSTM) networks, and ensemble models are among the most effective tools for analyzing environmental and agricultural data. For instance, a study proposed a 1D-CNN model for winter wheat yield prediction, which outperformed traditional linear models by reducing RMSE (root mean squared error) by 7–14% and increasing correlation coefficients by 4–50% (Srivastava et al., 2022). Similarly, LSTM networks have been used to capture temporal dependencies in weather and crop phenology data, demonstrating superior accuracy in yield forecasting (Bansal et al., 2024). An important benefit of utilizing AI in crop production prediction is its capacity to rapidly process and analyze vast quantities of data. This is particularly advantageous for farmers who need to monitor multiple variables that can affect crop production, including meteorological patterns, soil quality, and occurrences of pests and diseases (Avinash Khatri et al., 2023).

AI-driven crop management systems integrate remote sensing, weather forecasts, and soil data to optimize agricultural practices. Remote sensing data, such as the enhanced vegetation index (EVI), provides real-time insights into crop health, enabling early detection of stress factors (Joshi et al., 2024). For instance, a study demonstrated that combining remote sensing with weather forecasts improved wheat yield prediction accuracy by 16%, with optimal results achieved using 25 days of forecast data (Peng et al., 2024).

Precision agriculture is vital for optimizing wheat production amidst the challenges of changing climate patterns by addressing variability at the field level and efficiently aligning resource inputs with crop requirements. This approach utilizes technologies such as hyperspectral remote sensing, geographical information systems (GIS), and global positioning system (GPS) to improve agricultural practices, boost crop yields, and minimize negative environmental impacts (Finco et al., 2023). AI also facilitates precision agriculture by enabling variable rate applications of fertilizers, pesticides, and irrigation. For example, neural meta-models that combine multiple AI algorithms have been developed to predict wheat yields with a mean absolute error (MAE) of 0.82 tonnes/ha, providing actionable insights for farmers (Bansal et al., 2024). These applications not only enhance productivity but also promote sustainable resource use, aligning with global food security goals.

1.3.3 Climate Change

Climate change refers to long-term changes in the statistical characteristics of the climate system, including changes in temperature and precipitation averages over a 30-year period. These climate-driven alterations in temperature and precipitation patterns, and the frequency of extreme weather events are already affecting food yield, quality, and safety. Ensuring food safety and availability is a critical component of food security and maintaining an adequate standard of living (Karanth et al., 2023). The impact of climate change on food security varies across nations. In areas with lower hunger rates, the effects are less severe and may even improve over time. However, disparities in food access are expected to widen due to uneven climate change impacts. Vulnerable populations, in particular, face greater risks from extreme weather events, requiring urgent adaptation strategies to address rapid changes over the coming decades (Mushtaq et al., 2024).

Climate change has become one of the most pressing challenges in modern agriculture, with significant implications for global food security. The effects of changing climate conditions on crop yield, especially for staple crops such as wheat, have raised concerns about the future of food production (Iqbal et al., 2024). For instance, short periods of heat stress during flowering can cause sterility and grain abortion, leading to reduced grain numbers (Ferris et al., 1998). Similarly, heat stress during the grain-filling period can hasten leaf aging and shorten the grain-filling period, ultimately impacting final grain weight (Zhao et al., 2007). For every unit increase in the sum of daily heat degrees over 30 °C during anthesis and grain filling, grain yield was reduced by 1.0–1.6% (Bai et al., 2022; Li et al., 2016).

AI plays a pivotal role in assessing risks associated with wheat cultivation, particularly in the context of climate change. AI models leverage historical climate data and real-time monitoring to predict wheat yields and recommend adaptive strategies. The types of AI models that are most prevalent in climate-related studies include ML and DL. Machine learning, a branch of AI, helps with planning and running farms, and is often used to predict crop yields. Models such as multiple linear regression (MLR), ANNs (Fashoto et al., 2021), random forest regression (RFR) (Pang et al., 2022), and XGBoost (Sahbeni et al., 2023) are commonly used to estimate crop yields. Using ML models to predict wheat yield under climate change scenarios offers several advantages over traditional methods. These models provide enhanced predictive accuracy, which can help guide adaptation strategies and inform policy decisions (Iqbal et al., 2024). However, in crop yield estimation, it is essential to account for the effects of climate change and ecological differences on crop development (Tuğaç et al., 2022).

1.3.4 AI for Pest and Disease Management

Agricultural productivity is becoming more vulnerable due to the rising frequency and intensity of pest and disease outbreaks, which are worsened by climate change and changing environmental conditions (Sharada et al., 2025). These factors ultimately reduce total crop productivity to about 20–40% (Said Mohamed et al., 2021). The agriculture sector faces billions of dollars in losses each year due to plant diseases (Ahmad et al., 2024). Early detection and intervention are essential to reducing these losses. Implementing IoT systems for real-time crop monitoring can help minimize yield losses. AI-driven real-time disease detection in wheat fields has led to a 20% reduction in yield losses compared to conventional methods, underscoring the transformative potential of AI in improving crop health and productivity (Balaska et al., 2023).

Traditional methods of pest and disease control, while sometimes useful, are often ineffective when considering factors like time, cost, and environmental impact (Sharada et al., 2025). A potential solution to this issue is the timely identification of crop diseases. AI-driven image recognition and DL algorithms enable early and accurate detection of plant diseases, surpassing human visual assessment in both speed and accuracy (Khan et al., 2025). By integrating AI with Internet of things (IoT) devices and drones, farmers can monitor crop health in real time, enabling timely interventions to prevent disease outbreaks (Ahmad et al., 2024). AI-based diagnostics using leaf images have been shown to achieve an accuracy of 75%, outperforming alternative diagnostic methods (Barbedo, 2019; Martos et al., 2021).

Wheat is vulnerable to diseases such as rust, which can devastate yields and compromise food safety (Khan et al., 2025). Convolutional neural networks have been used to classify images of wheat leaves and diagnose diseases such as rust and powdery mildew with 85% accuracy (Hussain et al., 2018). AI has proven to be effective in disease identification. For instance, the agricultural expert system "Agpest" was developed for managing pests in wheat and rice crops using ANNs, genetic algorithms, and computer vision systems (Balleda et al., 2014). Various AI-based techniques have been established for identifying plant diseases and pests (Liu & Wang, 2021; Singh et al., 2023; Xu et al., 2022). However, the application of AI for plant disease detection requires large datasets that include both healthy and diseased plants. Although datasets like the plantVillage dataset exist, there is a need to develop more accessible datasets with diverse parameters. Additionally, significant challenges in deploying AI for disease detection include the need for greater computational resources, substantial financial investment, ongoing maintenance costs, and the availability of skilled personnel (Ahmad et al., 2024).

1.4 Conclusions

AI has emerged as a powerful tool for securing wheat cultivation, offering innovative solutions to enhance yield prediction, pest and disease management, and climate adaptation. By leveraging advanced algorithms and diverse datasets, AI enables precision agriculture and sustainable resource management, ensuring global food safety and security. However, addressing challenges such as data scarcity, high costs, and limited technical expertise will be crucial for widespread adoption. As AI continues to evolve, its integration with emerging technologies promises to revolutionize wheat cultivation, ensuring a sustainable and equitable food system for a growing global population.

References

Abioye, S. O., Oyedele, L. O., Akanbi, L., et al. (2021). Artificial intelligence in the construction industry: A review of present status, opportunities and future challenges. *Journal of Building Engineering, 44*, 103299. https://doi.org/10.1016/j.jobe.2021.103299

Acevedo, M., Zurn, J. D., Molero, G., et al. (2018). The role of wheat in global food security. In U. S. Nagothu (Ed.), *Agricultural development and sustainable intensification* (pp. 81–110). Taylor & Francis Group/Routledge.

Ahmad, A., Liew, A. X., Venturini, F., et al. (2024). AI can empower agriculture for global food security: challenges and prospects in developing nations. *Frontiers in Artificial Intelligence, 7*, 1328530.

An, D., Zhang, L., Liu, Z., et al. (2023). Advances in infrared spectroscopy and hyperspectral imaging combined with artificial intelligence for the detection of cereals quality. *Critical Reviews in Food Science and Nutrition, 63*(29), 9766–9796.

Avinash Khatri, K. C., Krishna, B. S., Logeshwaran, J., et al. (2023). Genetic algorithm based techno-economic optimization of an isolated hybrid energy system. *ICTACT Journal on Microelectronics, 8*(4), 1447–1450.

Ayres, L. B., Gomez, F. J. V., Linton, J. R., et al. (2021). Taking the leap between analytical chemistry and artificial intelligence: A tutorial review. *Analytica Chimica Acta, 1161*, 338403. https://doi.org/10.1016/j.aca.2021.338403

Bai, H., Xiao, D., Wang, B., et al. (2022). Simulation of wheat response to future climate change based on coupled model inter-comparison project phase 6 multi-model ensemble projections in the North China Plain. *Frontiers in Plant Science, 13*, 829580.

Balaska, V., Adamidou, Z., Vryzas, Z., et al. (2023). Sustainable crop protection via robotics and artificial intelligence solutions. *Machines, 11*, 774.

Balleda, K., Satyanvesh, D., & Sampath, N. (2014). Agpest: An efficient rule-based expert system to prevent pest diseases of rice & wheat crops. In *2014 IEEE 8th international conference on intelligent systems and control (ISCO), India, 10-11 January 2014.*

Bansal, Y., Lillis, D., & Kechadi, M. T. (2024). A neural meta model for predicting winter wheat crop yield. *Machine Learning, 113*(6), 3771–3788.

Barbedo, J. G. A. (2019). Plant disease identification from individual lesions and spots using deep learning. *Biosystems Engineering, 180*, 96–107. https://doi.org/10.1016/j.biosystemseng.2019.02.002

Ben Ayed, R., & Hanana, M. (2021). Artificial intelligence to improve the food and agriculture sector. *Journal of Food Quality, 2021*(1), 5584754.

Bhat, J. A., Feng, X., Mir, Z. A., et al. (2023). Recent advances in artificial intelligence, mechanistic models, and speed breeding offer exciting opportunities for precise and accelerated genomics-assisted breeding. *Physiologia Plantarum, 175*, e13969.

Chao, D., Wang, H., Wan, F., et al. (2025). MtCro: Multi-task deep learning framework improves multi-trait genomic prediction of crops. *Plant Methods, 21*(1), 12.

Chen, J., Zhang, M., Xu, B., et al. (2020). Artificial intelligence assisted technologies for controlling the drying of fruits and vegetables using physical fields: A review. *Trends in Food Science and Technology, 105*, 251–260. https://doi.org/10.1016/j.tifs.2020.08.015

Ciudad-Mulero, M., Barros, L., Fernandes, A., et al. (2020). Potential health claims of durum and bread wheat flours as functional ingredients. *Nutrients, 12*(2), 504.

Crossa, J., Montesinos-Lopez, O. A., Costa-Neto, G., et al. (2024). Machine learning algorithms translate big data into predictive breeding accuracy. *Trends in Plant Science, 2713*, 18.

Dietrich, O., Heun, M., Notroff, J., et al. (2012). The role of cult and feasting in the emergence of Neolithic communities. New evidence from Göbekli Tepe, south-eastern Turkey. *Antiquity, 86*, 674–695. https://doi.org/10.1017/S0003598X00047840

FAOSTAT. (2025). Food and Agriculture Organization of the United Nations. https://www.fao.org/faostat/en/#data/QCL (05.02.2025).

Fashoto, S. G., Mbunge, E., Ogunleye, G., et al. (2021). Implementation of machine learning for predicting maize crop yields using multiple linear regression and backward elimination. *Precision Agriculture, 6*, 679–697.

Ferris, R., Ellis, R., Wheeler, T. R., et al. (1998). Effect of high temperature stress at anthesis on grain yield and biomass of field-grown crops of wheat. *Annals of Botany, 82*, 631–639. https://doi.org/10.1006/anbo.1998.0740

Finco, A., Bentivoglio, D., Belletti, M., et al. (2023). Does precision technologies adoption contribute to the economic and agri-environmental sustainability of mediterranean wheat production? An Italian case study. *Agronomy, 13*, 1818.

Gardezi, M., Joshi, B., Rizzo, D. M., et al. (2023). A. Artificial intelligence in farming: Challenges and opportunities for building trust. *Agronomy Journal, 116*, 1217–1228.

Godfray, H. C. J., Beddington, J. R., Crute, I. R., et al. (2010). Food security: the challenge of feeding 9 billion people. *Science, 327*(5967), 812–818.

Goel, M., & Pandey, M. (2024). Crop yield prediction using AI: A review. In *2024 2nd International conference on disruptive technologies (ICDT) IEEE, India, 15–16 March 2024.*

Grote, U., Fasse, A., Nguyen, T. T., et al. (2021). Food security and the dynamics of wheat and maize value chains in Africa and Asia. *Frontiers in Sustainable Food Systems, 4*, 617009. https://doi.org/10.3389/fsufs.2020.617009

Hussain, A., Ahmad, M., Mughal, I. A., et al. (2018). Automatic disease detection in wheat crop using convolution neural network. In *The 4th International Conference on Next Generation Computing (ICNGC), Vung Tau, 20–23 December 2018*.

Iqbal, N., Shahzad, M. U., Sherif, E. S. M., et al. (2024). Analysis of wheat-yield prediction using machine learning models under climate change scenarios. *Sustainability, 16*(16), 6976.

Joshi, A., Pradhan, B., Chakraborty, S., et al. (2024). Deep-transfer-learning strategies for crop yield prediction using climate records and satellite ımage time-series data. *Remote Sensing, 16*(24), 4804.

Karanth, S., Benefo, E. O., Patra, D., et al. (2023). Importance of artificial intelligence in evaluating climate change and food safety risk. *Journal of Agriculture and Food Research, 11*, 100485.

Khan, G. E. L., Irshad, G., Ijaz, R., et al. (2025). Emboldening food security for global sustainability yoking artificial intelligence. *Discover Food, 5*(1), 2.

Köksel, H., & Cetiner, B. (2015). Future of grain science series: Grain science and industry in Turkey: Past, present, and future. *Cereal Foods World, 60*, 90–96. https://doi.org/10.1094/CFW-60-2-0090

Kumar, K., & Thakur, G. S. M. (2012). Advanced applications of neural networks and artificial intelligence: A review. *International Journal of Information Technology Computer Science, 4*, 57–68.

Li, K., Yang, X., & Tian, H. (2016). Effects of changing climate and cultivar on the phenology and yield of winter wheat in the North China plain. *International Journal of Biometeorology, 60*, 21–32. https://doi.org/10.1007/s00484-015-1002-1

Liu, J., & Wang, X. (2021). Plant diseases and pests detection based on deep learning: a review. *Plant Methods, 17*, 1–18. https://doi.org/10.1186/s13007-021-00722-9

Martos, V., Ahmad, A., Cartujo, P., et al. (2021). Ensuring agricultural sustainability through remote sensing in the era of agriculture 5.0. *Applied Sciences, 11*, 5911. https://doi.org/10.3390/app11135911

Montesinos-López, A., Crespo-Herrera, L., Dreisigacker, S., et al. (2024). Deep learning methods improve genomic prediction of wheat breeding. *Frontiers in Plant Science, 15*, 1324090.

Mushtaq, M. A., Ahmed, H.-D., & Zeng, Y. (2024). Applications of artificial intelligence in wheat breeding for sustainable food security. *Sustainability, 16*(13), 5688.

Özberk, İ., Atay, S., Altay, F., et al. (2016). *Türkiye'nin buğday atlası*. WWF-Türkiye (Doğal Hayatı Koruma Vakfı).

Pang, A., Chang, M. W., & Chen, Y. (2022). Evaluation of random forests (RF) for regional and local-scale wheat yield prediction in southeast Australia. *Sensors, 22*, 717.

Pantazi, X.-E., Moshou, D., & Bravo, C. (2016). Active learning system for weed species recognition based on hyperspectral sensing. *Biosystems Engineering, 146*, 193–202.

Peng, D., Cheng, E., Feng, X., et al. (2024). A deep-learning network for wheat yield prediction combining weather forecasts and remote sensing data. *Remote Sensing, 16*(19), 3613.

Ramesh, G., Logeshwaran, J., Kiruthiga, T., et al. (2023). Prediction of energy production level in large PV plants through AUTO-encoder based neural-network (AUTO-NN) with restricted Boltzmann feature extraction. *Future Internet, 15*(2), 46.

Ray, D. K., Mueller, N. D., West, P. C., et al. (2013). Yield trends are insufficient to double global crop production by 2050. *PLoS One, 8*(6), e66428.

Robles-Zazueta, C. A., Crespo-Herrera, L. A., Piñera-Chavez, F. J., et al. (2024). Climate change impacts on crop breeding: Targeting interacting biotic and abiotic stresses for wheat improvement. *Plant Genome, 17*, e20365.

Sahbeni, G., Székely, B., Musyimi, P. K., et al. (2023). Crop yield estimation using Sentinel-3 SLSTR, Soil data, and topographic features combined with machine learning modeling: A case study of Nepal. *AgriEngineering, 5*, 1766–1788.

Said Mohamed, E., Belal, A. A., Kotb Abd-Elmabod, S., et al. (2021). Smart farming for improving agricultural management. *Egyptian Journal of Remote Sensing and Space Science, 24*, 971–981.

Sharada, K., Choudhary, S. L., Harikrishna, T., et al. (2025). GeoAgriGuard: AI-driven pest and disease management with remote sensing for global food security. *Remote Sensing in Earth Systems Sciences, 8*, 1–14.

Sharma, R., Kukreja, V., & Gupta, R. (2023). Enhancing wheat crop resilience: An efficient deep learning framework for the detection and classification of rust disease. In *Proceedings of the 2023 4th international conference for emerging technology (INCET), Belgaum, India, 26–28 May 2023*.

Shiferaw, B., Smale, M., Braun, H. J., et al. (2013). Crops that feed the world 10. Past successes and future challenges to the role played by wheat in global food security. *Food Security, 5*, 291–317.

Singh, P., Kansal, M., Singh, M. K., et al. (2023). Crop disease prediction using deep learning algorithms. In A. Sharma, N. Chanderwal, & R. Khan (Eds.), *Convergence of cloud computing, AI, and agricultural science* (pp. 290–305). IGI Global. https://doi.org/10.4018/979-8-3693-0200-2.ch015

Srivastava, A. K., Safaei, N., Khaki, S., et al. (2022). Winter wheat yield prediction using convolutional neural networks from environmental and phenological data. *Scientific Reports, 12*(1), 3215.

Taneja, A., Nair, G., Joshi, M., et al. (2023). Artificial intelligence: Implications for the agri-food sector. *Agronomy, 13*(5), 1397.

Tuğaç, M. G., Özbayoğlu, A. M., Torunlar, H., et al. (2022). Wheat yield prediction with machine learning based on MODIS and landsat NDVI data at field scale. *International Journal of Environment and Geoinformatics, 9*(4), 172–184.

Umesha, S., Manukumar, H.-G., & Chandrasekhar, B. (2018). Sustainable agriculture and food security. In R. L. Sing & S. Mondal (Eds.), *Biotechnology for sustainable agriculture* (pp. 67–92). Elsevier. https://doi.org/10.1016/B978-0-12-812160-3.00003-9

United Nations. (2025). Department of Economic and Social Affairs, United Nations. https://www.un.org/en/desa/world-population-projected-reach-98-billion-2050-and-112-billion-2100. (10.01.2025).

Xu, M., Yoon, S., Jeong, Y., et al. (2022). Transfer learning for versatile plant disease recognition with limited data. *Frontiers in Plant Science, 13*, 1010981. https://doi.org/10.3389/fpls.2022.1010981

Yadav, S. P., & Yadav, S. (2019). Mathematical implementation of fusion of medical images in continuous wavelet domain. *Journal of Advanced Research in Dynamical and Control Systems, 10*(10), 45–54.

Zhao, H., Dai, T., Jing, Q., et al. (2007). Leaf senescence and grain filling affected by post-anthesis high temperatures in two different wheat cultivars. *Plant Growth Regulation, 51*, 149–158. https://doi.org/10.1007/s10725-006-9157-8

Chapter 2
Climate Change in Wheat (*Triticum* spp.) Under the Light of Remote Sensing, Digital Evolution, and Artificial Intelligence

Hakan Ulukan and Yusuf Ersoy Yildirim

Abstract As it is known, climate change and global warming caused by humanity threaten the agricultural sector in almost every field, and their effects are increasingly felt. Forest fires, melting of glaciers, excessive rainfall in some places, lack of rain in others for a long time, the transition of winters to warm instead of cold and freezing as usual, and a decrease in water resources are just some examples. On the other hand, although the fields devoted to plant production have reached their limits, lakes and seas cannot be used effectively, the world population's equal and healthy nutritional needs cannot be met adequately, food security cannot be ensured sufficiently, etc. It also brings with it various vital problems. Furthermore, and as mentioned above, within the framework of good agricultural practices (*GAP*), successful and appropriate irrigation and agricultural techniques, ensuring complete and adequate food safety, appropriate and sufficient storage, etc., require the process to be met in appropriate amounts and times, as well as the contribution and use of advanced technologies. With the support and collaboration of different branches of science, important contributions that provide valuable and long-term benefits have been achieved. In this study, the effects of climate change on wheat (*Triticum* spp.) cultivation are evaluated, and techniques that can be used to combat these effects are investigated. Climate change, in particular, is a growing concern in the agricultural sector and can also affect the cultivation of strategic crops such as wheat in various ways. Our study highlights how modern and advanced technologies such as remote sensing (*RS*), digital evolution (*DE*), and artificial intelligence (*AI*) can be used to combat climate change. While (*RS)* is a powerful tool for monitoring and analyzing changes in fields devoted to agriculture, *DE* and *AI* are important tools to process this

H. Ulukan
Department of Field Crops, The Ankara University, Ankara, Türkiye

Y. E. Yildirim (✉)
Department of Agricultural Structures and Irrigation, The Ankara University, Ankara, Türkiye
e-mail: yeyildirim@ankara.edu.tr

N. Zencirci et al. (eds.), *Empowering Wheat Cultivation with GIS, Digital Approaches and Artificial Intelligence*, https://doi.org/10.1007/978-3-031-99954-3_2

data in scenarios to predict future climate scenarios. However, the technological solutions they use may be different but complementary. Again, to develop sustainable and effective solutions for wheat agriculture, such approaches may need to be used, and effective social, socioeconomic, and political factors should be taken into account. In the future, with the use of advanced technologies such as *RS*, *DE*, and *AI*, the sustainability of wheat agriculture and resistance to diseases and pests will gain critical importance, and thanks to these technologies, smarter and more efficient production will be made, agricultural practices will be implemented, and especially the wheat plant will be able to grow under changing climate conditions. Its cultivation will become more sustainable, and as a result, a plant of strategic importance for human nutrition and shelter (i.e., wheat) will emerge.

Keywords Climate changes · Global warming · Good agricultural practices (*GAP*) · Remote sensing (*RS*) · Digital evolution (*DE*) · Artificial intelligence (*AI*)

Abbreviations

AI	Artificial ıntelligence
ANN	Artificial neural network
CO_2	Carbon dioxide
CH_4	Methane
CFCs	Chlorofluorocarbon
DE	Digital evolution
DL	Deep learning
FAO	Food and Agriculture Organization
GAP	Good agricultural practices
GHGs	Greenhouse gases
GPS	Global positioning system
IPCC	Intergovernmental Panel on Climate Change
IR	Infrared
ML	Machine learning
N	Nitrogen
N_2O	Nitrous oxide
RS	Remote sensing
SVM	Support vector machine
UN	United Nations
WUE	Water use efficiency

2.1 Introduction

2.1.1 What Is Climate? What Is Climate Changing? What Is Global Warming?

Modern agriculture, while facing the nutritional problems of an ever-increasing world population, also includes practices that take into account the use of limited resources and many negative environmental problems, especially in wheat agriculture with RS. Sensing technologies, including satellite- and drone-based systems (Fig. 2.15), are used to monitor and evaluate wheat agriculture. Similarly, RS enables real-time data collection, allowing farmers to make more informed decisions about irrigation, disease and pest control, and crop health. Wheat, which is a basic element of global nutrition, plays a critical role in ensuring food security for a significant part of the world population. On the other hand, as demand for wheat increases with population growth, the agricultural sector faces the challenge of optimizing production to meet these increasing needs. Again, this plant is still a staple food for a significant part of the world's population and has a critical role in ensuring global food security. As a matter of fact, due to the rapid increase in demand for wheat driven by global population growth, the agricultural sector faces the challenge of optimizing production in this regard. Fortunately, advanced technologies, including RS, DE, and AI, are changing the landscape of wheat farming in every aspect. In general, all cultivated plants need water, soil, nutrients, sunlight, and temperature to grow normally, develop, and provide the desired level of productivity. Parameters such as the genotype of the plant, environmental conditions, cultivation techniques, climate and soil requirements and conditions, and ecological factors are effective in all of these. According to the Food and Agriculture Organization (FAO), world food production must increase by 70.0% by 2050 to feed the increasing population (Uzhinskiy, 2023). It plays a central role in achieving the second Sustainable Development Goal prepared by the UN in securing the region's food supply and achieving zero hunger by 2030. In most countries where the expansion of cropland is almost impossible, agricultural automation has become the only option. At the same time, the advancement in these digital technologies has brought revolutionary changes in agriculture by providing intelligent systems with intelligence comparable to that of human experts that can monitor, control, and visualize various farm operations in real time (Subeesh & Mehta, 2021). Over the last few years, agriculture has been constantly facing climate change and other environmental problems, posing significant obstacles to increasing productivity. There are two possible options to overcome food shortage: farming on large areas or using them effectively. Technological support can be effective in this regard (Subeesh & Mehta, 2021). "*Climate*" refers to the average of temperature, humidity, atmospheric pressure, wind, precipitation, and other meteorological events over a certain period. In other words, many important parameters affect the climate, such as temperature, radiation, precipitation, soil moisture, and CO_2 concentration, and all these factors affect the agricultural activities carried out and the level of productivity

(Zhai & Zhuang, 2009). The climate has been changing naturally throughout world history until the Industrial Revolution in the mid-nineteenth century, and it is understood that human influence has had a significant share in the subsequent changes. However, "*global climate change*" is something else and it refers to "naturally occurring changes in climate" (Özdogan, 2011), as well as the changes that occur directly or indirectly in the chemical structure of the atmosphere but mostly occur with climate change, and almost all of them are caused by human activities (anthropogenic). However, "*Global World Temperature*" is the increase in world temperature due to the climate changes caused by these global climate changes and the effect of greenhouse gases (=GHGs) CO_2, CH_4, CFC, N_2O, and Water vapor, etc. (Anonymous, 2024b; IPCC, 2007). These gases prevent the reflection of sunlight and affect the interaction between the first and second atmospheric layers (troposphere and stratosphere) (Krupa, 1997). The world population is constantly increasing and is expected to reach 8.0–8.5 billion by the end of this century. This means that agricultural productivity (especially in wheat) will need to increase, although it depends on the applied techniques, technologies, and ecological parameters. While it can be affected by water scarcity, decreasing soil fertility, and various biotic (living) and abiotic (nonliving) stresses, drought (or lack of water), varying amounts of precipitation/water (oxidative stress) according to regions, and soil salinity are among the main stresses of global warming in wheat. The main climate parameters with which wheat plants interact negatively during global warming are undoubtedly temperature (Shanker & Venkateswarlu, 2011) and the CO_2 gas concentration of the environment. In this regard, wheat varieties show wide and diverse ecophysiological responses depending on location and orientation, genotype, and the duration and severity of the stress factor (Ackerly et al., 2000), causing stress in wheat production (Leung, 2008). Especially in this respect, the ion load and (K^+/Na^+) salinity in the plant-root region are extremely effective (Volkmar et al., 1998).

As mentioned above, increases in the amount of CO_2 gas in the environment will increase the growth rate, plant development, and water usage effectiveness (WUE) but will also reduce the efficiency of grain filling and nutrient use in plants. Changes in soil moisture due to rising temperature, changes in the physiology and morphology of diseases and pests, and their effects on agricultural production cause a decrease in fertilization and the expansion of cultivated areas, and also cause a decrease in (N) flow due to microbial activity; if the water in the soil is insufficient, photosynthetic activity and stomatal conductance will decrease, as well as some phenological changes in the plant, and the increase in CO_2 concentrations in the environment has undeniable and significant effects on the growth and productivity of various field crops (Reddy et al., 2010). According to IPCC (2014), CO_2 concentration is expected to reach 550 ppm by 2050. While this value was 315.0 ppm in 1960 (Tausz et al., 2013), it was measured as 422.80 ppm on January 1, 2024 (Anonymous, 2015a, b). GHGs are the second-largest industrial contributors, directly or indirectly, to climate change and global warming, of which CH_4 has approximately 300 times greater climate change impact and potential than CO_2 and approximately 20 times greater than N_2O. With research on rainfall, etc., it has been determined that the global average temperature on Earth will increase by 1.4–5.8 °C until 2100 due to weather events. There are many signs of global warming, such as the melting of glaciers and seas, icing, rising sea levels, longer flowering and ripening periods. Other factors include early development, shortening of vegetation periods, sudden maturation, loss of color in corals, migration of plants and animals toward the poles, and an increase in air temperature in Europe by 2.0–6.3 °C by 2100 (Fernandes et al., 2009). Animal husbandry, forestry, aquaculture, pastures, fallow lands, wetlands, etc. are in question. Various agricultural ecosystems, including agricultural ecosystems, will be affected at different levels by climate change and global warming; increasing CO_2 concentration will increase plant growth and WUE.

In general terms, this phenomenon will have varying effects on agricultural systems and on strategically important plants such as wheat, which is used as a grain product. In particular, temperature increase has increased by about 40.0% in the last 150 years, half of which has occurred in the last 30 years (IPCC, 2014). Agriculture has a large carbon footprint, accounting for more than 25.0% of global anthropogenic GHG emissions (IPCC, 2018). Specifically, soil organic matter decomposition and crop residue burning are the main sources of carbon dioxide (CO_2) emissions in agriculture (Zhao et al., 2023). Similarly, agricultural methane (CH_4) emissions come from flooded soils under rice cultivation, intestinal fermentation in livestock digestive systems, and the decomposition of feces and crop residues under moist conditions, and agricultural nitrous oxide (N_2O) emissions mainly come from the nitrogenous soil, manure, and compost. The pressure on the agriculture sector will increase with the continuing expansion of the human population, so agri-technology and precision farming have gained much importance in today's world. This is also termed digital agriculture (Qin et al., 2020), which means the use of high-tech computer systems to calculate different parameters such as weed detection, crop prediction, yield detection, crop quality, and many more machine learning techniques (Liakos et al., 2018). The increase in GHG concentration has wider impacts, such as more extreme weather events, deadly heat waves, and severe drought, posing serious threats to agricultural production (Leisner, 2020) and ultimately impacting net crop income (Hossain et al., 2019) and farmland value (Arshad et al., 2016). Scientific data show that global temperature will increase by 1.4–5.8 °C by the end of 2100 and that many plants, ecology, ecosystems, and related stress factor(s) can be affected. Particularly, climate change (and global warming) affects the physiology, growth, and development stages of wheat, the speed and capacity of photosynthesis, WUE values, harvest, and threshing. Particularly, local wheats, are very important sources of disease resistance and quality genes, parents in various hybridization programs, agronomy, tissue culture, genetic engineering studies, etc. for the agronomic limitations and bottlenecks which can be overcome to be studied by the multidisciplinary. Studies have shown that the most important cause of global warming is the emissions of CO_2, CH_4, and N_2O gases, which are the main GHGs, and many human activities such as burning fossil fuels, deforestation, industrialization, uncontrolled urbanization, and agricultural practices lead to the increase in greenhouse gas emissions in the atmosphere. Globally, the average temperature of the Earth's surface and water bodies has been increasing since 1861, averaging 0.8 °C in the twentieth century. During the period (1950–1993), maximum night time temperature and air temperature globally increased by 0.2 °C per decade, and only in the last 10 years maximum daytime temperature was realized 0.1 °C higher than the annual increase. The spread and lengthening of frost-free seasons in many middle and high latitudes is the effect of increasing temperatures (IPCC, 2001), and precipitation has increased significantly in the middle and high latitudes of the Northern Hemisphere in the last century. In regions with a Mediterranean climate, there was a decrease of 0.5–1.0% with an increase of 0.2–0.3% in precipitation. Additionally, the number of extremely rainy days in the Northern Hemisphere has increased by 2.0–4.0% (IPCC, 2001). These changes in the precipitation regime, due to global warming and climate change, have been one of the important factors affecting the agricultural sector.

Wheat, which is cultivated on 15.0% of agricultural areas globally and used in human nutrition, has a share of 20.0% in terms of calories and 25.0% in terms of optimum protein. It is considered the most important and strategic nutrient. It is a plant that can be grown almost everywhere in the world. In recent years, its importance has increased even more with the decrease in nutrition/production opportunities and food security coming to the fore, overshadowed by factors such as population growth and the COVID-19 epidemic. Statistics show the annual average for a healthy and balanced diet. It shows that 225.0 kg of wheat is needed (Anonymous, 2024a; Koç et al., 2009). The most important reason why wheat is at the forefront of global warming and food security discussions is its great importance in the nutrition of both human and animal populations, and its increased metabolic and enzymatic (RuBisCo enzyme) response to increasing temperature and CO_2 concentration. Generally, wheat cultivation in most parts of the world is done in insufficiently productive, mostly arid/semi-arid areas, and only dependent on rainfall, using dry farming techniques and without irrigation (Koç et al., 2009).

Climate change, extreme weather events, increasing weeds, diseases and pests, disease factors, etc., greatly affects agricultural/plant production (Taranto et al., 2018; Valizadeh et al., 2014). Studies show that for each degree Celsius increase in global average temperature, in wheat (*Triticum* spp.) and 7.4% in corn (*Zea mays*), 3.2% in rice (*Oryza sativa*), and 3.1% in soybean (*Glycine max*) yield level decrease, with average of (6.0–6.9%). It has been determined that the main factor in this decrease is the increase in GHGs and chlorofluorocarbon (CFC) concentrations due to human activities. These gases (GHGs) regulate the climate system and play a role in the absorption of light rays from the sun (Semenov et al., 2014; Tubiello et al., 2000; Zavarzin, 2001). However, due to the accumulation of these gases, the light rays from the sun cannot be reflected into space, and they warm the Earth like a blanket, thus increasing the temperature of the planet (IPCC, 2007, 2014). According to Fuhrer (2003), this situation will be felt differently in different parts of the world. The average temperature will increase between 1.4 and 5.8 °C until the year of 2100. The rate of CO_2 gas, which was 270 ppm before the Industrial

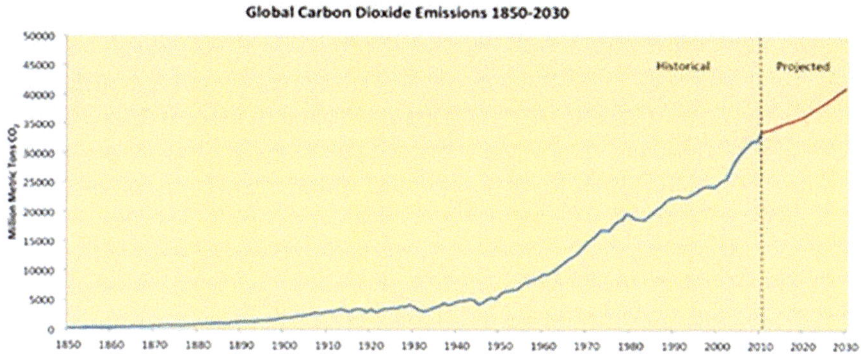

Fig. 2.1 Global CO_2 emissions trend between 1850 and 2030. (Anonymous, 2015c)

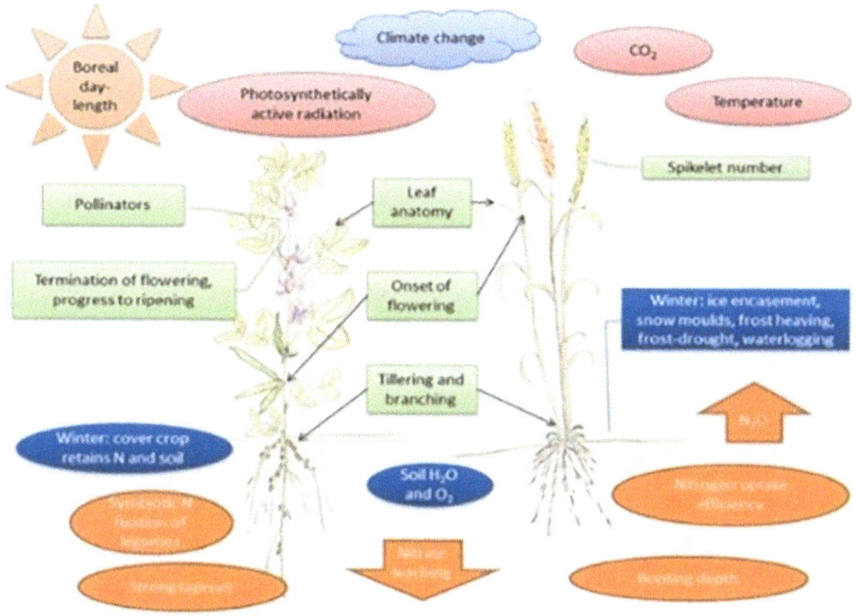

Fig. 2.2 Global Carbon Dioxide Emission in wheat agriculture progress. (Adapted from Frederick et al., 2011)

Revolution, has reached 355 ppm in modern times and is expected to reach 600 ppm in the twenty-first century (Rogers et al., 1994) (Figs. 2.1 and 2.2). GHGs from agricultural sources are mostly emitted from soil, automotive factories, petrochemical industries, aircraft and vehicle fleets, iron and steel factories, refineries, rice fields, etc. They are released into the atmosphere (Mei et al., 2007).

2.1.2 What Are the Effects of Climate Change on Wheat Agriculture?

Climate change has significant effects on almost every plant on Earth (Figs. 3, 6, and Table 2.2) The most important of these effects, in terms of the agricultural sector, is on wheat. As is well known, wheat is one of the most widely planted and consumed grain products worldwide. Determining the effects of climate change on wheat production and minimizing them is very important for the sustainability of the agricultural sector. In recent years, advanced technologies such as RS, DE, and AI have started to play an important role in the agricultural sector. These advanced technologies are used in many related fields, such as increasing efficiency in agricultural production, optimizing resource use, and minimizing or coping with the effects of negative factors such as climate change. These advanced technologies must be used for plants, especially wheat, to adapt to climate change and to grow and develop efficiently. In wheat farming, (RS) is a very good and useful technique for understanding the possible effects of climate change, and it can be used to monitor for flora, for temperature, for humidity and other relevant climate parameters with releted data which obtained from soil, and to understand the contributions and shares of effective factors much more clearly. Again, these data, determined in wheat production using advanced technologies such as DE and AI, constitute important tools for analysis and interpretation. All these technologies provide great benefits to wheat cultivation and agricultural sector with their ability to collect and analyze large volumes of relevant data. All these advanced digital technologies also increase efficiency and optimize resource use, such as sunlight, water, nutrients, etc., that the wheat plant needs. They can provide/present related advices or recommendations to producers/manufactures for keeping their crucial agricultural inputs at optimum levels. Therefore, they enable stakeholders in the agricultural sector and, in this regard, policymakers to both invest in technologies and take steps to use them. Similarly, revealing, measuring, and evaluating the effects of climate change on wheat agriculture is becoming increasingly important. Wheat is a basic agricultural product worldwide and the main source of food for many countries.

However, it is clear that climate change affects wheat production. As with other factors during production, changes in soil moisture negatively impact plant growth and phenology, alter precipitation patterns, and, depending on the frequency of meteorological events and temperature increases, can lead to negative developments in plant cultivation in some regions. When evaluated in general, all these factors affect the growth and development of wheat. Not only does it have a negative effect (especially the temperature increase), but it also accelerates the flowering process and grain formation (Fig. 2.5), and it negatively affects phenology, generative development, and plant physiology (Poggi et al., 2022; Rezaei et al., 2018; Richards et al., 2015). It has an accelerating effect on life cycles (Bosso et al., 2016; Liu et al., 2016) (especially for seed setting in wheat plants) (Figs. 2.3 and 2.5) and negatively affects drying by slowing down photosynthesis and causing water loss in its body. Again, dry matter yield and the formation of yield components (Asseng et al., 2011; Dias & Lian, 2009; Li et al., 2023; Liu et al., 2016; Porter & Gawith, 1999; Teixeira

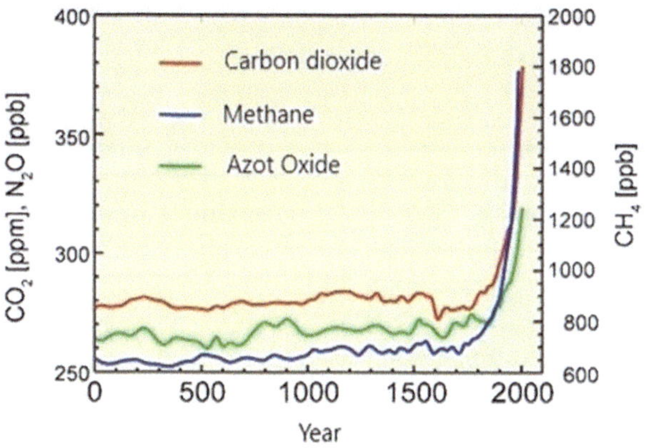

Fig. 2.3 GHGs concentration trends and variation at the global temperature. (Di Norcia, 2008)

Fig. 2.4 Major symptoms of wheat yellow rust (*Puccinia striiformis*) disease with significant yield losses. (Original)

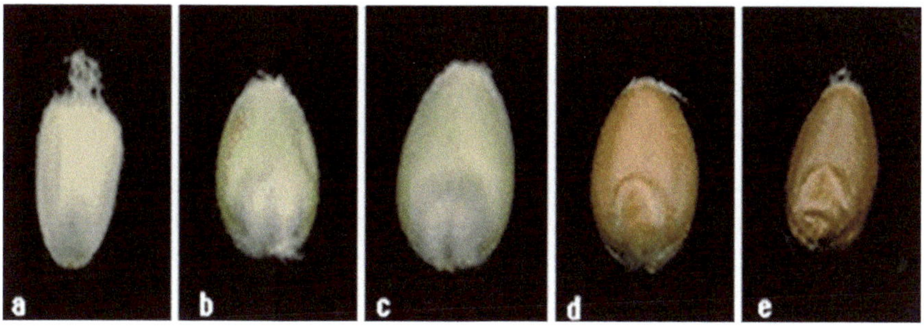

Fig. 2.5 Rapid seed ripening in wheat caused by significant climate changes (Original)

et al., 2013), especially rust diseases (yellow, black, brown rusts, etc.) (Figs. 2.4 and 2.5), excessive rainfall or excessive irrigation (Shen et al., 2021), while causing disease epidemics (epidemics), the plant (wheat) is more sensitive to drought stress (Shafr et al., 2021, Shafr et al., 2013; Diego et al., 2021; Ather et al., 2023) (Fig. 2.6).

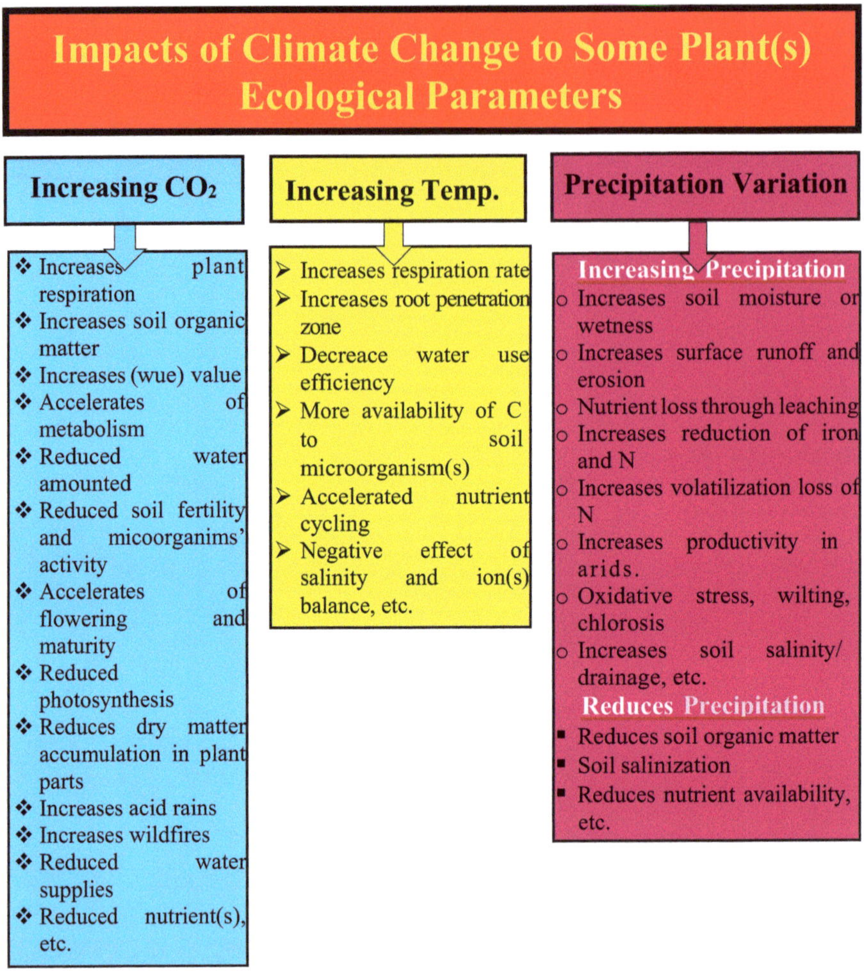

Fig. 2.6 Effect of global warming on wheat grain at various stages (Modified from Balasundram et al., 2023)

It is very important to accurately measure and analyze the climatic parameters affecting wheat production to determine and use future agricultural strategies. The use of advanced digital technologies such as *RS*, *DE*, *AI*, and drones (Fig. 2.15) in wheat farming not only save time, effort, and money, but also offers many advantages such as accuracy in the relevant decisions and long/short-term accuracy, and also provides high feasibility and rates. Especially with the application of digitalization in the plant production sector of agriculture, the use of genotypic and phenotypic combinations, computer-based models in plants, the combination and analysis of a large number (billions) of data with these and similar technologies, and the use of *AI* algorithms, the time-dependent changes in plant characteristics are clearly understood, and the most appropriate one among them is determined. It also makes it possible to select the plant trait(s) that provide the highest yield for production and are least affected by climate change. Research is being carried out in different parts of the world to develop wheat varieties with higher adaptation flexibility and resistance to adverse conditions. Scientific studies have revealed that countries, especially those in the Mediterranean basin, are much more likely to encounter global climate change than others.

Especially in this context, when the impact of climate change on wheat agriculture in Türkiye is evaluated, the temperature increase of (2.0–3.0) °C and its possible effects on the climate should be taken into consideration in light of agricultural policies for the next (10.0–20.0) years. For a temperature increase of (1.1–1.3) °C in this framework, there will be a 1.0% decrease in GDP between 2030 and 2034) and a 1.4% decrease during the 2040–2049 period. Similarly, the average productivity loss is estimated at (6.0–7.0)% for (2030–2039) and (8.0–9.0)% for (2040–2049). The variability in productivity levels, especially in wheat, will increase throughout the year. It has been observed that planting and harvesting will change by 4.0 days per 10 years, and an increase in field air temperature of 1.0 °C moves the harvest forward by 8.0 days (Karapınar et al., 2020; Şensoy et al., 2014). The main effect of the decrease in climate-related plant yield appears mostly in the temperature parameter (Fig. 2.6), which shortens and accelerates the time for vegetation and starch to fill the grain, reaching up to 125.0% and resulting in early ripening (Karapınar et al., 2020; Kumar & Tewari, 2013; Özdogan, 2011). According to previous scientific research, among the most important climatic parameters affecting the decrease in wheat plant productivity are maximum temperature at the time of planting and maximum rainfall during the flowering period (Özkan & Akcaoz, 2002) (Fig. 2.6). When the supply problems of the wheat production potential in the country during the period (1963–2012) and the risks arising from yield variability trends of wheat in terms of global climate change are examined (Dellal et al., 2015; Gençtan et al., 2015; Karapınar et al., 2020; Ulukan, 2021) (Fig. 2.7), it is seen that there are possible productivity losses. The negative impact on welfare will be at most 623.0 million dollars compared to 2012; according to another scenario, it is stated that this value will be 42.0% higher 267.0 million USD, but a 50.0% increase in the world wheat production level fluctuations due to effective climate-related factor(s)' and their interaction(s) in Türkiye (Anonymous, 2024a; Karapınar et al., 2020) (See Fig. 2.7).

2.1.3 *Significant Effects of Climate Change on the Agricultural Sector in General*

According to Dellal et al. (2015), Özgen et al. (2015), Ulukan (2018, 2021), and Anonymous (2024a), the main ones are; hotter and less rainy climate conditions; increase in extreme meteorological events; decrease in water resources; increase in the severity of drought; deterioration of water and soil quality and ecosystems,

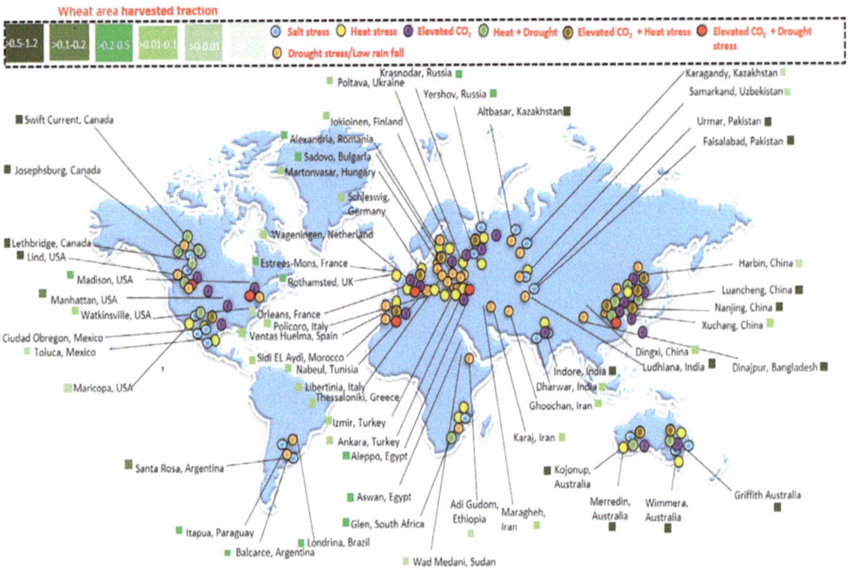

Fig. 2.7 World map showing the impact of climate change on the wheat (*Triticum* spp.) harvest areas in major wheat-growing countries in terms of drought, heat, CO_2, rainfall, and combined stresses. (Permitted by Zahra et al., 2023)

decrease in biodiversity; shrinkage of ecologies; decrease in agricultural production and quality; storage problems, increase in pests and diseases; fertilization and pesticide problems, threats to sustainable food security, etc. are the most important of these. Wheat, an essential and strategically important staple crop in nutrition globally, plays a critical role in ensuring food security for a significant portion of the world's population. As demand for wheat continues to increase with population growth, the agricultural sector also faces the necessity and difficulty of optimizing production to meet these increasing needs. However, recent developments in these technologies, especially in the fields of RS, DE, and AI, are reshaping the landscape of wheat agriculture. On the other hand, as is known, wheat agriculture has been an indispensable part of human and animal populations for thousands of years. However, as the world's population increases and resources become increasingly limited, finding innovative ways to increase efficiency and optimize resource use is of great and indispensable importance.

In recent years, RS, DE, and AI have emerged as promising and powerful tools in the field of agriculture (especially in wheat agriculture) and are being used increasingly widely, making it important and promising to benefit from these advanced technologies in wheat agriculture at an increasing rate in terms of both usability and price. RS, which is one of the digital methods that can be defined very generally as obtaining information about oneself or a phenomenon without making any physical contact with the object, is used to determine the products' field conditions, width, etc. It has been extremely effective in managing and monitoring easily, with minimal errors and in a short time, without any limits. Satellites equipped with advanced imaging sensors have allowed researchers and farmers to collect valuable and important data for wheat fields on a large scale. Among these, satellite-based RS allows continuous monitoring of crop health and helps identify diseases and pests by assessing the stress caused by water and nutrient content in the field. Again, by analyzing this data, making conscious decisions on irrigation, fertilization, and disease and pest control enables higher wheat yields to be obtained by using resources more efficiently. *DE* and *RS* have played, and continue to play, an important role in the transformation of the agricultural sector. The wheat plant is one of the oldest grains grown in the world and plays a vital role in global food security. Similarly, agriculture has been an integral part of humanity's nutrition and shelter for thousands of years, along with the subsectors with which it interacts (Vavilov, 1926; Wang & Engel, 1998). However, as the world's population increases and resources become limited, it is vital to increase efficiency and find innovative ways to optimize resource use. In recent years, RS, DE, and AI have emerged as effective and powerful tools in the field of agriculture, and their applications in wheat agriculture have shown great promise. Together with RS and DE, they play a very important role in transforming the agricultural sector. Additionally, digital technologies such as drones (Fig. 2.15) and unmanned aerial vehicles (UAVs) (Ballester et al., 2021; Yoshio, 2020) allow farmers to capture high-resolution images of wheat fields, providing detailed information on crop development and health. On the other hand, RS can be roughly defined as the science of obtaining information about an object or event without any physical contact; it has recorded notable, important, and remarkably promising results in terms of monitoring and management in crop production (especially wheat agriculture). In particular, satellites equipped with electronically advanced imaging sensors have allowed researchers and farmers to collect large-scale and valuable data for wheat fields. Satellite-based RS allows continuous monitoring of crop health, assesses water and nutrient stresses, if any, and detects disease and pest outbreaks, if any. Again, while all the data it determines can be analyzed, it makes decisions on irrigation, fertilization, disease, and pest control, which allows both more efficient use of resources and the achievement of high quality and quantity. These images can be further analyzed with machine learning (*ML*) algorithms to detect minor changes in plant physiology and identify potential problems before they become dominant. By integrating DE into wheat farming, farmers can proactively address their problems, minimize crop losses, and optimize resource allocation. AI is another advanced technology that is changing the entire application in wheat farming. With its algorithms, it can provide valuable information and recommendations to manufacturers by processing the large amounts of data generated through RS and DE; by analyzing previous years' yield values (data), weather patterns, soil properties, and other relevant factors, it can generate accurate growth forecasts, optimize irrigation programs, and adjust fertilization regimes according to crop, facilities, and timing.

Again, AI-powered farm management systems allow farmers to make data-based decisions, increasing efficiency, reducing costs, and, in parallel, improving sustainability. The integration of *RS*, *DE*, and *AI* in production holds tremendous importance and potential for the future of wheat agriculture. These advanced technologies enable producers to transition from traditional, one-size-fits-all approaches to site-specific and data-driven management strategies, as well as leverage the power of satellite imagery, drones, *ML*, and predictive analytics to address crop health, nutrient deficiencies, and yield-limiting factors. While transferring possible ideas into practice (Chen et al., 2023), environmental impact is minimized, and it enables the application of precision agriculture techniques with applications such as variable rates of fertilizer and pesticides for production optimization. At the same time, these technologies offer real-time monitoring and decision-making; with continuous data collection and analysis, farmers can identify sudden changes and prepare for or respond to situations such as possible disease outbreaks or adverse weather events; The ability to remotely monitor crops and obtain critical information on time is of great importance, especially in regions where access to agricultural resources and expertise is limited.

This information can help farmers make informed decisions on fertilization, irrigation, and disease and pest control, optimizing resource use and reducing environmental impact. DE refers to the process of harnessing the technological power of DE to encourage innovation and improvement in agriculture. Thanks to advances in data analytics, *ML*, and cloud computing, it will now be possible to analyze large amounts of data collected from wheat fields, identify patterns and trends, and make predictions about future crop performance. This allows farmers to make data-driven decisions and apply precision farming techniques, tailoring inputs and management practices to specific areas in a field. Additionally, training and capacity building are needed so that farmers can use these tools effectively and interpret the data obtained. Indeed, to build trust and use technology responsibly, concerns about data privacy and security also need to be addressed. With the integration of RS, DE, and AI applications into wheat farming, they have the potential to further revolutionize the sector by organizing the activity in question and transforming it as desired. So much so that these techniques provide real-time information and data that improve resource management, increase product durability, and offer a sustainable and efficient approach to production. Continued investment in research and development, as well as targeted capacity-building efforts, is vital to fully exploiting their potential and ensuring food security for future generations. They offer new and innovative ways to increase wheat production, optimize resource use, and reduce the effects of climate change. From the primitive plows and hand tools of the past to today's advanced machines and digital solutions, the agricultural sector has experienced a remarkable evolution. One of the important obstacles is the development of accurate models that can capture the complexity of wheat plant × product × climate parameters, as well as the need for robust and reliable data sources. Additionally, the adoption of these technologies by farmers, especially in developing countries, may require capacity building and investment in the necessary infrastructure. Ultimately, the combination of *RS*, *DE*, and *AI* represents a powerful approach to understanding and managing the impacts of climate change on wheat production. These technologies enable scientists and farmers to monitor and analyze climate-related variables, predict future scenarios, and design appropriate adaptation strategies. By utilizing these tools, sustainable agriculture can be encouraged, increasing food security and ensuring the resilience of wheat to a changing climate. Its impact on various crops, including wheat, has raised concern among researchers and farmers alike.

To achieve successful wheat production, the necessary care and attention must be paid to completing the following growth and development stages completely and correctly (Table 2.1).

Two factors are important for the germination of species-specific plantlets consisting of seeds and reaching the soil surface: soil temperature and soil moisture. However, increasing temperatures due to climate change have an accelerating effect on metabolic-physiological reactions in almost all plants. At the same time, increasing temperature affects the soil the most. This causes the plant to lose moisture and immediately enter to water stress. Having sufficient soil moisture at the appropriate stages ensures that the wheat completes its growth and development stages strongly and enters the other stages stronger and healthily completes them. Undoubtedly, weather conditions also affect this process. Additionally, CO_2 gas is effective in this regard. Studies have shown that the application of CO_2 gas in C3 plants, including wheat, increases the yield by approximately 34.0–36.0%, stating that this was due to the physiology of C3 plants (CO_2 gas fertilization) and that this increase begins to decrease after approximately 1 week to 10 days (Uzmen, 2007) (Figs. 2.2, 2.6, 2.7, 2.8, 2.10, and 2.11, and Tables 2.1, 2.2). Any error or deficiency in the growth and development stages of wheat, as listed above, will lead to irreparable situations in wheat agriculture. All C3 plants, especially wheat, respond differently to increased CO_2 and other climate factors. In general, scientific data show that the growth and development of plants decreases due to climate change, and these phenomena lead to negative consequences in terms of wheat plant productivity by time (Romanova, 2005). Among these effects, the increase in CO_2 increases the value of WUE, which is especially evident in C3 plants (Cutforth et al., 2007); especially during flowering and growth, stress due to increasing temperatures reduces the speed of germination from seed, disrupts nutritional balance, and negatively affects fruit, tuber, and grain formation and quality criteria. It has been observed that there are disruptions in pollen and meiosis in flowers, especially at temperatures in the range of 35–40 °C (Fuhrer, 2009).

The anatomical effects of climate change on plants, including local wheat, are generally due to the increase in the concentration of CO_2 gas and the increase in the parameters that support this, such as temperature and salinity, and their interaction with each other. With these factors and interactions, a shift occurs in various growth and development stages such as leaf thickness, stomatal aperture, chlorophyll number, tillering, and plant height, as well as a significant decrease in plant water intake; these changes occur due to the closure of stomata, especially due to the increase in chloroplasts in leaf cells (Mei et al., 2007; Romanova, 2005; Ulukan, 2008, 2009). Likewise, increasing temperature also reduces the moisture in the soil by increasing evaporation. Schlenker and Roberts (2009) reported that the productivity level of wheat at low CO_2 concentration varies between (36–40)% and (63–70)%; Hatfield and Prueger (2011) found it to range between 3.8% and 5.0%;

Table 2.1 Consequences and effects of some climate factors such as temperature (°C), humidity (%), sunlight density (Ft or Cd), and day length (hours) in wheat growth and development stages. (Photos are original)

	Wheat growth and development stages									
	Sowing	Emergency		Tillering		Jointing		Flowering and fertilization	Heading	Harvest and tresh
Climate factors	Opt.	Opt.	Max.	Opt.	Max.	Opt.	Ext.	Opt.	Max.	Ext.
Temperature	19–25	16–19	>35 <3.5	08–15	>27 <2.5	0.8–15	>24.5 <1.0 (tiller no decrease)	(10–(−18))	>24 <−1.0 (spike shorter)	>53 <20 (emergency/ seed setting decrease)
Humidity	0.4–1.0	1.0–4.0	>1.0 <0.4	1.0–4.0	>8.0 <0.4 (tiller no decrease)	1.0–4.0	<0.4 <8.0 (tiller no decrease)	(1.0–(−4.0))	<0.4 >80 (flower number decrease)	<0.4 >0.8 (spike partially dies)
Sunlight density		2500	>1.0	2500	<2000 (tiller no decrease)	2500	<2000 (tiller no decrease)	2500	<2000 (flower no decrease)	<2000 (flower no decrease)

Day length		12–14	>0.4	14–16	>8.0 <17		>18 <12		Low Quality
Outcome		☺	(decrease)	☺	(thin.)	☺		☺	☹
Possible climate change effect	The moisture of the soil decreases, and the seeds or plants cannot germinate sufficiently and come to the surface of the soil. For such reasons, the dry matter yield of wheat also decreases!... 			As soil moisture decreases, photosynthesis and tillering processes slow down, nutrient uptake slows down due to the lack of water in the environment, and eventually, dry matter production in the plant, which constitutes the yield and yield elements, decreases. 		During this period, plants begin to become erect, but for this they need water; they also need to produce increasing amounts of dry matter through photosynthesis and consume mineral matter in parallel. Global warming prevents all this. 		Care should be taken to protect against moisture or lack of water, especially during this period when low temperatures affect it extremely negatively. 	During this period, previous growth and development periods are valid, and according to satellite data, increasing temperatures will shorten the wheat harvest by approximately 2 weeks. 

Opt. Optimum, *Thin.* Thinning, *Max.* Maximum

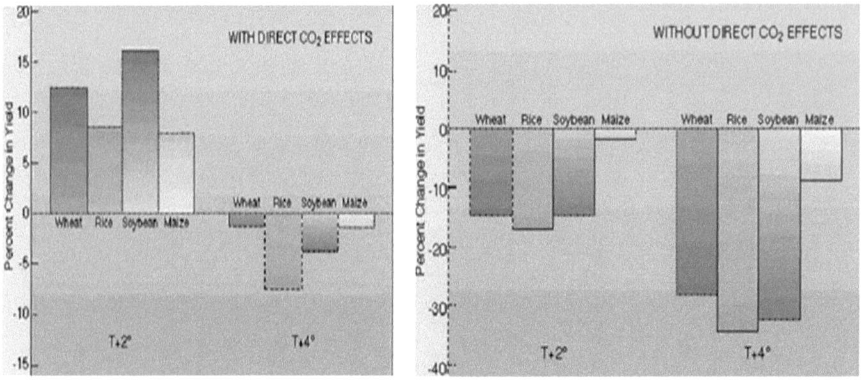

Fig. 2.8 Averaged crop yields of four major crops [wheat (*Triticum* spp.), rice (*Oryza sativa*), soybean (*Glycine max*) (= C3 plants), and maize (*Zea mays*) (= C4 plants)] and as a result of increases of 2.0 and 4.0 °C in average global surface temperature with (left) and without (right) the direct effects of CO_2 on plant growth and water use T = Temperature. (Modified from Rosenzweig & Hillel (1995), Mengü et al. (2008))

Table 2.2 Wheat physiological traits and responses to some climate factors

	Increased in some climate factors			
Traits	(CO_2)	(UV-A, B)	(O_3)	Response(S)
Roots	+	?	?	Root/stem parts
Photosynthesis	"	–	–	Yield and resp.
Leaf cond. and develop.	–	+	"	Leaf area
WUE	+	"	"	Stoma conduct. and aperture
Leaf area	Middle	–	"	PAR point
Leaf thickness	+	"	"	?
Maturity and tresh	"	Unaffected	?	Veget. and yield
Flowering (anthesis)	Acc.	Protects	–	"
Dry matter production	Doubled	Varied	Varied	Yield
Sensitivity in plants	Varied	"	"	"
Resistance to drought	Middle to sensitive			Wilt, dwarfness, and death
Nutrition	Less	Less	Sensitive	Dwarfness, death
Vernalization	+	?	?	Veget.
(CO_2/O_2)	?	"	"	Photos.
Respiration rate	"	"	"	Biomass
Maturation stage	"	"	"	Yield
Biomass production	+	"	"	"
Internodi number	?	"	"	Biomass and plant height
Weed distribution	"	"	"	+
Seed germination	+	"	"	Varies
Rhizomes	"	"	"	"
Seed longevity	"	"	"	?
DNA and sterility	"	"	"	Mutation and death
Ecological factors	"	+	+	Stress

Krupa (1997), Dhakhawa and Campbell (1998), modified from Tubiello and Ewert (2002)
+ positive, – negative, ? unknown, *P.A.R.* photosynthetic activity radiation, *Acc.* accelerates, *Resp.* respiration, *Veget.* vegetation, *Photos* photosynthesis, *Conduct.* conductivity

Fuhrer (2009) stated that a 1.0 °C increase in temperature can cause a (7.0–124.0)% decrease in efficiency.

Unit area yield depends primarily on the level of technology, climate, soil conditions, plant diseases, and the density and severity of pests and weeds. It is predicted that continents, oceans, and coasts will be more affected by this situation, and the poles will warm more faster than the equatorial regions. As atmospheric concentrations of GHGs continue to increase at unprecedented rates, they could have a major impact on wheat yield levels, and research is being conducted around the world to understand their effects on wheat production. Global climate change is a global problem of increasing importance, and its negative effects are already being felt acutely in some regions of the world.

When evaluated in this context, it would be a rational approach to develop methods and practices that will eliminate or minimize the negative effects of global warming; especially in a field crop like wheat, whose production and consumption concern a large number of living organisms in terms of quality and quantity, eliminating or minimizing the negative effects of global warming is among the important targets that should be worked on. In this respect, the most rational strategy is to prioritize the causes of global warming, use the relevant agricultural methods and inputs more effectively and appropriately, and turn to careful but long-term savings-oriented solutions. That is why the development of binding agreements containing regulation(s) regarding CO_2 emissions and/or the use of water resources according to their contribution to "*global climate change*" should be adopted. Excessive nitrogen fertilization should be avoided (to avoid polluting groundwater) and measures should be taken to reduce emissions of N_2O. GHGs continue to increase, and scientific research is being carried out in many countries and research institutions around the world to reveal the effects of this situation on crop production.

We can list our suggestions on this subject as follows (Anonymous, 2024a; Karapınar et al., 2020; Yavaş & Ünay, 2010): an urgent mobilization should be declared for "Adaptation to Climate Change" in agriculture, and relevant policy alternatives should be developed; the "*Adaptation Fund*" for climate change and the "*Institute for Research and Application of Climate Change in Agriculture*" should be established immediately; all other universities, institutions, and organizations related to this institute should work in a coordinated and effective manner; scientific research should be conducted at the national or international level; there should be no moisture loss, and water harvesting practices should be included through direct planting of wheat; wheat producers with low income levels should be supported financially in sufficient amount and time under the climate change adaptation package; it should be aimed to use organic or nature-friendly wheat production methods on a certain proportion of the lands allocated for agriculture; 100% pressurized irrigation in irrigation systems; national and international agricultural trade policies that prioritize risk management should be developed; wheat production should be increased, producers should be supported to use advanced technologies such as RS, DE, and AI; and agricultural extension should be given importance in this and similar aspects and relevant measures should be taken (Table 2.2, Figs. 2.8 and 2.9).

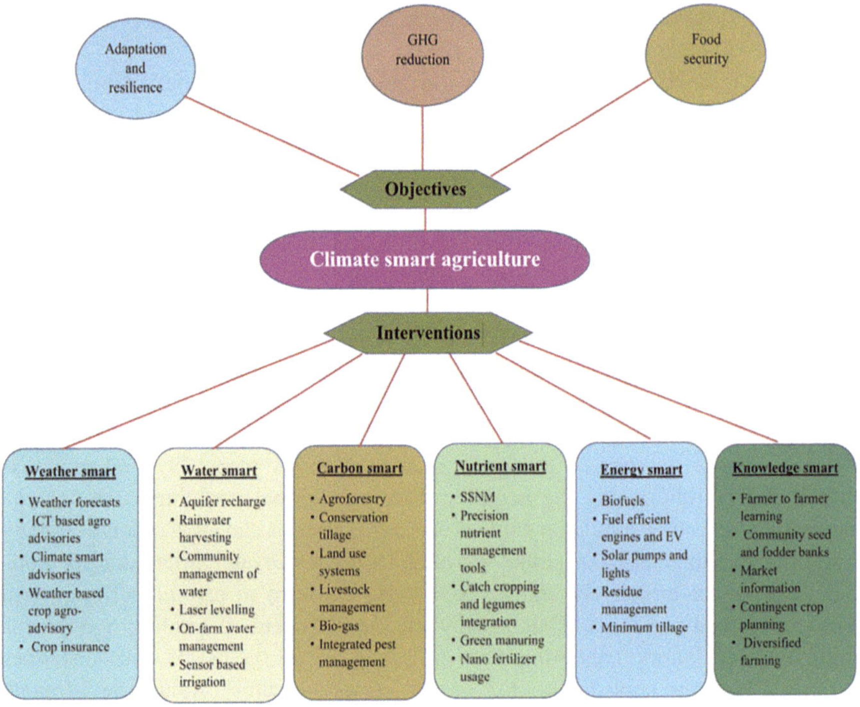

Fig. 2.9 Climate-smart agriculture. (Permitted by Balasundram et al., 2023)

One of the main problems in wheat production is the suitability of the ear for separation into its constituent parts (grains, etc.), or in other words, for blending (Fig. 2.10). To avoid these and similar negativities, wheat varieties with the mentioned characteristics should be selected and planted at the appropriate time for that region; as the restrictions caused by the decrease or limitation in resources increase, attention should be paid to growing wheat varieties that are resistant to drought, have low water consumption, yet have high efficiency and quality, and are ecologically suitable.

Additionally, crop rotation models that include animal feed and edible legumes should be adopted due to their ability to deliver products to storage organs (Nemecek et al., 2008; Ulukan, 2009); alternative energy sources should be used; minimum soil tillage techniques should be applied (Olesen & Bindi, 2004; Ulukan, 2008); the planting area of field crops such as bread wheat (*Triticum aestivum*) and corn should be narrowed, but the planting of durum wheat (*Triticum durum*) should be reduced (Thomson et al., 2005); organic farming should be practiced and the practice of burning agricultural wastes should be avoided. For this reason, due to global climate change and the resulting global warming, there is a significant decrease in water resources on our planet, and there are changes in the precipitation regime across regions and countries (IPCC, 2001; Kapur et al., 2012; IPCC, 2014, 2018). Care should be taken to grow grains such as wheat and barley (*Hordeum* spp.), which are

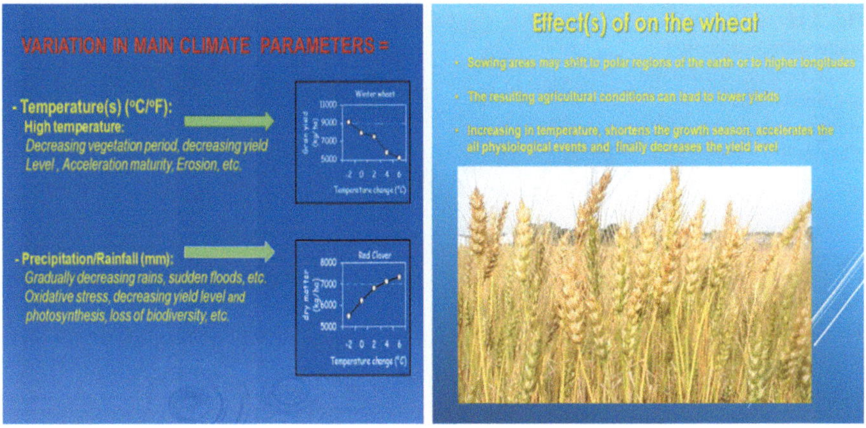

Fig. 2.10 Important climate parameters and their effects on a wheat plant. (Original)

C3 plants (due to metabolic and enzymatic reasons such as RuBisCo enzyme); practices with mechanisms such as insurance, credit, and financing for product losses resulting from *Global Climate Change* should be implemented; policies that reduce the agricultural inputs used by producers in wheat production and ensure optimum costs should be followed, and appropriate and effective incentives and supports should be provided; agricultural tools and equipment such as tractor, plow, seeder, combine harvester, etc., should be replaced as they lose their optimum function and become worn out; correct and scientific water use methods should be applied; productivity-based smart agricultural practices should be disseminated and implemented as soon as possible; it must be ensured that the solutions are within the limits determined by ecology and genotype, according to the needs of countries, cities, and regions; the agricultural planning process should be carried out together with the stakeholders; production capacity should be increased; to give an example from our country, since temperature anomalies in the Mediterranean, Aegean and Central Anatolia regions have been determined to cause low yields in wheat, temperature changes/fluctuations in wheat growing should be carefully avoided; climate-friendly agricultural practices should be followed, as wheat yields decrease due to the negative effects of temperatures on accelerating development; and photoperiodic sensitivity emerges with slow growth, but resistant varieties should be used (Kapur et al., 2012) (Fig. 2.11).

Wheat agronomy continues to be affected by the negative effects of climate change and global warming. In light of the techniques, information, and findings presented so far regarding the increasing severity of climate change and global warming felt on a global scale (Figs. 2.10, 2.11, 2.12, and 2.13), the following visual sheds light on the relationship between the origins of climate change and global warming in wheat farming and global agricultural production:

Fig. 2.11 Sacked wheat grains after production which is being tried to be sustained under the influence of climate change. (Original)

Fig. 2.12 Harvested wheat seeds under the climate change effect. (Original)

2.2 Why AI and RS … and Their Roles in Agriculture

As is well known, agricultural production [especially crop production (wheat production branch)] is exposed to many biotic and abiotic stresses. From this perspective, the efficiency level in agricultural production is very important. In addition, it is affected by many natural parameters nature and is mostly made by humans, that is, subjectivity. It is essential that such activities, where the phenomenon is at the forefront, be carried out in a way that provides maximum performance and efficiency with the least errors and highest accuracy. In this context, it is essential to use many advanced technologies such as RS, robots and drones (Fig. 2.15), satellite photographs, spectral cameras, etc. Their use saves both time and effort and provides an

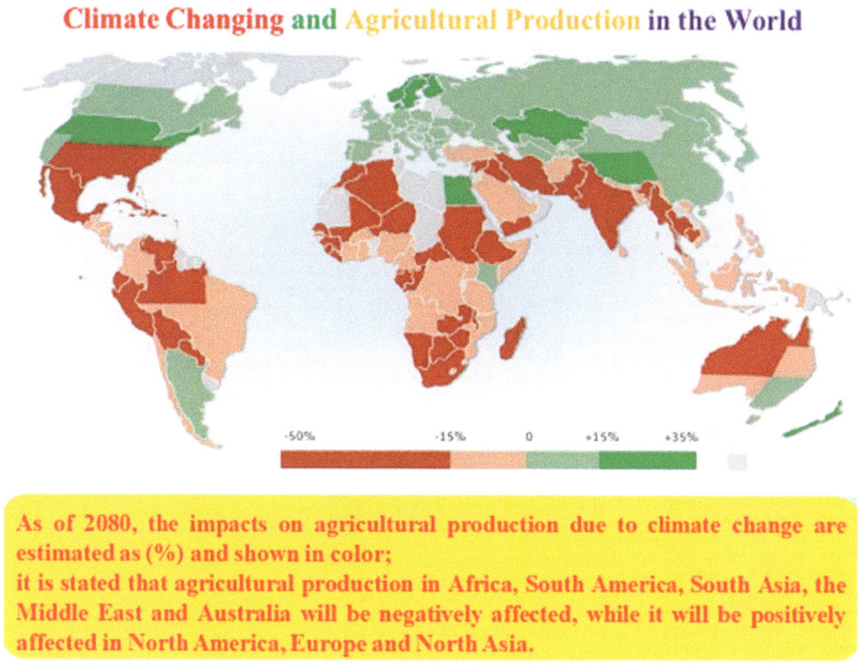

Fig. 2.13 Climate change effect and agricultural production in the world. (Anonymous, 2016)

extremely accurate and appropriate application. In addition, it is one of the most powerful and effective digital techniques in terms of its numerous benefits in agriculture. Therefore, the application of DE in the field of agriculture, and its adaptation in wheat agriculture, has led to very important developments. On the other hand, the use of all these applications based on RS, DE, and AI has resulted in many agricultural production advantages that make it possible to achieve extremely high productivity levels in the shortest time. Their implementation will lead to the elimination of almost all agricultural problems that arise in this regard in the long term, such as human-based determination of physical and plant phenology. For this reason, in this section, the place and importance of artificial intelligence (*AI*) such as *RS*, *DE* etc. in wheat agronomy is mentioned, and other appropriate agricultural techniques are also mentioned in the related section title.

AI emerged in the 1950s and is defined as the ability of machines to perform some operations as skillfully as humans. Machine learning (ML) was introduced in the 1980s and became even more popular with the use of data mining (Javid et al., 2023).

Deep learning (DL) began to be used in the 2010s and can perform machine learning calculations across many layers simultaneously (Fig. 2.14). It consists of artificial neuron networks (ANNs) consisting of one or more hidden layers and similar ML algorithms. In other words, it involves using at least one "ANN" and

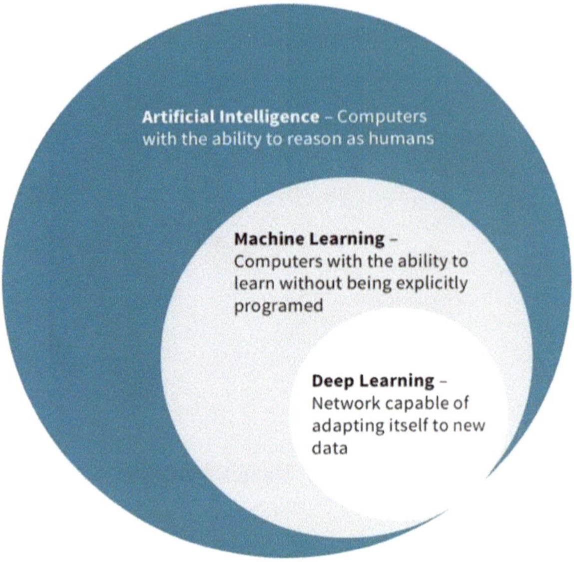

Fig. 2.14 AI and its various layers or stages. (Anonymous, 2017)

many algorithms to obtain new data from existing data. DL is one of the different methods used in machine learning. Specifically, the DL method is inspired by the learning and decision-making methods of the human brain (Kaya et al., 2019), that is, it uses ANNs by simulating the movements of neurons. "ANN" is an information processing technology inspired by the human brain's ability to process information. "ANNs" mimic the way the biological nervous system works. In other words, they are digital models of biological neuron cells and the synaptic connection of these cells with each other. "ANNs" are computer systems developed to automatically perform functions such as obtaining new information through learning, creating new information, and discovering new information, which are unique to the human brain, without the need for any assistance.

ML is a science concerned with designing and developing algorithms that enable computers to learn based on types of data, such as sensor data or databases. The focus of ML research is to give computers the ability to detect complex patterns and make rational decisions based on data. This type of learning is a system that investigates the working and construction units of algorithms that can learn as a structural function and make predictions based on data. It is involved in every stage throughout the entire growing season. It starts with planting the seed in the soil, soil preparation, planting method, fertilizer and pesticide applications, irrigation applications, and maturity determination from computer-aided imaging of robots and continues until the moment of harvest, and these are evaluated under subheadings: (a) supervised learning, (b) unsupervised learning, (c) semisupervised learning, (d) reinforced learning, and (e) intensive learning.

Big Data are making data meaningful. Data are analyzed, classified, and processed, and thus transformed into meaningful and usable information sources rather than just a pile of data. So, we can say that "Big Data" is the process of making sense of a pile of data. From the beginning of agriculture to the present day, farmers have had to take into account countless variables when determining how to cultivate their fields: Which seeds should be planted, when they should be planted, and what fertilizer should be used? Throughout history, farmers have carefully collected information to get the best answers to these questions. However, nowadays it is much easier than before to collect this data and make choices based on it. These questions, which in the past depended on the knowledge and experience of growers/farmers, can now be accessed through specialized machines and drones. The system in the field can detect problems such as when to fertilize or not, and when to plant, by collecting different data with various types of sensors used in daily agricultural work. Values obtained from climate and soil moisture stations in the field are widely used to determine irrigation time and amount.

2.2.1 AI and Its Future in Agriculture

As predicted, AI holds great potential for transformation and use in the agricultural sector (Mohanty et al., 2016). In particular, ML algorithms can perform very important tasks such as detecting and controlling diseases and pests, and determining yield levels by analyzing agricultural data (Kamilaris & Prenafeta-Boldú, 2018), while DL models can be used in the diagnosis of problematic plant diseases (Kamilaris & Prenafeta-Boldú, 2018), such as rusts (*Puccinia* spp.) (Fuentes et al., 2017). The integration of technologies such as RS, DE, and AI into the agricultural sector has the potential to transform future agricultural practices. Effective use of these technologies can increase agricultural productivity, enable more efficient use of natural resources, and improve food security. However, realizing this potential requires overcoming challenges such as infrastructure, education, and financing (Horton et al., 2017), therefore, it is very important and essential to make investments in the agricultural sector in order to develop relevant policies as well as technological innovations and advancements in order to protect and ensure agricultural sustainability, food security and globalization in this sense. AI is a branch of computer science that uses ML and DM algorithms, among others, to gain insights from and interpolate data to mimic human intelligence. These algorithms make predictions by dynamically connecting input and output variables. While all industries will achieve significant productivity gains with AI systems and ML capabilities, it is predicted that agriculture will also experience a digital transformation. However, these predictions can help create various solutions to both simple and complex situations, and AI will carry agriculture, one of the oldest sectors, into the future. There are a surprising number of agricultural applications of AI, with many studies using it to monitor and predict the yield of wheat, one of the most important food resources in the world (Kaya, 2017; Peker et al., 2021) (Figs. 2.9, 2.10, 2.11, and 2.13).

2.2.2 Artificial Intelligence (AI) in Wheat Growing and Wheat Growing Fields

The overcome some important factors such as feeding of the growing population (in terms of plant production), stress and to ensure that the population is fed and sheltered for a sufficient time (FAO, 2020); so much so that the use of advanced technologies such as *RS*, *DE*, and *AI* is of great importance in terms of artificial intelligence (*AI*) and has an undeniable contribution in terms of efficiency and sustainability of production (Nabiyev, 2016; Liu et al., 2019). Of these, *RS* is a widely used technology for monitoring soil properties, plant growth, and water (moisture) status in agricultural fields, for example, *Landsat and Sentinel Satellites* (Jackson & Huete, 1991; Wu et al., 2019). In this method, agricultural productivity is estimated based on biophysical parameters such as vegetation indices (Thenkabail et al., 2000). DE, another important parameter, refers to the digitalization and technological integration of the agricultural sector in general. It has an effect that makes traditional agricultural practices more efficient (Reed, 2018). In this regard, some examples can be given (agricultural machines and satellite-connected soil cultivation tools). It reduces labor costs by optimizing heavy agricultural operations such as planting and harvesting.

According to Zhang and Chaisattapagon (1995), Moshou et al. (2004, 2005), Guevara-Hernandez and Gomez-Gil (2011), Tian et al. (2012), Moshou et al. (2014), Majumdar et al. (2015), Jiang et al. (2016), Mondal and Kole (2016), Olgun et al. (2016). Pantazi et al. (2017a, 2017b), Shi et al. (2017), Zhang et al. (2017), Incicas et al. (2020), Taşan et al. (2022), and Cemek et al. (2023), they can be mentioned as follows:

- *Detection and Monitoring of Plant Health and Productivity Levels*

  Monitoring plant health and growth efficiency in wheat farming is critical to increasing crop productivity. Factors such as plant stress, chlorophyll content, and growth rate can be monitored through remote sensing techniques and sensors.
- *Water Management*

  It is critical to combat pressures on water resources, such as drought, and to use water resources effectively. *RS* techniques can be used with sensors to monitor data such as soil moisture content and plant water stress.
- *Timing of Harvest and Planning of Crop Production*

  Harvesting at the right time is critical for product quality and efficiency level. Data analysis with *AI* can be used to predict the timing of harvest and optimize its management. This technique allows farmers to bring their products to market at the right time and minimize losses.
- *Identification of Diseases and Pests*

  Diseases and pests can cause crop losses at varying rates and frequencies before and after harvest. *RS* and image analysis algorithms can ensure the early detection of diseases and pests, their severity and spread, and timely control, thus preserving the quality and quantity of the product.

- *Soil Fertility and Fertilization Management*
  Soil fertility is (directly) related to the availability of plant nutrients in the soil in the right amount(s) and at the right time. Sensors and soil analysis can be used to assess soil fertility and optimize fertilization strategies. Thus, productivity increases and environmental impacts are minimized.
- *Optimization of Fertilization and Spraying Dosage*
  Fertilizers and pesticides are critical for plant health and productivity and must be applied in the correct dosages. With *AI* and data analysis, we can optimize fertilizer and pesticide dosages by analyzing soil and plant data, which reduces agricultural input costs and again minimizes environmental impacts.
- *Field (Area) Management and Productivity Analysis*
  Using agricultural lands most efficiently is of great importance in field management. Data analysis with digital maps and analysis of site morphology and usage patterns can provide efficiency analyses. This can lead farmers to optimize their farming practices and adopt sustainable farming methods.
- *Machine and Robot Usage*
  Agricultural mechanization and robots are in an important position in terms of the automation of agricultural operations. *AI*-supported agricultural machines can perform automatic operations such as planting, irrigation, and harvesting.
- *Data Analysis Processes and Decision Support Phenomenon*
  This application allows producers to accurately and quickly calculate and interpret agricultural practices and reach conclusions in a very short time.
- *Market Analysis and Estimating Potential Demand*
  Market analysis and demand forecasting provide farmers with important information for marketing their products. Data analysis and AI allow farmers to increase their income by marketing their products more effectively through market trend analysis and demand-target audience prediction.
- *Agricultural Extension and Consultancy*
  The complexity of applied agricultural technologies and equipment and the constantly changing technological environment require continuous training and consultancy for producers and those other than the experts (farmers) who use this equipment.
- *Breeding of Wheat Genotypes with Genetic Resistance to Diseases and Pests*
  Developing wheat varieties resistant to diseases and pests that are problems in the region through AI and genomic analyses can reduce product losses and also contribute to reducing the amount of pesticides needed for chemical control.
- *Adaptation to Climate Change.*
  Adaptation strategies should be developed to cope with the effects of climate change. Combined analysis of remote sensing and climate data can help identify adaptive farming practices and thus minimize the impact of climate change on wheat cultivation.
- *Identification of Pests and Diseases*
  Early detection and follow-up of any disease requires early diagnosis and accurate detection. In general, the diagnosis of plant diseases is made mostly by visual methods. This process is inefficient and prone to human error. After

scanning hundreds of thousands of photographs, a computer that is suitable for the purpose can identify diseased plant photographs by special Machine Learning (*ML*) algorithm can be easily developed to answer questions such as the type of disease on the plant, how effective it may be, what its role will be in the formation of yield and yield components, and how much yield loss it may cause. In agriculture, *ML* allows for a more accurate disease diagnosis. During this period, it helps prevent waste of energy and resource consumption due to incorrect diagnoses. Producers can also diagnose plant diseases and develop appropriate pesticide plans by uploading images from satellites, *UAVs*, robots, smartphones, etc., with the help of relevant software.

- *Identification of Weeds*

  Various mobile robots, robotic methods, UAVs are being used to clear weeds, particularly around airport runways, large and extensive tourist facilities, museums, golf courses, and parks. In countries like the Netherlands, Austria, Canada, the United States, and China, unwanted plants/weeds spread over millions of hectares are being cleared. Indeed, a famous e-commerce site uses drones for product delivery, and AI-powered robotic dogs and uav are being explored for various situations, such as earthquakes, fires, forest fires, floods, and cave/mine collapse. Weeds are an important threat known to cause (0–100)% yield and production losses in any crop production. On the other hand, the biggest problem in weed control is the difficulty of separating them from the parent plant. All weeds depend on the main plant for various essential growth factors such as water and plant nutrients, etc., and they (weeds) exploit the main plant to provide them. For healthy plant production, weeds in the environment must be separated from the main plant immediately. "Computer vision" and "*ML* algorithms" are used to detect weeds subject to this process at low cost and without causing environmental problems. In the future, robots will also be used to remove weeds from the environment, minimize the need for herbicides, and save time.

- *Seed/Seedling, Fertilizer, Agronomic Inputs,* etc.

  ML is involved at every stage throughout the entire growing and harvesting cycle. A successful production of wheat begins with exceeding the standards of seed quality and quantity criterion (germination rate and emergency speed etc.), sowing time, a well-cultivated seedbed, soil fertility, etc. Then the production period ends when the chlorophyll breaks down, all the plants turn to yellow, and they (plants) reach harvest maturity. that latest stages of the wheat cultivation/growing. In other words, it ends with the completion of a cycle ranging from monitoring the growth of the planted plant, regularly measuring the amount of water and nutrients it consumes, and determining its maturity with the computer vision of the robots, to harvesting and storing the product. Ensuring healthy germination and emergence is vital for "seed quality," which is an important determinant of efficiency and quality in production. The germination rate of seeds is determined in the laboratory by experts, which is a tedious and subjective process and therefore prone to error. Therefore, researchers have developed various *ML* and image recognition techniques to automate the process of seed sorting and evaluation, using IR image recognition and *thermal infrared (IR)*

Fig. 2.15 A drone equipped with a spectral camera is moving around a wheat field covered with weeds (left); a drone is spraying a wheat field/fields dense with weeds (on right)

processes. In this context, "*hyperspectral*" imaging has been widely used in rating system along with *ML*.

- *Monitoring and Evaluation of Climate Change Impacts*

  RS techniques are an important tool for monitoring and evaluating the impact (s) of climate change on wheat cultivation. RS techniques, which include the detection and analysis of data obtained from satellite images and aircraft, can be used to monitor and evaluate the effects of climate change in agricultural areas. It is important to understand the effects of factors such as drought, water stress, temperature changes, and other climate parameters on agricultural productivity.
- *Harvest forecasting and risk management*

  AI and RS techniques can be used for harvest forecasting and risk management. When integrated with meteorological data, crop growth models, and other data, AI can help farmers predict future harvests, plan harvest periods, and take measures to reduce some losses and risks.
- *Water Resources Management and Irrigation Strategies*

  DE and *AI* techniques can be used to optimize water resources management and irrigation strategies. Irrigation programs created using artificial intelligence (*AI*) algorithms are made more efficient and useful, increasing the sustainability of water resources.
- *Precision Agriculture Applications*

  AI and *RS* techniques can be used to support precision farming practices. Again, data provided by drones (Fig. 2.15), satellite imagery, and other *RS* tools enable detailed mapping and analysis of agricultural fields, allowing farmers to carry out a variety of agricultural practices such as optimizing plant growth, fertilization, disease, and pest control.
- *Wheat Genetics and Variety Breeding*

  AI contributes to both plant genetics and various plant developmental processes. In particular, analyzing phenotypic and genetic data of the plant with big data analytics and *ML* techniques can contribute to the development of new wheat varieties with the desired characteristics. Such wheat varieties may be more resistant to "climate change" conditions, which are the subject of this section, and more efficient in utilizing water and nutrients.

Moreover, all these impacts are sustainability, biotechnology, harvesting robots, air and soil quality monitoring tools, digital trading platforms, artificial intelligence AI-supported meteorological forecasts, savings and increased efficiency in energy use, reduction of carbon/water footprint, food security, food security monitoring, the farm-totable production process, interactive education and knowledge sharing, precision agriculture practices, irrigation systems, climate control systems, fertilization, integrated pest control, crop rotation, GPS-supported tractors, local seed protection, research funds, and agricultural insurance.

2.2.3 Overview of Some Important Limitations

- *AI* is lacking solutions that can be transferred to simple agricultural applications.
- Farmers and growers lack the digital skills to individually implement *AI* solutions and experience difficulties in integrating them.
- Since the techniques do not work outside their programs, accurate prediction machines need a large amount of data to be trained and to make accurate predictions, and they cannot be fully trusted. Although spatial data, especially regarding agricultural areas, can be easily obtained, there are difficulties in providing temporal data.
- Since plant data can only be obtained once a year during the growing season (i.e., vegetation period), it takes a long time to establish the data infrastructure and, therefore, to create the ML model.

2.2.4 The Future of Technologies in Agriculture and Wheat Agriculture

From controlling diseases and pests in crops to weather forecasting, it produces rational, logical, quick, and easily applicable solutions for almost every place and stage, from determining soil moisture and nutrient distribution to assistance in farm work. In the coming years and quite possibly with the contribution of other sciences, perhaps with the contribution of *AI* itself, safe and beneficial advances will be made that are directly related to agriculture and that the parties can successfully implement at every stage of production. Of these, AI companies are working in farm environments, military, pharmaceutical delivery, cargo and transportation, explosive detection, caves, historical excavations, cave-ins, etc.

They are developing robots that can perform tasks in any desired place and manner/way. Robots designed to harvest crops faster and better than humans can detect the amount of calcite (calcium carbonate crystallizing as the chemical formula CaCO3) in plants. It is a primary building block of almost all living things. With the contributions of all releted techniques, robots, drones, etc. have already entered an indispensable and essential learning process for sustainable and successful wheat production Precautions can be taken in this regard, as AI algorithms will use relevant (satellite) photographs and historical data to determine whether an insect is damaging the plant and which insect is effective in this regard. Again, by sending notification(s) to the mobile phones of its subscribers, it will ensure that the necessary precautions are taken and that the relevant pest(s) are controlled. At the same time, this will make significant contributions to producers in terms of time, effort, planning, and return. Future technical advances will help businesses interested in developing AI-based products and services, such as automated machine production with drones using educational data for agriculture (e.g., Farmlabs-Agro4, R&D products), to overcome some of the challenges associated with providing food to people.

2.3 Conclusion

The use of technology, which has significantly contributed to increasing productivity in agriculture, has gained a different dimension, especially with the advent of digital technologies. The fastest developing among these technologies has been agricultural early warning systems. Current research areas continued with the regular measurement of climate data and soil moisture levels, the use of cameras to monitor diseases and pests, regular monitoring of plant growth using thermal and spectral cameras, the use of multispectral and hyperspectral cameras, and later thermal cameras on drones (Fig. 2.15). Spectral libraries have begun to be created for each plant, especially with images taken by drones; today, studies on the creation of spectral libraries for plants grown in different countries around the world have gained momentum. By combining satellite images with these spectral libraries and images taken from drones, more reliable results have begun to be obtained, especially in estimating plant yields. Similarly, in addition to plant yield estimates, similar techniques have begun to be used to obtain results about the distribution of diseases and pests in plants, weed invasion, and water stress; and in this process, these studies started with "*AAN*" and continued with *ML* under the umbrella of *AI* has begun to be used intensively. AI in agricultural applications will always be used to improve the entire agricultural process.

References

Ackerly, D. D., Dudley, S. A., Sultan, E. S., Schmitt, J., Coleman, S. J., Linder, C. R., Sandquist, D. R., Geber, M. A., Evans, S. A., Dawson, E. T., & Lechowicz, M. J. (2000). The evolution of plant ecophysiological traits: Recent advances and future directions. *Bioscience, 50*, 979–995.

Anonymous. (2015a). *Earth' CO_2, Home page.* https://www.co2.earth/. Accessed 8 May 2015.

Anonymous. (2015b). *National Oceanic and Atmospheric Administration.* https://www.noaa.gov/. Accessed 1 Feb 2024.

Anonymous. (2015c). *C2ES, Center for Climate and Energy Solutions, International Emissions.* http://www.c2es.org/facts-figures/. Accessed 16 Feb 2024.

Anonymous. (2016). *Research Program on Climate, Agriculture and Food Security, 2016 Report.*

Anonymous. (2017). *The artificial use of the term "Intelligence".* https://becominghuman.ai/the-artificial-use-of-the-term-intelligence-c82878447dfd. Accessed 24 May 2024.

Anonymous. (2024a). T.C. Tarım ve Orman Bakanlığı, Tarım Reformu Genel Müdürlüğü, Tarımsal Çevre ve Doğal Kaynakları Koruma Daire Başkanlığı, "İklim Değişikliği ve Tarım Raporu". Pdf [Erişim Tarihi 30.01.2024]", 1–20, Ankara. [In Turkish].

Anonymous. (2024b). https://climate.nasa.gov/faq/19/what-is-the-greenhouse- effect/#:~:tExtremet=Greenhouse%20gases%20consist%20of%20carbon,that%20initially%20caused%20the%20warming. Accessed 12 June 2024.

Anonymous. (2024c). *Smarter-wheat monitoring with artificial intelligence.* https://dug.com/smarter-wheat-monitoring-with-artificial-intelligence/. Accessed 26 June 2024.

Arshad, M., Kächele, H., Krupnik, T. J., Amjath-Babu, T. S., Aravindakshan, S., Abbas, A., Mehmood, Y., & Müller, K. (2016). Climate variability, farmland value, and farmers perceptions of climate change: Implications for adaptation in rural Pakistan. *The International Journal of Sustainable Development World Ecology, 11*, 532–544.

Asseng, S., Faster, L., & Turner, C. N. (2011). The impact of temperature variability on wheat yields. *Global Change Biology, 17*(2), 997–1012.

Ather, T., Pandey, A., Khan, K. M., Hamurcu, M., Rustagni, A., Gezgin, S., Singhal, K. R., & Youssaf, H. (2023). Climate change triggering abiotic stresses and losses in wheat production and quality, chapter 26. In *Abiotic stresses in wheat* (pp. 413–425). Elsevier Inc. https://doi.org/10.1016/B978-0-323-95368-9.00003-5

Balasundram, K. S., Shamsiri, R. R., Sridhnara, S., & Rizan, N. (2023). The role of digital agriculture in mitigating climate change and ensuring food security: An overview. *Sustainability, 15*, 5325. https://doi.org/10.3390/su15065325. 1–23.

Ballester, C., Torres-Sánchez, J., García-Ferrer, A., et al. (2021). A review of the use of unmanned aerial vehicles and artificial intelligence in precision agriculture. *Remote Sensing, 13*(2), 348.

Bosso, B., Liu, L., & Ritche, T. S. (2016). A comprehensive review of the CERES-wheat, maize, and rice models performances. *Advances in Agronomy, 136*, 27–132., Chapter 20. https://doi.org/10.1016/bs.agron.2015.11.004

Cemek, B., Tasan, S., Canturk, A., Tasan, M., & Simsek, H. (2023). Machine learning techniques in estimation of eggplant crop evapotranspiration. *Applied Water Science, 13*(6), 136.

Chen, J., Zhou, J., Li, Q., Li, H., Xia, Y., Jackson, R., Sun, G., Zhou, G., Deakin, G., Jiang, D., & Zhou, J. (2023). CropQuant-Air: An AI-powred system to enable a phenotypic analysis of yield and performance-related traits using wheat canopy imagery collected by low-cost drones. *Frontiers in Plant Science, 14*, 1219983. https://doi.org/10.3389/fpls.2023.1219983

Cutforth, H. W., McGinn, S. M., McPhee, E., & Miller, P. R. (2007). Adaptation of pulse crops to the changing climate of the Northern Great Plains. *Agronomy Journal, 99*, 1684–1699.

De Maman, W. T. A., da Silva, G. A. J., Binelo, O. M., Scremin, B. D., Kraisig, R. A., Carvalho, R. I., Preira, M. L., Berlezi, D. J., & Argenta, V. C. (2019). Artificial intelligence simulating grain productivity during the wheat development considering biological and environmental indicators. *Journal of Agricultural Studies, 7*(3), 197–212., Macro Think Institute, ISSN 2166-0379. https://doi.org/10.5296/jas.v7i3.15153

Dellal, İ., Engürülü, B., Ulukan, H., Özveren, A. Ş. and Ünal, M. (2015). İklim değişikliğinin tarım sektörüne ekonomik yansımaları/Economical Reflection of Climate Change to Agricultural Sector, VII. Technical Congress of T.Z.M.O., (12–16) January 2015, Çankaya Municipality Contemporary Arts Center, pp. 62–80, Ankara, Türkiye [In Turkish].

Dhakhawa, G. B., & Campbell, L. C. (1998). Potential effects of differential day-night warming in global climate change on crop production. *Climate Change, 40*, 647–667.

Di Norcia, V. (2008). Global warming is man-made: Key points in the international panel on climate change 2007 report. In *Hard like water-ethics in business* (pp. 1–4). Oxford University Press.

Dias, S. A., & Lian, F. F. (2009). Evaluation of grain filling rate and duration in bread and durum wheat, under heat stress after anthesis. *Journal of Agronomy and Crop Science, 195*(2), 137–147.

Diego, N. L. P., Hernandez-Ocka, M. I., Reynolds, M., Sander, K., Milan, M. A., Robertson, D. R., Lopes, S. M., Xiong, W., Kropff, M., & Asseng, S. (2021). Climate impact and adaptation to heat and drought stress of regional and global wheat production. *Environmental Research Letters, 16*(2021), 054070. https://doi.org/10.1088/1748-9326/abd970

FAO (Food and Agriculture Organization of the United Nations). (2020). *The future of food and agriculture—Alternative pathways to 2050.*

Fernandes, I., Uzun, B., Pascoal, C., & Cassio, F. (2009). Responses of aquatic fungal communities on leaf litter to temperature-change events. *International Review of Hydrobiology, 94*, 410–418.

Frederick, L. S., Mäkelä, P. S. A., & Puhakainen, T. (2011). In J. Blanco (Ed.), *Adaptation of boreal field crop production to climate change, climate change research and technology for adaptation and mitigation*. InTech. https://doi.org/10.5772/25258. Available from: http://www.intechopen.com/books/climate-change-research-and-technology-for-adaptation-and-mitigation/adaptation-of-boreal-field-crop-production-to-climate-changeagriculture

Fuentes, A., Yoon, S., Kim, S., et al. (2017). A robust deep-learning-based detector for real-time tomato plant diseases and pests recognition. *Sensors, 17*(9), 2022.

Fuhrer, J. (2003). Agroecosystem responses to the combination of elevated CO_2, ozone, and global climate change. Agriculture, Ecosystems & Environment, 97, Pg. 20.

Fuhrer, J. (2009). *Climate change-risks and opportunities for agriculture, 1st Int. Congress on Global Climate Change and Agriculture.* May (28–30) Tekirdağ University Agriculture Faculty, Tekirdağ, Türkiye, pp: 81–87.

Gençtan, T., Öktem, A., Birsin, M. A., Bilgin, O., Ulukan, H., Balkan, A., Sürek, H. and Özseven, İ., (2015). Yurdumuzun tahıl üretim potansiyeli, sorunları ve çözüm önerileri/Cereal production potential of country, its problems and solution offers/VII. Technical Congress of T.Z.M.O., (12–16) January 2015, Çankaya Municipality Contemporary Arts Center, pp. 352–386, Ankara, Türkiye [In Turkish].

Guevara-Hernandez, F., & Gomez-Gil, J. (2011). A machine vision system for classification of wheat and barley grain kernels. *Spanish Journal of Agricultural Research, 9*(3), 672–680.

Hatfield, J. L., & Prueger, J. H. (2011). In S. S. Yadav, R. J. Redden, J. L. Hatfield, H. Lotze-Campen, & A. E. Hall (Eds.), *Agroecology: Implications for plant response to climate change* (pp. 27–43). Wiley-Blackwell.

Horton, D. E., Johnson, N. C., Singh, D., et al. (2017). Contribution of changes in atmospheric circulation patterns to extreme temperature trends. *Nature, 522*(7557), 465–469.

Hossain, M. S., Lu, Q., Arshad, M., Shahid, S., Fahad, S., & Akhter, J. (2019). Climate change and crop farming in Bangladesh: An analysis of economic impacts. *International Journal of Climate Change Strategies and Management, 7*, 424–440.

Incicas, R., Cotuna, O., & Sala, F. (2020). Mobile application for detecting some wheat pathogens using AI. *Research Journal of Agricultural Science, 52*(3), 56–65.

IPCC. (2001). *Climate change: Impacts, adaptation and vulnerability.* Contribution of Working Group II to the Third Assessment Report of the Intergovernmental Panel on Climate Change (pp. 398–400).

IPCC. (2007). New assessment methods and the characterization of future conditions. In *Climate change 2007: Impacts, adaptation, and vulnerability. Contribution of working group II to the fourth assessment report of the intergovernmental panel on climate change* (p. 976). Cambridge University Press.

IPCC. (2014). *Climate change 2014: Synthesis report.* Contribution of Working Groups I, II, and III to the fifth assessment report of the Intergovernmental Panel on Climate Change [Core Writing Team, R.K. Pachauri and L.A. Meyer (eds.)] (151 pp). IPCC.

IPCC. (2018). *Summary for policymakers of IPCC special report on global warming of 1.5°C approved by governments.* Cambridge University Press.

Jackson, R. D., & Huete, A. R. (1991). Interpreting vegetation indices. *Preventive Veterinary Medicine, 11*(3–4), 185–200.

Javid, M., Haleem, A., Khan, H. I., & Suman, R. (2023). Understanding the potential applications of artificial intelligence in the agricultural sector. *Advanced Agrochem, 2*, 15–30.

Jiang, G., Wang, X., Wang, Z., & Liu, H. (2016). Wheat row detection at the early growth stage based on Hough transform and vanishing point. *Computers and Electronics in Agriculture, 123*, 211–223.

Kamilaris, A., & Prenafeta-Boldú, F. X. (2018). Deep learning in agriculture: A survey. *Computers and Electronics in Agriculture, 147*, 70–90.

Karapınar, B., Özertan, G., Tanaka, T. An, N. and Turp, T. M. (2020). İklim Değişikliği Etkisinde Tarımsal Arzın Sürdürülebilirliği, TÜSİAD 2020 Yılı Tarım ve Gıda Raporu, 65 sf. [In Turkish].

Kaya, S. (2017). Konvansiyonel ve otomatik meteoroloji istasyon verilerinden hesaplanan referans evapotranspirasyon değerlerinin kıyaslanması. *Turkish Doğa ve Fen Derg., 6*(2), 83–87. [In Turkish].

Kaya, U., Yılmaz, A. and Courville, A. (2019). *Deep learning*, Available from www.deeplearningbook.org/. Accessed 15 Feb 2024.

Koç, M., Barutçular, C., & Kapur, B. (2009). Buğday, Küresel Isınma ile Baş Edebilecek Mi. *Türk Tarım Dergisi, 171*, 64–71. [In Turkish].

Krupa, S. V. (1997). Global climate change: Processes and products overview. *Environmental Monitoring and Assessment, 46*, 73–88.

Kumar, S., & Tewari, M. L. (2013). *Effects of climate change on plant phenology* (Vol. 9, pp. 26–28). Oaks. ISSN:0975–5918.

Leisner, C. P. (2020). Review: Climate change impacts on food security on perennial cropping systems and nutritional value. *Plant Science, 7*, 110412.

Leung, H. (2008). Stressed genomics-bringing relief to rice fields. *Current Opinion in Plant Biology, 11*, 201–208.

Li, Y., Liu, H., Wei, J., Ma, X., Zhang, G., & Xi, L. (2023). Research on winter wheat growth stages recognition based on mobile edge computing. *Agriculture, 13*(3), 2–16. https://doi.org/10.3390/agriculture13030534

Liakos, K. G., Busato, P., Moshou, D., Pearson, S., & Bochtis, D. (2018). Machine earning in agriculture: A review. *Sensors (Basel), 18*(2674), 1–29.

Liu, B., Asseng, S., Liu, L., Cao, W., & Zhu, Y. (2016). Testing the responses of four wheat crop models to heat the stress of anthesis and grain filling. *Global Change Biology, 22*(5), 1890–1903.

Majumdar, D., Ghosh, A., Kole, D.K., Chakraborty, A., and Majumder, D. D. (2015). *Application of fuzzy C-means clustering method to classify wheat leaf images based on the presence of rust disease.* In: Paper presented at the Proceedings of the 3rd International Conference on Frontiers of Intelligent Computing: Theory and Applications (FICTA) (2014), pp. 277–284.

Mei, H., Chengjun, J. I., Wenyun, Z., & Jinsheng, H. E. (2007). Interactive effects of elevated CO_2 and temperature on the anatomical characteristics of leaves in eleven species. *Acta Ecologica Sinica, 26*, 326–333.

Mengü, G. P., Şensoy, S., & Akkuzu, E. (2008). *Effects of global climate change on agriculture and water resources* (pp. 1–10). Balwois. (27–31) May 2008.

Mohanty, S. P., Hughes, D. P., & Salathé, M. (2016). Using deep learning for image-based plant disease detection. *Frontiers in Plant Science, 7*, 1419.
Mondal, D., & Kole, D. K. (2016). A time-efficient leaf rust disease detection technique of wheat leaf images using Pearson correlation coefficient and rough fuzzy C-means. In *Information systems design and intelligent applications* (pp. 609–618). Springer.
Moshou, D., Bravo, C., West, J., Wahlen, S., McCartney, A., & Ramon, H. (2004). Automatic detection of "yellow rust" in wheat using reflectance measurements and neural networks. *Computers and Electronics in Agriculture, 44*, 173–188.
Moshou, D., Bravo, C., Oberti, R., West, J., Bodria, L., McCartney, A., & Ramon, H. (2005). Plant disease detection based on data fusion of hyperspectral and multi-spectral fluorescence imaging using Kohonenmaps. *Real-Time Imaging, 11*, 75–83.
Moshou, D., Pantazi, X.-E., Kateris, D., & Gravalos, I. (2014). Water stress detection based on optical multi-sensor fusion with a least squares support vector machine classifier. *Biosystems Engineering, 117*, 15–22.
Nabiyev, V. V. (2016). Yapay Zeka [Artificial Intelligence]. Maximum Yayınları, Türkiye, Sf. 25 [In Turkish]
Nemecek, T., von Richthofen, J. S., Dubois, G., Casta, P., Charles, R., & Pahl, H. (2008). Environmental impacts of introducing grain legumes into European crop rotations. *European Journal of Agronomy, 28*, 380–393.
Olesen, J. E., & Bindi, M. (2004). Agricultural impacts and adaptations to climate change in Europe. *Farm Policy Journal, 1*, 36–46.
Olgun, M., Onarcan, A. O., Ozkan, K., Işik, Ş., Sezer, O., Ozgişi, K., & Koyuncu, O. (2016). Wheat grain classification by using dense SIFT features with an SVM classifier. *Computers and Electronics in Agriculture, 122*, 185–190.
Özdogan, M. (2011). Modeling the impacts of climate change on wheat yields in Northwestern Türkiye. *Agriculture, Ecosystems and Environment, 141*(1–2), 1–12.
Özgen, M. A., Adak, S. M., Ulukan, H., Benlioğlu, B., Peşkircioğlu, M., Koyuncu, N., Yıldız, A., and Tuna, E. D. (2015). İklim Değişikliği ve Bitkisel Gen kaynakları/Climate Change and Plant Genetic Resources, VII. Technical Congress of TZMO, (12–16) January 2015, Çankaya Municipality Contemporary Arts Center, pp. 184–211, Ankara, Türkiye [In Turkish].
Özkan, B., & Akcaoz, H. (2002). Impacts of climate factors on yields for selected crops in southern Türkiye. *Mitigation and Adaptation Strategies for Global Change, 7*(4), 367–380.
Pantazi, X.-E., Moshou, D., Alexandridis, T. K., Whetton, R. L., & Mouazen, A. M. (2016). Wheat yield prediction using machine learning and advanced sensing techniques. *Computers and Electronics in Agriculture, 121*, 57–65.
Pantazi, X. E., Tamouridou, A. A., Alexandridis, T. K., Lagopodi, A. L., Kontouris, G., & Moshou, D. (2017a). Detection of *Silybum marianum* infection with *Microbotyum silybum* using VNIR field spectroscopy. *Computers and Electronics in Agriculture, 137*, 130–137.
Pantazi, X. E., Tamouridou, A. A., Alexandridis, T. K., Lagopodi, A. L., Kashefi, J., & Moshou, D. (2017b). Evaluation of hierarchical self-organizing maps for weed mapping using UAS multispectral imagery. *Computers and Electronics in Agriculture, 139*, 224–230.
Peker, F., Bayhan, H., & Akkoyunlu, A. (2021). Global gerçek evapotranspirasyon (ETa) haritalarından arazi kullanım sınıflarına ait (ETa) kayıplarının tahmini. *European Journal of Engineering and Applied Sciences, 4*(1), 18–26. [In Turkish].
Poggi, G. M., Aloisi, I., Corneti, S., Esposito, E., Naldi, M., Fiori, J., Piana, S., & Ventura, F. (2022). Climate change effects on bread wheat phenology and grain quality: A case study in the north of Italy. *Frontiers in Plant Science, 13*, 936991. https://doi.org/10.3389/fpls.2022.936991
Porter, R. J., & Gawith, M. (1999). Temperatures and the growth and development of wheat: A review. *European Journal of Agronomy, 10*(1), 23–36.
Qin, Z., Zhang, P., Cheng, X., et al. (2020). Digital agriculture: The untapped potential of emerging technologies. *Trends in Plant Science, 25*(11), 1093–1095.

Reddy, A. R., Rasineni, G. K., & Raghavendra, A. S. (2010). The impact of global elevated CO_2 concentration on photosynthesis and plant productivity. *Current Science, 99*(1–10), 46–57.

Reed, M. (2018). Digital transformation in agriculture: Industry 4.0 and digital farming. *Journal of Agribusiness in Developing and Emerging Economies, 8*(1), 98–114.

Rezaei, E. E., Siebert, S., Hüging, H., & Ewert, F. (2018). Climate change's effect on wheat phenology depends on cultivar change. *Science Reports, 8*, 4891. https://doi.org/10.1038/s41598-018-23101-2. 1–10.

Richards, R. A., Condon, A. G., & Rebetzke, G. J. (2015). Chapter 88–100. In M. P. Reynolds, J. I. Ortiz-Monasterio, & A. McNab (Eds.), *Application of physiology in wheat breeding* (p. 219). CIMMYT.

Rogers, H. H., Runion, G. B., & Krupa, S. V. (1994). Plant responses to atmospheric enrichment with emphasis on roots and the rhizosphere. *Environmental Pollution, 83*, 155–189.

Romanova, A. K. (2005). Physiological and biochemical aspects and molecular mechanisms of plant adaptation to the elevated concentration of atmospheric CO_2. *Russian Journal of Plant Physiology, 52*, 112–126.

Rosenzweigh, C., & Hillel, D. (1995). Potential impact of climate change on agriculture and food supply. *Consequences, 1*, 24–32.

Ruby, U. A., Chandran, C. G., Chaitan, N., Jain, S. T. J., & Patıl, R. (2024). A wheat leaf diseases classification using a modified RasNet 50 convolutional neural network model. *Multimedia Tools and Applications, 2024*, 1–19. https://doi.org/10.1007/s11042-023-18049-z

Schlenker, W., & Roberts, M. J. (2009). Nonlinear temperature effects indicate severe damage to U.S. crop yields under climate change. *Proceedings of the National Academy of Sciences, 106*, 15594–15598.

Semenov, M. A., Stratonovitch, P., Alghabari, F., & Gooding, M. J. (2014). Adapting wheat in Europe to climate change. *Journal of Cereal Science, 59*(3), 245–256.

Şensoy, S., Türkoğlu, N. and Çiçek, İ. (2014). Phenological effects of climate change in Türkiye. In *Climate change and climate dynamics international conference*, (8–10) September 2014, 83–87, Istanbul, Türkiye.

Shafr, U., Mumtaz, R., Qureshi, M. D. M., Mahmood, Z., Tanveer, K. S., Haq, U. I., & Zaidi, M. H. S. (2013). *Embedded AI for wheat yellow rust infection type classification* (Vol. 11, pp. 23726–23738). EEE Access. https://doi.org/10.1109/ACCESS.2023.3254430

Shanker, A. K., & Venkateswarlu, B. (2011). *Abiotic stress in plants-mechanisms & adaptations* (p. 140). In Tech.

Shen, H., Jiang, K., Sun, W., Xu, Y., & Ma, X. (2021). Irrigation decision method for the winter wheat growth period in supplementary irrigation based on a support vector, machine algorithm. *Computer and Electronics in Agriculture, 182*(1–9), 0168–1699. https://doi.org/10.1016/j.compag.2021.106032. Elsevier B. V.

Shi, Y., Huang, W., Luo, J., Huang, L., & Zhou, X. (2017). Detection and discrimination of pests and diseases in winter wheat based on spectral indices and kernel discriminant analysis. *Computers and Electronics in Agriculture, 141*, 171–180.

Shivalal, N., Mukti, R. P., Binod, P., Ayali, D., Harkala, P., & Radhakrishna, B. (2024). Drought stress effect, tolerance, and management in wheat: A review. *Cogent Food & Agriculture, 10*(1), 2296094. https://doi.org/10.1080/23311932.2023.2296094

Subeesh, A., & Mehta, C. R. (2021). Automation and digitization of agriculture using artificial intelligence and the Internet of Things. *Artificial Intelligence in Agriculture, 5*, 278–291.

Taranto, F., Nicolia, A., Pavan, S., De-Vita, P., & D'Agostino, N. (2018). Biotechnological and digital revolution for climate-smart plant breeding. *Agronomy, 8*, 277. https://doi.org/10.3390/agronomy8120277. 1–20.

Taşan, S., Cemek, B., Taşan, M., & Cantürk, A. (2022). Estimation of eggplant yield with machine learning methods using spectral vegetation indices. *Computers and Electronics in Agriculture, 202*, 107367.

Tausz, M., Tausz-Posch, S., Norton, R. M., Fitzgerald, G. J., Nicholas, M. E., & Seneweera, S. (2013). Understanding crop physiology to select breeding targets and improve crop

management under increasing atmospheric CO_2 concentrations. *Environmental and Experimental Botany, 88*, 71–80.
Teixeira, I. E., Fisher, G., Veltwizen, van, H., Walter, C., & Ewert, F. (2013). Global hot spots of heat stress on crops due to climate change. *Agricultural and Forest Meteorology, 170*, 206–215.
Thenkabail, P. S., Lyon, J. G., & Huete, A. (2000). Advances in hyperspectral remote sensing of vegetation and agricultural croplands. In *Remote sensing of agriculture and vegetation* (pp. 3–56). Springer.
Thomson, A. M., Brown, R. A., Rosenberg, N. J., Lazuralde, R. C., & Benson, V. (2005). Climate change impacts for the conterminous USA: An integrated assessment = part 3. Dryland production of grain and forage crops. *Climate Change, 69*, 43–65.
Tian, Y., Zhao, C., Lu, S. and Guo, X. (2012). SVM-based multiple classifier system for recognition of wheat leaf diseases. In: World Automation Congress (WAC), 2012 (pp. 189–193). IEEE.
Tubiello, F. N., & Ewert, F. (2002). Simulating the effects of elevated CO_2 on crops: Approaches and applications for climate change. *European Journal of Agronomy, 18*, 57–74.
Tubiello, F. J., Donatelli, M., Rosenzweig, C., & Stockle, C. O. (2000). Effects of climate change and elevated CO_2 on cropping systems: Model predictions at two Italian locations. *European Journal of Agronomy, 13*, 179–189.
Türkoğlu, N., Çiçek, İ., and Şensoy, S. (2014). Türkiye'de iklim değişikliğinin meyve ağaçları ve tarla bitkilerinin fenolojik dönemlerine etkileri/Climate changes and trends in phenology of fruit trees and field crop in Türkiye, TÜCAUM - VIII. Coğrafya Sempozyumu, (23–24) Ekim 2014, 60–71, Ankara, Türkiye [In Turkish].
Ulukan, H. (2008). Agronomic adaptation of some field crops: A general approach. *Journal of Agronomy and Crop Science, 194*, 169–179.
Ulukan, H. (2009). The evolution of cultivated plant species: Classical plant breeding versus genetic engineering. *Plant Systematics and Evolution, 280*, 133–142.
Ulukan, H. (2018). Climate change (CC) impacts on field crops: A general approach, 3rd Int. Sci., Symposium, 725B, New Horizons in Science, (5–8) September, Priştina, Kosovo, Ecological Life Sci., ISSN 130B, pp. 163–170, https://doi.org/10.12739/NWSA.2018.13.4.5A0105.
Ulukan, H. (2021). Chapter 9: Climate change and global warming effect(s) on wheat landraces: A general approach. In N. Zencirci, F. S. Balosh, E. Habyarima, & C. Gyuhwa (Eds.), Wheat landraces (pp. 168–191). ISBN 978–3–030–77378, ISBN 978–3–030–77377–3 (eBook), https://doi.org/10.1007/978-3-030-77388-5_9. Springer Nature, Switzerland AG, pp. 1–267.
Uzhinskiy, A. (2023). Advanced technologies and artificial intelligence in agriculture. *Applied Mathematics, 3*(4), 799–813. https://doi.org/10.3390/appliedmath3040043
Uzmen, R. (2007). Küresel Isınma ve İklim Değişikliği, İnsanlığın sonu mu? İş Bankası Kültür Yayınları. [Global Warming and Climate Change, Is it a catastrophe for humanity?, İş Bank Culture Publ.]. No 221, pg. 176, Ankara, Türkiye [in Turkish].
Valizadeh, J., Ziaei, S. M., & Mazloumzadeh, S. M. (2014). Assessing climate change impacts on wheat production (a case study). *Journal of the Saudi Society of Agricultural Sciences, 13*, 107–115.
Vavilov, N. I. (1926). *Studies on the origin of cultivated plants* (p. 248). Bulletin of Applied Botany and Plant Breeding.
Volkmar, K. M., Hu, Y., & Steppuhn, H. (1998). Physiological responses of plants to salinity: A review. *Canadian Journal of Plant Science, 78*, 19–22.
Wang, E., & Engel, T. (1998). Simulation of phenological development of wheat crops. *Agricultural Systems, 58*(1), 1–24.
Wu, C., Niu, Z., Tang, Q., Zhang, X., Zhou, Q., Huang, W., et al. (2019). Use of remote sensing for crop management. *Advances in Agronomy, 157*, 123–158.
Yavaş, İ., & Ünay, A. (2010). Ege Bölgesi Ekmeklik Buğdaylarında Kurağa Dayanıklılık Özelliklerinin Saptanması. Doktora Tezi [Determination of drought resistance characteristics of Aegean region of bread wheat, Ph. D Thesis]. University of Adnan Menderes (ADÜ), Pp 103, Aydın, Türkiye [In Turkish].

Yoshio, I. (2020). Satellite-and-drone-based remote sensing of crops and soils for smart farming—A review. *Soil Science and Plant Nutrition, 66*(6), 798–810. https://doi.org/10.1080/00380768.2020.1738899

Zahra, N., Hafeez, B. M., Wahid, A., Al-Masruri, M. H., Ullah, A., Siddique, M. H. K., & Farooq, M. (2023). Impact of climate change on wheat grain composition and quality. *Journal of the Science of Food and Agriculture, 103*(6), 2745–2751.

Zavarzin, A. G. (2001). The role of biota in global climate change. *Russian Journal of Plant Physiology, 48*(2), 265–272.

Zhai, F., & Zhuang, J. (2009). Agricultural impact of climate change: a general equilibrium analysis with special reference to Southeast Asia, ADBI Working Paper 131, Tokyo: Asian Development Bank Institute. pp. 1–21. Available: http://www.adbi.org/workingpaper/2009/02/23/2887agricultural.impact.climate.change/. Accessed 16 June 2024.

Zhang, N., & Chaisattapagon, C. (1995). Effective criteria for weed identification in wheat fields using machine vision. *Transactions of ASAE, 38*(3), 965–974.

Zhang, S., Wu, X., Yu, Z., & Zhang, L. (2017). Leaf image-based cucumber disease recognition using sparse representation classification. *Computers and Electronics in Agriculture, 134*, 135–141.

Zhang, H., Tang, Y., Chandio, A. A., Sargani, R. G., & Twumani, A. M. (2022). Measuring the effects of climate change on wheat production evidence from Northern China. *International Journal of Environmental Research and Public Health, 19*(19), 12341. https://doi.org/10.3390/ijerph191912341

Zhao, J., Liu, D., & Huang, R. A. (2023). Review of climate-smart agriculture: recent advancements, challenges, and future directions. *Sustainability, 15*, 3404. https://doi.org/10.3390/su15043404

Chapter 3
Geographic Information Systems, Remote Sensing, Agroecosystems, Global Databases, and Agricultural Planning

Hakan Mete Doğan

Abstract In this chapter, geographic information system (GIS) and remote sensing (RS) tools, which are gaining importance today, are briefly mentioned. The characteristics of agroecosystems are summarized, and the importance of topographic, climatic, soil, and land use/cover databases in modeling studies to be conducted for these systems is emphasized. The importance of freely downloadable, reliable, accessible, and continuous databases is emphasized, and some important global or international databases from which the mentioned data can be obtained are summarized. In this context, the Global Digital Elevation Model (ASTGTM) version 3 topographic databases provided by EARTDATA of the National Aeronautics and Space Administration (NASA), soil databases provided by the International Soil Reference and Information Center (ISRIC), climate databases provided by the High-Resolution Climatologies for World Land Surface Areas (CHELSA), and Coordination of Environmental Information (CORINE) databases provided by the European Environment Agency (EEA) are introduced. The roles of GIS and RS technologies and methods and the use of these databases in agricultural planning are discussed.

Keywords Agricultural planning · Agroecosystems · GIS · Global databases · Modeling · Remote sensing

Abbreviations

ASTER	Advanced Spaceborne Thermal Emission and Reflection Radiometer
ASTGTM	Global Digital Elevation Model Version 3
CEC	Cation Exchange Capacity
CHELSA	High-Resolution Climatologies for World Land Surface Areas

H. M. Doğan (✉)
Faculty of Arts and Sciences, Geography Department, Tokat Gaziosmanpaşa University, Tokat, Türkiye

N. Zencirci et al. (eds.), *Empowering Wheat Cultivation with GIS, Digital Approaches and Artificial Intelligence*, https://doi.org/10.1007/978-3-031-99954-3_3

CORINE	Coordination of Information on the Environment
DEM	Digital Elevation Model
EEA	European Environment Agency
ESRI	Environmental Systems Research Institute
GFDL	Geophysical Fluid Dynamics Laboratory
GIS	Geographic Information System
IPSL	Institut Pierre Simon Laplace
ISRIC	International Soil Reference and Information Center
MPI	Max Planck Institute
MRI	Meteorological Research Institute
NASA	National Aeronautics and Space Administration
RS	Remote Sensing
WDS	World Data System
WSI	World Soil Information
WSL	Swiss Federal Institute for Forest, Snow and Landscape Research

3.1 Introduction

Recently, the importance of geography in problem-solving and management stage has been recognized by many people because every decision in our daily life is affected, limited, or controlled by geographic facts. Increasing population coupled with decreasing natural resources causes irreversible damage on the Earth (Holdren & Ehrlich, 1974). Careful management and planning of natural resources came to the top of the agenda (Dandy et al., 2014). For these reasons, the use of geographic information systems (GIS), which date back to the 1960s (Goodchild, 2011), started to become widespread in the 1990s (Maguire, 1991).

GIS operates the intersection of many disciplines such as cartography, photogrammetry, RS, statistics, and computer science (Gold, 2006). It has been effectively used to solve complicated problems in many fields (Wei et al., 2011). Today, GIS, which has developed considerably, serves many disciplines and is a very useful tool for identifying and solving environmental problems. Consequently, it is possible to see many GIS users among market researchers, land and natural resource managers, planners, tax officials, private enterprises, and public service personnel. RS is an important data source for GIS because it contains broad areas and supplies very detailed information in a short time (Champbell and Wynne, 2011; Jensen, 2015). GIS contains efficient mechanisms that can utilize the satellite images (Liu & Mason, 2016). With the rapid development of GIS technology and related units, integration between vector-based data and grid data from satellite images or aerial photographs has become possible (Hild & Fritsch, 1998). The query capability based on the relational database structure and complex modeling process, combined with the high cartographic quality of modern vector GIS, creates the most effective tool for spatial research and applications (Allen & Terry, 2005).

One of the most important areas of use of GIS and RS is agriculture. Since modeling agroecosystems is very complex, reliable, systematically updated, continuous (regularly archived), and easily accessible data is the key to obtaining the necessary benefits from GIS and RS techniques. For successful agricultural modeling and planning studies, topographic, soil, and climate databases are very important and indispensable. Today, there are organizations that provide such kinds of databases on a global scale (CHELSA, 2025; EARTHDATA, 2025; EEA, 2024; ISRIC, 2025a). This chapter aims to inform the audience to benefit from global databases in agricultural planning.

3.2 Geographic Information Systems and Remote Sensing

Today, information is defined as the results obtained by correlating the data resulting from an observation or process with each other. Information Systems are generally systems that store data in predetermined formats to be used with instant methods in order to obtain information. GIS are systems of hardware, software, personnel, and methods designed to solve complex planning and management problems covering the collection, management, processing, analysis, modeling, and visual presentation of geographically located data (DeMers, 2008). The concept of organization has recently been added to this definition. As can be understood from these definitions, databases do not only constitute information on their own. The processing of these databases and their transformation into results produces information. The main theme of GIS, which has important functions in the creation of databases, is to solve complex planning and management problems. GIS technology enables the creation of databases for the purpose of inquiry for the benefit of humanity through mapping, the production of information using the power of spatial analysis, and the classification of this information in accordance with the objectives. GIS describes objects and events, predicts results, and stands out in strategic planning. Thanks to these features, GIS has surpassed other information systems and its applications are rapidly spreading worldwide (DeMers, 2008; ESRI, 2004). The use of GIS in many areas such as urban planning, natural resource management, economic development, education, agriculture, and forestry has increased rapidly in recent years. It is a powerful tool that performs spatial and environmental analysis that assists planning processes. GIS applications are tools that allow users to create interactive queries, analyze spatial information by organizing map data, and present activity results (Bolstad, 2019). Determining environmental boundaries in sustainable land use is an important step in determining the land suitability potential (Malczewski, 2004).

RS is a technique for obtaining and evaluating information about natural and artificial objects on Earth without coming into physical contact with the objects, through measurement instruments placed on platforms that move in the atmosphere or space at a certain distance from the Earth (Campbell & Wynne, 2011). In other words, it can be defined as the science and art of obtaining information about objects through measurements made from a certain distance without coming into physical contact with the objects. RS can also be defined as tele-detection or earth observation in short. RS is the entirety of methods, techniques, and tools used in the observation of the Earth's surface from long distances, whether stationary or mobile, by creating images (Jensen, 2015). One of the most important methods used in the collection of geographical data on the Earth's surface is the RS technique. This technique is widely used to obtain data directly in the planning of very large areas. In today's UA technology, areal data is recorded digitally, and information can be produced from images with the help of image interpretation and analysis processing elements (Lillesand et al., 2015). For example, many natural resource maps are made utilizing UA today (Chuvieco, 2020). Satellite images are used as the basic data in many forest, geology, topographic, climatic, land use, and soil maps.

3.3 Agroecosystems

The term agroecosystem is a shortened form of the term agricultural ecosystem. According to Odum and Barrett (2004), agroecosystems differ from natural or seminatural ecosystems (e.g., lakes and forests) that are completely dependent on solar energy in three fundamental ways. First, in addition to solar energy input (as a supplement), auxiliary energy sources under human control are also used in agroecosystems. These auxiliary energies include human and animal power, fertilizers, pesticides, irrigation, and fuel-powered agricultural machinery. Second, the diversity of living species and products in agroecosystems has been significantly reduced by humans. The aim here is to maximize the yield of certain products. Third, the determination of the plants and animals grown intensively in agroecosystems is based on artificial selection rather than natural selection. In other words, the main purpose of agroecosystems is to ensure that solar energy and other auxiliary energies are converted into food or other marketable products to the extent possible.

In order to achieve this goal, Odum and Barrett (2004) reported two methods that are used in the operation of agroecosystems. (1) The establishment of order and continuity in natural ecosystems is achieved through solar energy. In agroecosystems, auxiliary (support) energy is used; thus, a greater proportion of solar energy is directly converted into food. (2) Plants and animals from which food is obtained are subjected to genetic selection in a way that will give the best product under special care and in an energy-supported environment. As a result of these, productivity is increased; however, as in all intensive and specialized agricultural practices, in addition to the benefits provided by such practices, there are also many harms. Negative effects such as soil erosion, water and environmental pollution caused by pesticides and fertilizers, high cost of fuels, decrease in biodiversity, loss of resistance of genetically improved living species (becoming more sensitive to weather conditions and pests) are the disadvantages of monoculture and specialized agriculture (Odum & Barrett, 2004).

In a highly simplified form, Odum and Barrett (2004) defined agroecosystems by three broad groups. These are (1) preindustrial agriculture, (2) industrial agriculture

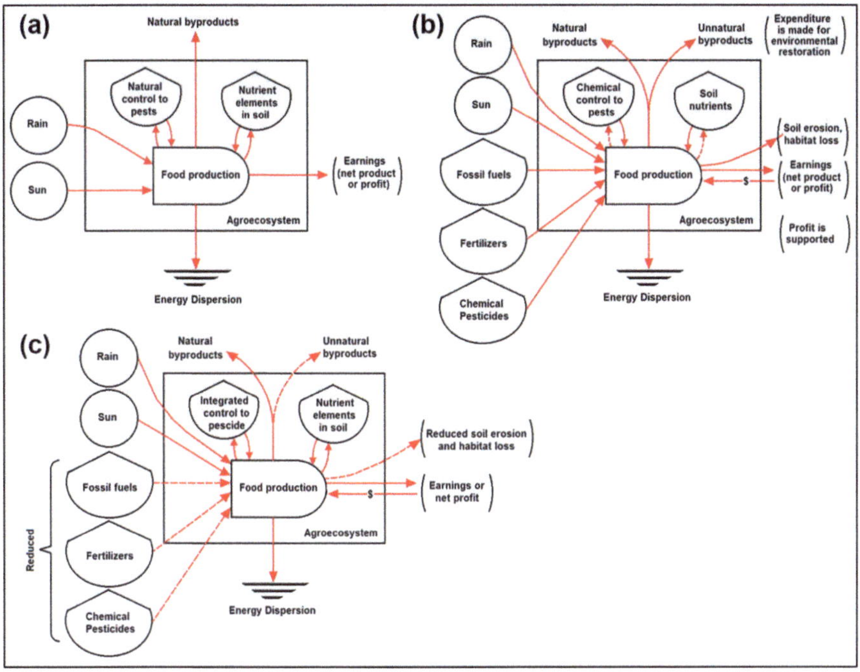

Fig. 3.1 (**a**) Preindustrialization agroecosystem diagram: it is a natural, unsubsidized (without subsidy support), solar energy-based system (Barrett et al., 1990), (**b**) industrial agroecosystem diagram: it is a system that uses human-supported energy as well as solar energy (Barrett et al., 1990), and (**c**) alternative agriculture diagram: It is modified versions of industrial agroecosystem in order to reduce some expenses (Odum & Barrett, 2004)

(today's traditional agriculture), and (3) alternative agriculture. Preindustrial agriculture is a self-sufficient, labor-intensive form of agriculture (auxiliary energy support is provided by animal and human labor). Production is not high and only enough food is produced for the farmer and his family in general. Enough product can be obtained to sell or barter in markets; however, not enough production is produced for export. The preindustrialization agroecosystem diagram developed by Barrett et al. (1990) is shown in Fig. 3.1a. In contrast, industrial agriculture (conventional agriculture today) is a heavily mechanized, fuel-supported form of agriculture (auxiliary energy support is provided by machines and chemicals). Product yields can generally increase in parallel with increased support. Production is greater than local needs and is used for export and commercial purposes. The industrial agriculture agroecosystem diagram developed by Barrett et al. (1990) is shown in Fig. 3.1b.

Finally, alternative agriculture is a low-input sustainable agriculture (National Research Council, 1989). According to Odum and Barrett (2004) the target of alternative agriculture is to reduce soil erosion and increase biodiversity. They stated that it is important to reduce supporting inputs such as fossil fuels, pesticides and fertilizers while maintaining crop yields and income (Fig. 3.1c).

When agroecosystems summarized in three large groups are examined carefully (Fig. 3.1), it is seen that the basic inputs of the systems are: precipitation, solar energy, fossil fuel energy, fertilizer, and chemical pesticides, and the outputs are natural and unnatural by-products, profit or gain obtained from food production and energy distribution (loss). As a result, any modeling study related to agroecosystems should definitely take into account the data on these inputs and outputs. In this context, topographic data, climate data, and soil data naturally come to the fore in agroecological modeling studies. It is undoubtedly important that these data should have certain standards such as reliability, continuity, accessibility.

3.4 Global Databases

Global databases have become increasingly important by providing very useful, continuous, and reliable spatial data and have increased the potential for effective use of GIS. Nowadays, four reliable major data providers, especially important for biology, ecology, agriculture, geography, and environmental sciences, have made topographic, soil, climate, and land use/cover data accessible, usable, and continuous. These global or international data providers can be summarized as the United States' National Aeronautics and Space Administration (NASA), International Soil Reference and Information Center (ISRIC), Climatologies at high resolution for the Earth's land surface areas (CHELSA), and European Environment Agency (EEA).

As a global data provider, NASA's EARTHDATA supplies freely downloadable topographic data that can be used and evaluated with tools such as GIS without restriction. One of the important EARTHDATA sources is the Terra Advanced Spaceborne Thermal Emission and Reflection Radiometer (ASTER) Global Digital Elevation Model version 3 (ASTGTM). This database is available to users from NASA's official website (EARTHDATA, 2025) and can be downloaded from there (Fig. 3.2). Basically, ASTGTM is a reliable digital elevation model (DEM) that covers land surfaces between 83°N and 83°S with a spatial resolution of 1 arc sec (approximately 30 m horizontal posting at the equator).

ISRIC is an independent foundation that supplies soil information at global, national, and subnational levels for application into sustainable management of soil and land (ISRIC, 2025a) via World Soil Information (WSI). Due to its longstanding reputation for providing reliable global information products about soils, ISRIC's WSI is an accredited organization as the member of International Science Council (ISC) World Data System (WDS) (ISRIC, 2025b). Basically, WSI provides valuable layers of soil with 250 m spatial resolution (Fig. 3.3) that are freely available on the web (https://soilgrids.org/). ISRIC provides soil databases prepared for five different depths of five physical (bulk density, coarse fragments, sand, silt, clay) and four chemical (Cation Exchange Capacity, Nitrogen, Soil Organic Carbon, and pH water) soil variables (Table 3.1).

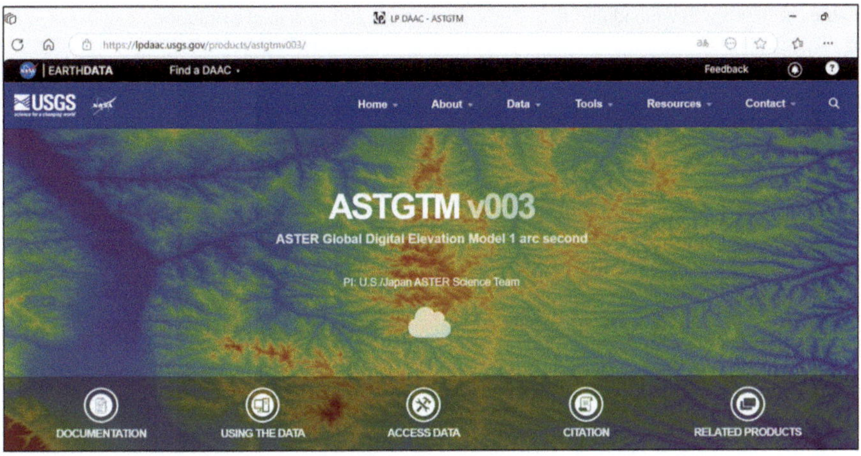

Fig. 3.2 ASTER global digital elevation model (ASTGTM version 3). (EARTHDATA, 2025)

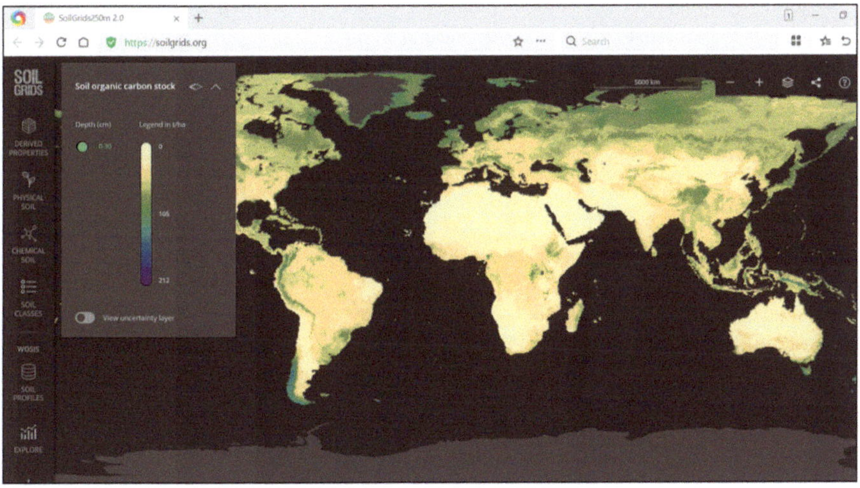

Fig. 3.3 Available ISRIC soil layers on the web (https://soilgrids.org/)

CHELSA is hosted by the Swiss Federal Forest, Snow and Landscape Research Institute WSL (WSL, 2025) and established to provide free access to high-resolution climate data including climate layers for various time periods and variables, ranging from the Last Glacial Maximum to the present and future (CHELSA, 2025). CHELSA has a mission to serve the international community as a reliable database provider and supplies high-resolution (30 arc sec, ~1 km) global climate data that is constantly updated and refined for research and application. Basically, the CHELSA supplies database originated from the five research centers around the World including GFDL-ESM 4, IPSL-CM6A-LR, MPI-ESM 1–2-HR, MRI-ESM 2–0, and

Table 3.1 Freely downloadable IRSIC soil layers with 250-m spatial resolution

Physical soil properties	*Unit*	*Available soil depth (cm)*
Bulk density	cg/cm^3	0–5, 5–15, 15–30, 30–60, 60–100
Coarse fragments	cm^3/dm^3	0–5, 5–15, 15–30, 30–60, 60–100
Sand	g/kg	0–5, 5–15, 15–30, 30–60, 60–100
Silt	g/kg	0–5, 5–15, 15–30, 30–60, 60–100
Clay	g/kg	0–5, 5–15, 15–30, 30–60, 60–100
Chemical soil properties	*Unit*	*Available soil depths (cm)*
Cation exchange capacity	mmol(c)/kg	0–5, 5–15, 15–30, 30–60, 60–100
Nitrogen	cg/kg	0–5, 5–15, 15–30, 30–60, 60–100
Soil organic carbon	dg/kg	0–5, 5–15, 15–30, 30–60, 60–100
pH water	pH*10	0–5, 5–15, 15–30, 30–60, 60–100

Table 3.2 Information about the CHELSA Centers and the database they produced (Karger & Zimermann, 2021)

Title	Model	Institution	Native resolution	Ensemble member	Priority
GFDL-ESM4	gfdl-esm4	National Oceanic and Atmospheric Administration, Geophysical Fluid Dynamics Laboratory, Princeton, NJ 08540, USA	288 × 180	r1i1p1f1	1
UKESM1–0-LL	ukesm1–0-ll	Met Office Hadley Centre, Fitzroy Road, Exeter, Devon, EX1 3 PB, UK	192 × 144	r1i1p1f2	2
MPI-ESM1-2-HR	mpi-esm1-2-hr	Max Planck Institute for Meteorology, Hamburg 20,146, Germany	384 × 192	r1i1p1f1	3
IPSL-CM6A-LR	ipsl-cm6a-lr	Institut Pierre Simon Laplace, Paris 75,252, France	144 × 143	r1i1p1f1	4
MRI-ESM2-0	mri-esm2-0	Meteorological Research Institute, Tsukuba, Ibaraki 305–0052, Japan	320 × 160	r1i1p1f1	5

UKESM1–0-LL (Karger & Zimermann, 2021). Information about these centers were given in Table 3.2. Although, the databases produced by these centers are close to each other, GFDL-ESM4 center is the first in order of priority. CHELSA version 2.1 layers (Figure 3.4) produced by the GFDL-ESM4 center can be preferably downloaded from the relevant website (https://chelsa-climate.org/downloads/).

All CHELSA variables are time-based and include daily, monthly, or annual averages, but not mixtures of these. Instantaneous parameters are not provided. Monthly time-averaged files are typically based on synoptic times. Monthly files represent averages of calendar months, taking leap years into account. For monthly averages, each file contains a single month. For annual means or accumulations, files contain a single year (Beck et al., 2020; Karger et al., 2017, 2020, 2021; Karger & Zimermann, 2021).

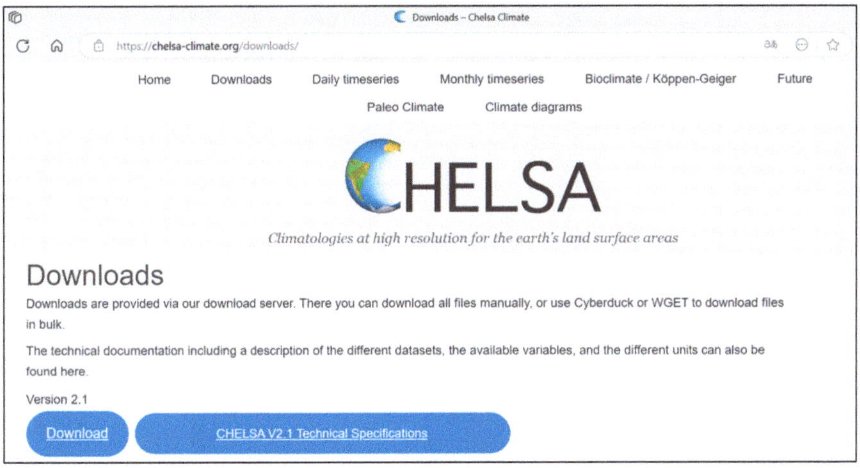

Fig. 3.4 Freely downloadable CHELSA version 2.1. climatic layers

The CHELSA database (version: 2.1) contains EUR-11 for European Countries and GLOBAL for the World. The GLOBAL database provides data under six main headings: (1) annual, (2) climatologies, (3) daily, (4) daily normal, (5) input, and (6) monthly. In climatologies, a file contains averages for a specific period, for example, 1981–2010 (Beck et al., 2020; Karger et al., 2017, 2020, 2021; Karger & Zimermann, 2021). Databases are available for four time periods (1981–2010, 2011–2040, 2041–2070, 2071–2100) under Climatologies headings in CHELSA, and the climate parameters in Table 3.3 can be downloaded. For future periods (2011–2040, 2041–2070, 2071–2100), climate maps were produced according to scenarios from the most optimistic to pessimistic.

Because the downloadable raster database has raw data characteristics, it needs to be processed by using the transformation equations supplied by CHELSA documentation (Table 3.3) after download. For example, to convert mean annual air temperature (bio1) values to degrees centigrade, it is necessary to multiply the grid values by the scale value and add the offset value to it (Table 3.3). This conversion can be easily performed with a GIS software and spatial analysis map calculation process. Accordingly, the downloaded database can be transformed to real units of climatic variables by utilizing the supplied equations (scale and offset values) in Table 3.3. More information on how to benefit from CHELSA databases can be found in the resources provided above.

The Coordination of Information on the Environment (CORINE) Land Cover Project of the European Environment Agency (EEA) was initiated in 1985 in order to collect the same basic data and create a standard database for the purposes of (1) determining environmental changes in the land, (2) rationally managing natural resources, and (3) determining environmental policies in all member countries of the EEA (EEA, 2024). In line with the criteria and classification determined by the EEA, data sets for the years 1990, 2000, 2006, 2012, 2018, and 2000 including Turkey

Table 3.3 Climatic layers and their transformation equations (scales and offset values) supplied by CHELSA (2025)

Shortname	Longname	Unit	Scale	Offset	Explanation
bio1	Mean annual air temperature	°C	0.1	−273.15	Mean annual daily mean air temperatures averaged over 1 year
bio8	Mean daily mean air temperatures	°C	0.1	−273.15	The wettest quarter of the year is determined (to the nearest month)
bio9	Mean daily mean air temperatures of the driest quarter	°C	0.1	−273.15	The driest quarter of the year is determined (to the nearest month)
bio10	Mean daily mean air temperatures of the warmest quarter	°C	0.1	−273.15	The warmest quarter of the year is determined (to the nearest month)
bio11	Mean daily mean air temperatures of the coldest quarter	°C	0.1	−273.15	The coldest quarter of the year is determined (to the nearest month)
bio12	Annual precipitation amount	kg m^{-2}	0.1	0	Accumulated precipitation amount over 1 year
bio13	Precipitation amount of the wettest month	kg m^{-2}	0.1	0	The precipitation of the wettest month.
bio14	Precipitation amount of the driest month	kg m^{-2}	0.1	0	The precipitation of the driest month.
bio16	Mean monthly precipitation amount of the wettest quarter	kg m^{-2}	0.1	0	The wettest quarter of the year is determined (to the nearest month)
bio17	Mean monthly precipitation amount of the driest quarter	kg m^{-2}	0.1	0	The driest quarter of the year is determined (to the nearest month)
bio18	Mean monthly precipitation amount of the warmest quarter	kg m^{-2}	0.1	0	The warmest quarter of the year is determined (to the nearest month)
bio19	Mean monthly precipitation amount of the coldest quarter	kg m^{-2}	0.1	0	The coldest quarter of the year is determined (to the nearest month)
pr	Precipitation amount	kg m^{-2}	0.1	0	"Amount" means mass per unit area. "Precipitation" in the earth's atmosphere means precipitation of water in all phases.
Tasmax	Mean daily maximum 2 m air temperature	°C	0.1	−273.15	Daily maximum air temperatures at 2 metres from hourly ERA5 data
Tas	Mean daily air temperature	°C	0.1	−273.15	Daily mean air temperatures at 2 metres from hourly ERA5 data
Tasmin	Mean daily minimum air temperature	°C	0.1	−273.15	Daily minimum air temperatures at 2 metres from hourly ERA5 data

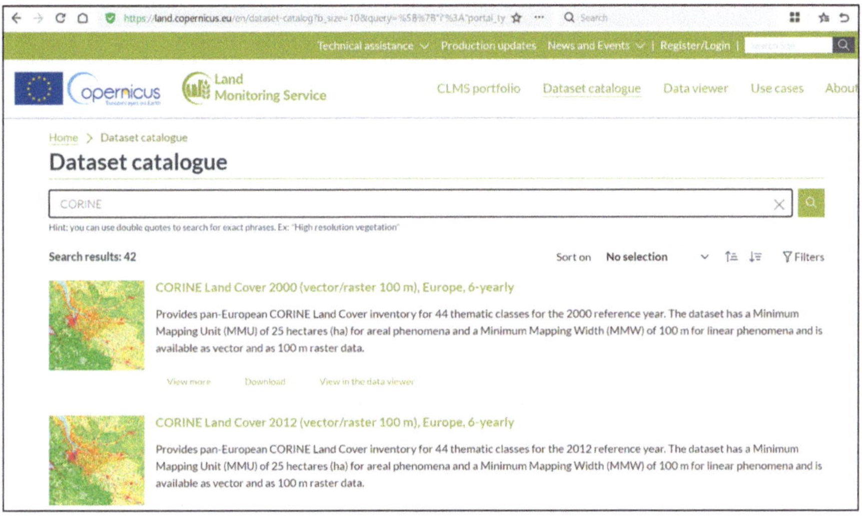

Fig. 3.5 CORINE databases in both raster and vector bases

have been created to date. CORINE databases in both raster and vector bases (Fig. 3.5) are freely available to users from the official website Copernicus of the European Union (https://land.copernicus.eu/en/dataset-catalog). The CORINE land cover-use classification system consists of three levels, including Level-I, Level-II, and Level-III. These level classes are as shown in Table 3.4.

As an example of the mentioned global databases, ASTGTM digital elevation (m), ISRIC 0-5 cm soil depth cation exchange capacity (mmol(c)/kg) and CHELSA (2011-2040) January average precipitation (mm) raster maps were downloaded for Tuz Gölü and its surroundings (32-34 East longitudes and 38-40 North latitudes), one of the most well-known regions of Turkey, are given in Figure 3.6a-c. CORINE Level-3 land use/cover vector and raster maps of the same area are shown in Fig. 3.6d, e.

3.5 Agricultural Planning

One of the most important application areas of GIS and RS is agriculture. Today, GIS and RS techniques are successfully used in monitoring and tracking agricultural lands at regular intervals throughout the season, product yield estimates, detection of problem areas and agricultural planning. Using RS techniques in a GIS environment allows planning of large areas and decision makers to implement their decisions effectively in the shortest time possible (Nguyen et al., 2020).

The suitability of the areas to be cultivated, especially for the products planned to be cultivated, should be taken into consideration. While making calculations such as

Table 3.4 The CORINE land cover-use classification system (EEA, 2024)

Clc Code	Level-1	Level-2	Level-3
111	Artificial surfaces	Urban fabric	Continuous urban fabric
112	Artificial surfaces	Urban fabric	Discontinuous urban fabric
121	Artificial surfaces	Industrial, commercial and transport units	Industrial or commercial units
122	Artificial surfaces	Industrial, commercial and transport units	Road and rail networks and associated land
123	Artificial surfaces	Industrial, commercial and transport units	Port areas
124	Artificial surfaces	Industrial, commercial and transport units	Airports
131	Artificial surfaces	Mine, dump and construction sites	Mineral extraction sites
132	Artificial surfaces	Mine, dump and construction sites	Dump sites
133	Artificial surfaces	Mine, dump and construction sites	Construction sites
141	Artificial surfaces	Artificial, non-agricultural vegetated areas	Green urban areas
142	Artificial surfaces	Artificial, non-agricultural vegetated areas	Sport and leisure facilities
211	Agricultural areas	Arable land	Non-irrigated arable land
212	Agricultural areas	Arable land	Permanently irrigated land
213	Agricultural areas	Arable land	Rice fields
221	Agricultural areas	Permanent crops	Vineyards
222	Agricultural areas	Permanent crops	Fruit trees and berry plantations
223	Agricultural areas	Permanent crops	Olive groves
231	Agricultural areas	Pastures	Pastures
241	Agricultural areas	Heterogeneous agricultural areas	Annual crops associated with permanent crops
242	Agricultural areas	Heterogeneous agricultural areas	Complex cultivation patterns
243	Agricultural areas	Heterogeneous agricultural areas	Principally agriculture with natural vegetation
244	Agricultural areas	Heterogeneous agricultural areas	Agro-forestry areas
311	Forest and semi natural areas	Forests	Broad-leaved forest
312	Forest and semi natural areas	Forests	Coniferous forest
313	Forest and semi natural areas	Forests	Mixed forest
321	Forest and semi natural areas	Scrub and/or herbaceous vegetation associations	Natural grasslands

(continued)

Table 3.4 (continued)

Clc Code	Level-1	Level-2	Level-3
322	Forest and semi natural areas	Scrub and/or herbaceous vegetation associations	Moors and heathland
323	Forest and semi natural areas	Scrub and/or herbaceous vegetation associations	Sclerophyllous vegetation
324	Forest and semi natural areas	Scrub and/or herbaceous vegetation associations	Transitional woodland-shrub
331	Forest and semi natural areas	Open spaces with little or no vegetation	Beaches, dunes, sands
332	Forest and semi natural areas	Open spaces with little or no vegetation	Bare rocks
333	Forest and semi natural areas	Open spaces with little or no vegetation	Sparsely vegetated areas
334	Forest and semi natural areas	Open spaces with little or no vegetation	Burnt areas
335	Forest and semi natural areas	Open spaces with little or no vegetation	Glaciers and perpetual snow
411	Wetlands	Inland wetlands	Inland marshes
412	Wetlands	Inland wetlands	Peat bogs
421	Wetlands	Maritime wetlands	Salt marshes
422	Wetlands	Maritime wetlands	Salines
423	Wetlands	Maritime wetlands	Intertidal flats
511	Water bodies	Inland waters	Water courses
512	Water bodies	Inland waters	Water bodies
521	Water bodies	Marine waters	Coastal lagoons
522	Water bodies	Marine waters	Estuaries
523	Water bodies	Marine waters	Sea and ocean

economic, cultural, labor, and investment costs in feasibility studies, ecological factors should also be kept in the foreground. Selecting the product to be cultivated according to the ecological characteristics of the region and making plans in this direction constitute the basis of agricultural investments. Reducing some of the costs indicated in the alternative agriculture diagram by Odum and Barrett (2004) can only be achieved in this way. Because determining the most suitable planting areas for plants by considering their topographic, soil, and climate requirements will enable the rational planning of the correct use of especially arable lands from the very beginning, and thus it will be possible to reduce fertilization, irrigation, pesticide, and even soil processing costs. At this point, GIS and RS today offer the opportunity to analyze and present all the characteristics of the regions together. The opportunity to evaluate the climate characteristics, soil characteristics, water resources, and other factors of the region together can be easily done with today's GIS and RS technologies (ESRI, 2004). In countries that are agriculturally and technologically developed, the mapping processes of agricultural areas, with their characteristics in the foreground, have been completed and made accessible.

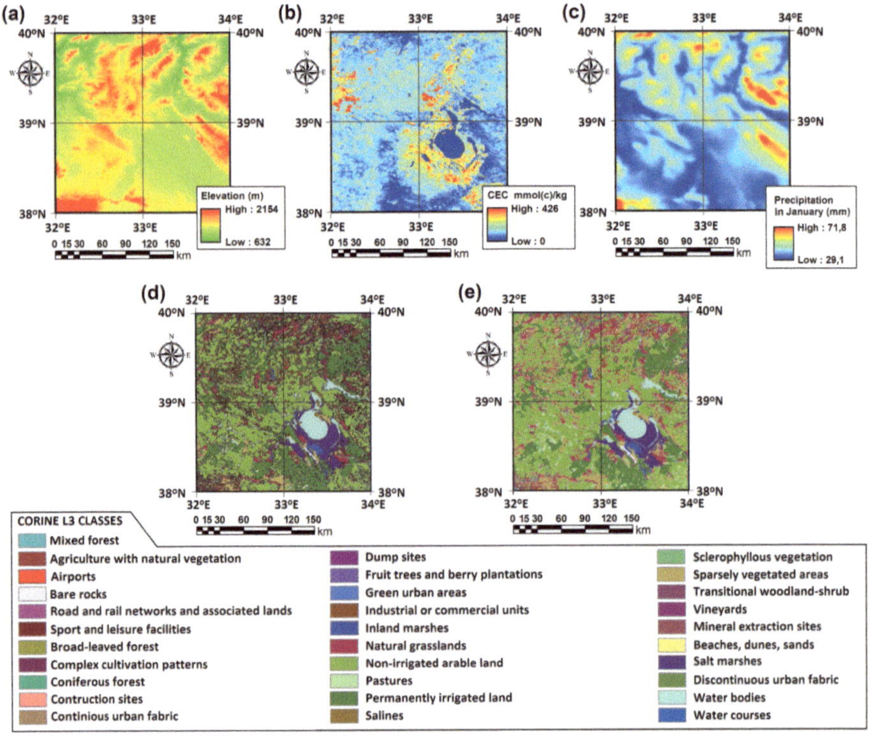

Fig. 3.6 Maps of Tuz Gölü and its surroundings (32–34° East longitudes and 38–40° North latitudes): (**a**) EARTHDATA ASTGTM digital elevation raster map (m), (**b**) ISRIC 0–5 cm soil depth cation exchange capacity (mmol(c)/kg) raster map, and (**c**) CHELSA climatologies (2011–2040) average precipitation (mm) in January raster map, (**d**) vector CORINE Level-3 map, and (**e**) raster CORINE Level-3 map

The most important components of environmental factors are climate, topography, and soil. Many scientific studies conducted to date have shown that different plant species have their own environmental requirements. If any plant is to be produced for agricultural purposes, first of all, it is necessary to determine the appropriate growing environments based on the environmental preferences of this plant species. This location determination will minimize the risks that may occur in the production of the plant, reduce costs and increase profitability. In addition, more conscious fertilization and maintenance activities will be carried out for the selected areas. Today, with the use of GIS and RS technologies, themed (thematic) maps for each environmental feature can be produced. The obtained thematic maps can be used to identify and map the most suitable areas using environmental requirements of target plant species and spatial analysis methods. Thus, the desired quality and yield for target plant species in the identified areas are achieved and they are grown in the most economical way.

3.6 Conclusion and Future Outlooks

Especially in developing countries, which have not fully systematized agricultural policies, many agricultural investments are made without sufficient feasibility studies, in line with traditional accumulations and habits, or as a result of the guidance of the market economy. This situation unfortunately causes investments to face setbacks from the very beginning and ultimately reflects on the economy as time, labor, and financial loss.

Effective and correct use of resources is an important element in ensuring development. In today's world, where needs are increasing and diversifying, there is great pressure on resources. The wrong land use resulting from this pressure causes the destruction of land resources, and the increase of poverty and other social problems. In this context, in countries, where the population is increasing, it is imperative that land is used in accordance with its potential. Because the sustainability of agriculture can only be achieved by determining the natural and cultural potential and implementing land use appropriate to the ecological structure. The effect of environmental factors on the yield and quality of agricultural products is important.

In order to carry out agricultural activities more economically, determining the most suitable growing environments for important target agricultural products with high and stable market value using GIS and RS techniques is important for public and private sector organizations, especially farmers operating in this sector. As stated at the beginning of this book chapter, information is defined as the results obtained by correlating the data resulting from an observation or process with each other. Therefore, in order to produce good quality information, first of all, reliable geographical or spatial data is needed. Reliable, systematically updated, continuous (regularly archived), and easily accessible data is the key to obtaining the necessary benefits from GIS and RS techniques. In short, the quality of the data used affects the quality of the outputs to be obtained. Institutionalized global and international databases, which are becoming increasingly widespread and whose quality standards are increasing, have come to meet the needs in this field. There is no doubt that these databases will increase the success of future modeling studies and the efficiency of economic activities, especially agriculture.

References

Allen, S. L., & Terry, E. (2005). *Beginning relational data modeling* (2nd ed.). Apress.

Barrett, G. W., Rodenhouse, N., & Bohlen, P. J. (1990). Role of sustainable agriculture in rural escapes. In C. A. Edwards, R. Lai, P. Madden, R. H. Miller, & G. House (Eds.), *Sustainable agricultural systems* (pp. 624–636). Soil and Water Conservation Society.

Beck, H. E., Wood, E. F., Mcvicar, T. R., Zambrano-Bigiarini, M., Alvarez-Garreton, C., Baezvillanueva, O. M., Sheffield, J., & Karger, D. N. (2020). Bias correction of global

high-resolution precipition climatologies using streamflow observations from 9372 catchments. *Journal of Climate, 22*(3), 1299–1315.
Bolstad, P. (2019). *GIS fundamentals: A first text on geographic information systems* (6th ed.). XanEdu Publishing.
Campbell, J. B., & Wynne, R. H. (2011). *Introduction to remote sensing* (5th ed.). The Guilford Press.
CHELSA. (2025). *Climatologies at high resolution for the earth's land surface areas*. Retrieved January 24, 2025, from: https://chelsa-climate.org/
Chuvieco, E. (2020). *Fundamentals of satellite remote sensing an environmental approach* (3rd ed.). CRC Press.
Dandy, N., Fiorini, S., & Davies, A. L. (2014). Agenda-setting and power in collaborative natural resource management. *Environmental Conservation, 41*(4), 311–320.
Demers, M. N. (2008). *Fundamentals of geographical information systems* (4th ed.). Wiley.
EARTHDATA. (2025). *Advanced Spaceborne Thermal Emission and Reflection Radiometer (ASTER) Global Terrain Model 1 arc second, Version: 3 (ASTGTM v003)*. National Aeronautics and Space Administration (NASA). Retrieved January 19, 2025, from: https://lpdaac.usgs.gov/products/astgtmv003/
EEA. (2024). *What is CORINE Land Cover?* European Environment Agency (EEA). Retrieved December 28, 2024, from: https://www.eea.europa.eu/en/about/contact-us/faqs/what-is-corine-land-cover
ESRI. (2004). *ArcGIS 9, Geoprocessing in ArcGIS*. Environmental Systems Research Institute.
Gold, C. M. (2006). What is GIS and what is not? *Transactions in GIS, 10*(4), 505–519.
Goodchild, M. F. (2011). Spatial thinking and the GIS user interface. *Procedia Social and Behavioral Sciences, 21*, 3–9.
Hild, H., & Fritsch, D. (1998). Integration of vector data and satellite imagery for geocoding. In D. Fritsch, M. Englich, & M. Sester (Eds.), *GIS-between visions and applications* (Vol. 32, Part 4, pp. 246–251). International Society For Photogrammetry and Remote Sensing (IAPRS).
Holdren, J. P., & Ehrlich, P. R. (1974). Human population and the global environment. *American Scientist, 62*(3), 282–292.
ISRIC. (2025a). *World soil information, purpose and strategy*. International Soil Reference and Information Center (ISRIC). Retrieved December 25, 2024, from: https://www.isric.org/about
ISRIC. (2025b). *World Data Centre for Soils (WDC-Soils)*. International Soil Reference and Information Center. Retrieved December 28, 2024, from: https://www.isric.org/about/world-data-centre-soils-wdc-soils
Jensen, J. R. (2015). *Introductory digital image processing: A remote sensing perspective* (4th ed.). Pearson Education.
Karger, D. N., & Zimermann, N. E. (2021). *Climatologies at high resolution for the earth land surface areas CHELSA V2.1: Technical specification*. Swiss Federal Research Institute WSL. Retrieved January 7, 2025, from https://chelsa-climate.org/wp-admin/download-page/CHELSA_tech_specification_V2.pdf
Karger, D. N., Conrad, O., Böhner, J., Kawohl, T., Kreft, H., Soria-Auza, R. W., et al. (2017). Climatologies at high resolution for the Earth's land surface areas. *Scientific Data, 4*, Article e170122. https://doi.org/10.1038/sdata.2017.122
Karger, D. N., Schmatz, D. R., Dettling, G., & Zimmermann, N. E. (2020). High-resolution monthly precipitation and temperature time series from 2006 to 2100. *Scientific Data, 7*, Article e248. https://doi.org/10.1038/s41597-020-00587-y
Karger, D. N., Wilson, A. M., Mahony, C., Zimmermann, N. E., & Jetz, W. (2021). Global daily 1 km land surface precipitation based on cloud cover-informed downscaling. *Scientific Data, 8*, Article: e307. https://doi.org/10.1038/s41597-021-01084-6
Lillesand, T., Kiefer, R. W., & Chipman, J. (2015). *Remote sensing and image interpretation* (7th ed.). Wiley.
Liu, J. G., & Mason, P. J. (2016). *Image processing and GIS for remote sensing: Techniques and applications*. Wiley.

Maguire, D. J. (1991). An overview and definition of GIS. In D. J. Maguire, M. F. Goodchild, & D. W. Rhind (Eds.), *Geographical information systems: Principles and applications* (Vol. 1, pp. 9–20). Wiley.

Malczewski, J. (2004). GIS-based land-use suitability analysis: a critical overview. *Progress in Planning, 62*(1), 3–65.

National Research Council. (1989). *Alternative agriculture*. National Academy Press.

Nguyen, T. T., Hoang, T. D., Pham, M. T., Vu, T. T., Nguyen, T. H., Quyet-Thang, H., & Jun, J. (2020). Monitoring agriculture areas with satellite images and deep learning. *Applied Soft Computing, 95*, Article e106565. https://doi.org/10.1016/j.asoc.2020.106565

Odum, E., & Barrett, G. W. (2004). *Fundamentals of ecology* (5th ed.). Cengage Learning.

Wei, H., Xu, Q., & Tang, X. (2011). A knowledge-based problem solving method in GIS application. *Knowledge-Based Systems, 24*, 542–553.

WSL. (2025). Swiss Federal Institute for Forest, Snow and Landscape Research. Retrieved January 14, 2025, from: https://www.wsl.ch/en/

Chapter 4
Big Data Utilisation in Wheat Genetic Resources

Ayşe Yıldız, Kürşad Özbek, and Nusret Zencirci

Abstract For many years, the decrease in plant genetic resources has been noticed by mankind and many studies have been initiated to protect species that are considered important. In genebank where biological diversity is preserved and protected for a long time, registered varieties, local varieties, and wild relatives are collected, stored, recorded, and made available to researchers. In addition, in order for end users to use the material efficiently, it must be characterised and documented in accordance with standards and the documents must be made available to end users. A strong information management system is required for the effective and secure operation of a genebank. This system should cover material acquisition, registration, storage conditions, collection types (basic collection, active collection, secure copy), monitoring the viability of accessions before and after storage, regeneration, characterisation, and evaluation processes. In addition, it is important to regularly record data on germplasm health tests, material distribution, and the number of subsamples and seed quantities of each accession held in the genebank. In order to adapt to the ever-increasing diversity and big data-driven nature of the modern research environment, it is of great importance that genebanks' information systems are integrated with other data sources and made accessible over networks. Such integrations provide new information on plant genetic resources, thus increasing the usability of these resources and strengthening the scientific reputation of genebanks. For this big dataset, comprehensive studies in accordance with international standards are needed in the collection, preservation, characterisation, documentation, and use of genetic resources

A. Yıldız
Republic of Türkiye Ministry of Agriculture and Forestry Soil, Fertilizer and Water Resources Central Research Institute, Ankara, Türkiye

K. Özbek
Republic of Türkiye Ministry of Agriculture and Forestry Directorate of National Botanical Garden of Türkiye, Ankara, Türkiye

N. Zencirci (✉)
Bolu Abant İzzet Baysal University, Art and Sciences Faculty, Biology Department, Gölköy, Bolu, Türkiye

N. Zencirci et al. (eds.), *Empowering Wheat Cultivation with GIS, Digital Approaches and Artificial Intelligence*, https://doi.org/10.1007/978-3-031-99954-3_4

Keywords Plant genetic resources · Genebanks · Biological diversity · Local varieties · Wild relatives · Secure copy

Abbreviations

CBD	Convention on biological diversity
CGIAR	Consultative Group on International Agricultural Research
CIMMYT	International Maize and Wheat Improvement Centre
ECPGR	European Cooperative Programme on Genetic Resources
EURISCO	European Search Catalogue for Plant Genetic Resources
FAO	Food and Agriculture Organization
FAO/IPGRI	Genebank Standards for Plant Genetic Resources for Food and Agriculture
GAA	Germplasm Acquisition Agreement
GBS	Genotyping by Sequencing
GMO	Genetically Modified Organisms
GRIN	Germplasm Resources Information Network
ICARDA	International Centre for Agricultural Research in Dry Areas
ITPGRFA	International Treaty on Plant Genetic Resources for Food and Agriculture
MLS	Multilateral System of Access and Benefit Sharing
NBPGR	National Bureau of Plant Genetics Resources
NCBI	National Center for Biotechnology Information
NGS	Next-Generation Sequencing
NSGC	National Small Grains Collection
OECD	Organisation for Economic Co-operation and Development
SMTA	Standard Material Transfer Agreement
VIR	Vavilov Research Institute
WheatIS	Wheat Information System

4.1 Genetic Resources and Importance

Humanity has recognised the decline in plant genetic resources for many years, and many studies have been initiated to conserve essential species. This protected material also constitutes a vital gene pool for registering new varieties containing resistance genes against the stress factors encountered. In addition, the evolution process in nature can be understood more clearly through studies on material collected over different years. Humanity has been struggling for many years with the difficulty of feeding its ever-increasing population. Considering the limited agricultural areas, the primary priority is to increase yield per unit area. For this purpose, genetic resources that have developed resistance to different biotic and abiotic stress conditions over many years will always have an increasing importance due to the genes they store in them against the stress factors we may encounter. In other words, human beings will need genetic resources to continue their generation. In addition to all these, genetic resources have an essential value in many fields, especially in the pharmaceutical industry, apart from feeding humanity.

Genetic resources belonging to wild species can continue their lineage by struggling with stress factors in their natural cycles. Local varieties, which constitute the most essential part of genetic resources, have faced human intervention since the beginning of agriculture in the world. In other words, not only natural conditions or stress factors but also human intervention has played a role in the evolution of these species. Farmers have shaped the evolution of these species with their cultivation methods. The long-standing cultivation of ancient wheat varieties like *Triticum monococcum ssp. monococcum* L. (Siyez) and *Triticum dicoccum Schuebler* (Gernik) serves as a prime example of this practice. The output of primitive wheat varieties such as *Triticum monococcum* ssp. *monococcum* L. (Siyez) and *Triticum dicoccum* Schuebler (Gernik) over the years are the best examples of this (Karagöz et al., 2010).

As is known, wild and primitive populations enrich the gene content of modern cultivars. In breeding programmes, yield is the main priority; therefore, the varieties developed often lack resistance to biotic and abiotic stresses. Therefore, plant genetic resources are utilised either directly or as bridge species in breeding programmes as gene sources for quantitative and qualitative characters (disease resistance, etc.).

In genebanks where biological diversity is preserved and protected for a long time, registered varieties, local varieties, and wild relatives are collected, stored, recorded, and made available to researchers. During all these processes, the most important issues are the preservation and reproduction of the viability of genetic resources. Because even under the best conditions materials may lose their vitality over time, new seeds should be obtained by regeneration in specific periods. In addition, for end users to use the material efficiently, it should be characterised and documented by standards, and the documents should be open to end users. There are seed banks, field genebanks, cryobanks, vegetative banks, and DNA banks operating worldwide to carry out all these procedures (Anonymous, 2025a).

4.2 Why Wheat Genetic Resources

Wheat is one of the most important genetic resources that constitutes the basis for adaptation, diversification, and survival in human evolution, the guarantee of our food security, the basis of agricultural development in the world, and the insurance of our future. For this reason, the development of new varieties of wheat is of vital importance for economic development. Unfortunately, wheat biodiversity is gradually decreasing due to faulty agricultural policies, uniformity of breeding programmes, population pressure, and changing climatic conditions. Therefore, it becomes essential to protect our gene resources, which consist of wild relatives, local varieties, modern and old varieties, and breeding lines (Anonymous, 2025b).

Wheat is an annual plant species that can be used as a feed plant, and it is a significant source of protein and carbohydrates for human beings. It can be used as a feed plant; high-yielding quality flour can be obtained, and it has high adaptability to different conditions. For many years, wheat has been used to prepare healthy and easy-to-preserve foods rich in dietary fibre and nutrients, containing high amounts of minerals and beneficial phytochemicals. Wheat is still being used and is a species suitable for modern urban life. In addition, wheat, which is used in producing products such as animal nutrition, starch, malt, and alcohol, is an important animal nutrient. Different studies on wheat have been carried out in many research centres worldwide, and wheat genetic resources remain indispensable in many fields.

Today, 95% of the wheat, of which there are around 25,000 cultivated varieties, is bread wheat (*T. aestivum*), and the remaining 5% is *T. turgidum* wheat used in pasta or semolina production, which we call durum or hard wheat (Özbek et al., 2024).

There are more than 80 wheat collections worldwide, including the International Maize and Wheat Improvement Centre (CIMMYT), the National Small Grains Collection (NSGC), the Centre for Genetic Resources (CGN), the Vavilov Research Institute (VIR), the International Centre for Agricultural Research in Dry Areas (ICARDA) and the National Bureau of Plant Genetics Resources (NBPGR). Further research and big data are needed on duplications that may be present in these collections, which are dominated by local or regional material consisting of breeders' working collections (Anonymous, 2025b).

Full-scale studies by international standards in the collection, conservation, characterisation, documentation, and utilisation of genetic resources are needed for this big data.

4.3 Collection of Genetic Resources

The first data on the collected plant genetic resources are obtained using survey information to be carried out in the collected region, descriptions made by systematics experts, and a literature review. Since this information is the primary source of data on genetic resources, it will affect all future studies. It should be done as carefully as possible, confirmed, and within the framework of international standards. However, no errors should be made when entering these data into databases.

4.4 Conservation of Genetic Resources

Many genebanks conserve and share plant genetic resources with users (Özbek et al., 2024). These banks aim to preserve the genetic heritage of plants for future generations by housing extensive collections of genes that contribute to solving global food security problems and adapting agriculture to changing climates (Ghamkar et al., 2024). However, despite their important position, many genebanks face difficulties characterising their extensive genetic material and making it available for practical use (Weise et al., 2020). In this context, the international standards "Genebank Standards for Plant Genetic Resources for Food and Agriculture (FAO/IPGRI)", first published in 1994, provide a guiding framework for genebank operations. The material stored in genebanks is of inestimable value, the product of long-lasting efforts of researchers, high-budget projects, and breeding and collection programmes. Therefore, security measures should be taken at the highest level where these valuable genetic resources are preserved. Facilities where the material is kept should be effectively protected against natural disasters and artificial threats. The conditions of the genebanks should be regularly checked, and all data should be meticulously recorded.

Equipment in cold rooms should be maintained without interruption, and critical spare parts should be kept ready for emergencies. In order to prevent possible power outages, it is essential to keep powerful generators ready to operate at all times. In addition, the operation of drying and cold rooms by international standards should be checked and recorded using sensors at regular intervals. In case of exceeding the standards set in the room conditions, the system should send warnings to the relevant persons via mobile phone and e-mail and inform the surrounding personnel with audible warnings. In addition, these housing conditions should be easily monitored online. Personnel control in the facilities is also of great importance. Access to cold rooms should be limited as much as possible, and only authorised people should be allowed. Entry and exit procedures should be recorded with a digital card system, and this process should be monitored separately. All areas, such as the herbarium, biotechnology, and physiology laboratories in the seed genebank, should also be monitored and recorded with a camera system. In addition to all these measures, it should also be ensured that security backups of the genetic material are created in different geographical locations. This approach minimises possible risk factors, and priceless genetic resources can be secured (Özbek et al., 2024).

Security replication contributes to the prevention of potential losses caused by disasters. Through this process, accessions from one genebank are included in the core collections of another genebank, and for the safety of the collection, inaccessible black box samples from one genebank are stored in different genebanks. During security duplication, details such as accession number, estimated seed quantity, weight of the sample (g), date of renewal, year of security duplication, and information on the institution holding the duplicate should be recorded in full (Anonymous, 2025b).

Effective management of genebanks brings along the need for well-trained personnel. In this respect, it is essential to distribute responsibilities balanced among specialised personnel. A genebank should establish a strategic plan and budget for the minimum workforce that can fulfil the necessary tasks related to the provision, conservation, and distribution of germplasm. It is also desirable for each genebank to have staff with different specialisations in line with its purpose and tasks. The seeds' health and usefulness depend on the genebank's safety and reliability. Arrangements such as having energy backups or building durable structures in earthquake-prone areas can be considered in this context. In addition, the genebank should promote and implement a systematic risk management approach to protect its collections and data from potential physical or biological risks. Certified training programmes and/or vocational training should be provided to ensure that staff are adequately equipped, and training needs should be analysed in detail. Genebank staff should be made aware of the risks to which germplasm may be exposed and safety procedures aimed at minimising these risks. Genebank facilities should be designed in a structure resistant to natural disasters such as storms, tornadoes, earthquakes, or floods that may occur in the geography where they are located. Temperature monitoring devices should be available in the drying and storage rooms to monitor the changing temperature values over time. In addition, fire alarm systems and fire response equipment should be mandatory in the genebank. Considering that most of the fires are caused by faulty electrical circuits, compliance with safety standards should be checked periodically. Fire extinguishing equipment should include extinguishers and fire blankets. In areas where thunderstorms are common, the genebank should have a properly installed lightning rod (Özbek et al., 2024).

4.5 Characterisation of Genetic Resources

For the long-term conservation of plant genetic resources and to make the conserved samples usable, these samples must be accurately identified and characterised. Information at various levels must be documented (Weise et al., 2020). Basic data on plant genetic resources are generally included in passport information. Passport data aims to clarify the identity of the material and includes the accession number, scientific name, origin of the material, and acquisition information (e.g., acquisition year, donor institution, collection task, and location). Furthermore, documenting information on regeneration and how and where the material is stored (active or base collections, cold storage, shelf locations, or secure propagation areas) is critical for effective genebank management (FAO, 2014). The geographical origin of a genebank entry, in combination with environmental data, can provide information on possible adaptations to biotic or abiotic stressors. In addition to this data, the dataset should be completed with details such as the status of the material (e.g., wild form or local breed). Other important information is the phenotypic characterisation of the material, including morphological and agronomic traits. This basic information serves as a guide to describe the characteristics of plant genetic resources better and determine their potential for use in breeding programmes. The expansion of such basic data is essential for the development of genebank collections. Following material descriptions, it is possible to determine to what extent a genebank represents the gene pool of a species or which species or subspecies are missing.

Furthermore, geographical information can be used to more precisely identify which regions are missing material and need to be collected (Fig. 4.1) This detailed

Fig. 4.1 Phenological characterisation of genetic material and an image of the materials in the genebank. (Original photos by Kürşad Özbek)

knowledge of the collection composition is necessary for gap analyses. Comparing the data provides information about possible duplicate specimens within the same collection (Weise et al., 2020).

Combining and analysing data from different genebanks makes it possible to determine whether specific accessions are present in other collections and, if necessary, to obtain them from them. Such comparisons provide an effective way of targeting the expansion of genetic material through exchange with other banks. In addition, this method contributes to the more efficient functioning of genebanks (Engels & Maggioni, 2018).

International agreements govern the conservation, sustainable use, right of access, and benefit sharing of plant genetic resources. These include the Convention on Biological Diversity (CBD) and its additional text, the Nagoya Protocol. Under these regulations, it is crucial to document the geographical origin of accessions in genebanks, when they were included in the collection, and their current legal authorisation for collection.

Entering into force in 2004, the International Treaty on Plant Genetic Resources for Food and Agriculture (ITPGRFA) by the Convention on Biological Diversity (CBD) includes the obligation of signatory states to conserve, identify, utilise, and ensure the sustainable use of plant genetic resources. The Multilateral System of Access and Benefit Sharing (MLS), which is the core component of the Agreement, regulates the provision of facilitated access to these resources and the equitable sharing of the benefits derived. In this context, the standard material transfer agreement (SMTA) has been developed and covers the reporting obligations usually carried out by the documentation units of Genebanks. The FAO Genebank Standards recommend that genetic material be backed up in other genebanks in geographically diverse and remote areas to avoid the loss of entire collections or access to genetic resources in the event of war or disaster (FAO, 2014).

Many genebanks can only provide end users with limited passport information and basic characterisation data based on morphological traits associated with standard international identifiers. The number of genetic materials with agronomic and quality traits data is minimal. Although there are about 1750 independent genebanks worldwide that conserve about 7.4 million agricultural genetic materials, only 10% of these materials have comprehensive phenotypic and genotypic characterisation, and many materials are not accessible to researchers or breeding programmes. The material in genebanks cannot be effectively utilised for breeding work due to incomplete resources, temporal limitations, and insufficient comprehensiveness of available data (Nguyen & Norton, 2020).

High throughput phenotyping (HTP) is a promising, efficient, cost-effective method for collecting phenotypic data for various traits in large-scobale experiments. This approach is particularly valuable when combined with genomic data. For example, it has been successfully applied in genomic selection processes in wheat (Araus & Kefauver, 2018; Nguyen & Norton, 2020). Phenotyping is an essential but costly part of product development programmes to understand the interactions between genotypes and environments and the genetic basis of these processes. Using HTP tools and techniques by genebanks will not only speed up the phenotyping process but also contribute to reducing costs by increasing the efficiency of the processes. It will also improve selection accuracy, especially in large-scale breeding projects. Missing or insufficient passport, characterisation, or evaluation data are the main problems that cause inefficient use of genetic resources. Keeping complete records of morphological traits and flowering, fruit, and seed qualities is important. However, due to financial constraints or high workloads, many genebanks ignore determining physiological and agronomic traits in their standard curation procedures. However, such information is vital for correctly evaluating genetic material and selecting valuable germplasm for priority use in breeding programmes. Core collections can be constructed using phenotypic data of agronomic traits, which allows for a more efficient evaluation of available germplasm (Nguyen & Norton, 2020).

In addition, DNA fingerprinting data are indispensable tools for genotyping and plant breeding (Zhao et al., 2021). New applications of modern molecular biology techniques, such as next-generation sequencing (NGS) and genotyping by sequencing (GBS), provide scientists with the opportunity to improve the quality, efficiency, and cost-effectiveness of genebank operations, as well as deepen the scientific knowledge of genebank collections. These advances also guide the shaping of effective conservation management. Molecular tools contribute to long-term, efficient conservation by enabling informed decisions to reduce unnecessary duplication in crop collections. Validation of passport data provides important information on possible duplicates in crop collections, which can be verified by molecular marker techniques to more clearly reveal genetic differences between materials. Thanks to advances in molecular biology and genomics, DNA from nuclei, mitochondria, and chloroplasts is increasingly being used to assess patterns of intraspecific and interspecific genetic variation, to map and characterise reproductively desirable traits or important genes, to carry out taxonomic studies that distinguish species from one another, and for evolutionary analyses of genome structure in plants. The GBS method dramatically simplifies the process by combining genotyping and genome-wide molecular marker detection in a single step. This method allows the discovery of new germplasm or species that have not been previously characterised without identifying and studying polymorphisms through molecular marker studies. Furthermore, thanks to advances in bioinformatics, the development of new reference genomes, and the availability of increasing volumes of sequence data, GBS datasets can then be re-analysed to reveal additional information, such as new polymorphisms or annotated genes (Engels & Ebert, 2021).

4.6 Documentation and Utilisation of Genetic Resources

The documentation process marks the crucial first step upon the receipt of any sample in a genebank. Genebank collections serve as invaluable genetic reservoirs for both contemporary and future breeding programmes, while also functioning as a dependable source of material for researchers and other stakeholders. To maintain their utility, each specimen must be precisely and comprehensively documented upon its entry into the genebank. Upon arrival, every seed or plant material sample is registered with a unique entry number to ensure its distinct identification. This step enables precise tracking, fosters accurate inventory management for conservation and distribution, and facilitates overall germplasm management (Fig. 4.2).

New samples entering the genebank are typically procured from collectors, other genebanks, or authorised sources under material transfer agreements, ensuring compliance with national and international conservation and distribution regulations. Accompanying every specimen is vital passport information to confirm it is not already present in the collection. Among this information are details like the cultivar name, collector number, common crop name or genus and species, collection number, collection site location, country of origin, collection date, phenology, collection source, and the number of plants sampled. The primary expectation is that new specimens should exhibit genetic differences from those already catalogued.

Interestingly, specimens with identical or similar names and grain characteristics might be genetically distinct, while others with differing names could share genetic similarities. To reliably identify duplicates, morphological, biochemical, and molecular approaches can be utilised based on the facilities and resources at hand. Additionally, all seeds and plant materials undergo a meticulous visual inspection

Fig. 4.2 Images of documentation of genetic material. (Original photos by Kürşad Özbek)

under a stereoscopic microscope to ensure they are free from pathogens, fungal growth, bacterial or viral infections, and pests such as insects. Only material that meets these rigorous quality standards and is available in quantities sufficient for storage can be approved for registration and subsequently assigned a unique accession number (Anonymous, 2025b).

An effective information management system should be established for each genebank. This system should be usable by genebank curators and staff to search and access specific information through various queries. The system should keep records of genebank data and cover aspects such as storage space, stocks, monitoring processes, health tests, and distribution status. It should also include the management of germ plasma orders, shipment information, and tracking of contact details of the genebanks. Barcoding is a valuable tool that complements this information management system. Each genetic material must have certain documents to be accepted in genebanks. These documents should include the Germplasm Acquisition Agreement (GAA), Standard Material Transfer Agreement (SMTA), phytosanitary certificate, import permit, non-GMO certificate, package list of the material and seed source, available passport data, and characterisation data. While accepting the incoming material with these documents, the accuracy of the shipments should be checked. In this context, the staff in charge should compare all incoming packages with the accompanying list, prepare a new list if the list is incomplete or there is no match, and send the documents to the donor or provider for approval to obtain the correct information. They should also check the seeds for insects, inspect incoming products for phytosanitary conformity, and isolate infected or contaminated samples. Seeds should be cleaned using fumigation before storage (Anonymous, 2025b).

The lack of high-quality phenotype-related data is recognised as one of the major obstacles limiting the efficient use and evaluation of genetic resources in functional genomics and modern breeding processes (Engels & Ebert, 2021). DNA or digital sequences are of no real value in themselves without concrete information about the samples from which they originate. In order to manage the vast amount of genetic and phenotypic information and make such valuable information on germplasm accessions accessible to users worldwide, a collaborative platform that supports data collection, analysis, and sharing is needed. In addition, information networks must be unified, globally accessible, and continuously updated in line with new research findings. In this context, some international databases already contribute to genetic resources. For example, platforms such as Genesys, EURISCO, OECD, and NCBI are notable resources in this field (Engels & Ebert, 2021).

In addition, there are several websites providing information on international ex situ base collections—World Information and Early Warning System on Plant Genetic Resources for Food and Agriculture-WIEWS—GeneBank Platform-CGIAR—Multilateral System of International Agreement—Svalbard Global Seed Vault—European Cooperative Programme on Genetic Resources-ECPGR—Global-GRIN based on the Germplasm Resources Information Network (GRIN) plant database (Engels & Ebert, 2021).

In addition, GrainGenes, one of the major databases in the USA, provides services for cereal groups such as wheat, barley, rye, and oats. GrainGenes (https://wheat.pw.usda.gov) is active among biological data communities, collaborating with curators and genome sequencing teams. This database also contributes to the AgBioData Consortium and the International Wheat Initiative through the Wheat Information System (WheatIS). The database is available at https://wheat.pw.usda.gov (Yao et al., 2022).

A genebank's efficient and secure operation requires a strong information management system. This system should cover material acquisition, registration, storage conditions, collection types (essential collection, active collection, secure copy), monitoring of the viability of accessions before and after storage, regeneration, characterisation, and evaluation. It is also important to regularly record data on germplasm health tests, material distribution, subsample numbers, and seed quantity of each accession held in the genebank. Although genebanks are encouraged to record molecular data generated by genomics, proteomics, metabolomics, phenomics, and bioinformatics, these data are often generated by collaborative projects with other research teams or specialised research institutes rather than by genebank staff. Due to the complexity of omics data and their exponential increase in volume over time, such information is often stored in specialised public databases, made accessible or allocated to specific systems (Engels & Ebert, 2021).

The advent of high-throughput sequencing technologies has ushered in the era of big data. In managing big data, all data are divided into six subunits:

Project, sample, experiment, run, compilation, and variation: These subheadings should include the project name, data type, submitter or institution, funding information, publication details, sample type, name, taxonomy ID, collection information, isolate, tissue type, location, phenotype and disease information, file type, sequencing platform, library strategy, source and layout, file name and MD5 value, molecule type, scope, sequencing technology, assembly method, variation itself and detection method. A project may be associated with more than one project; likewise, a sample may be associated with more than one experiment, assembly, or variation. Experiments are usually associated with one or more runs. All data are assigned unique access numbers for ease of access and reference. Users wishing to submit data must first register, log in, and fill in the submitter information. A typical data submission sequence follows: create a project, add sample information, and complete the process with relevant experiments, runs, assemblies, and variations. Submitted projects are first evaluated by ethical guidelines and published after approval by the curators. Favouring standard file formats such as FASTQ, BAM, FASTA, and VCF is recommended to increase data reusability. Genebank databases are important repositories for archiving raw sequencing data and analysis results. For the database to serve a broader user base, new data types associated with omics data, such as sequence, protein, metabolism, expression, clinical information, and images, should be gradually added over time (Guo et al., 2020). Rather than storing big data, the critical point is the timely processing of these data and obtaining meaningful results (Ghamkar et al., 2024). Significantly, when the data size exceeds a single machine's capacity, and extensive analyses are required, the problems related to big data analytics become more prominent. Despite all these challenges, the increasing use of big data technologies has started to create a new job field for "data scientists", computer scientists, and mathematicians who are competent in big data analytics, statistics, data mining, and computer programming (O'Driscoll et al., 2013).

4.7 Problems and Recommendations

Conservation methods based on exceptional standards are being developed to protect genetic resources and provide material for scientific studies. The most critical problem encountered in this process is ensuring the continuity of the studies. In addition, the correct selection and successful implementation of the scientific methods to be applied are also important for success. For the security of collections and their documents or databases, all kinds of measures, especially backup, should be taken by utilising up-to-date technologies.

Scientific studies on genetic resources are compiled by international standards and organised according to the type of conservation. Adherence to these standards in the design and implementation of storage environments is essential to ensure the future and usability of collections.

These standards manage many collections around the world and thus gain international recognition (Özbek et al., 2024).

Although genebanks have a great capacity to phenotype and characterise genetic resources, they face several challenges. One of these challenges is financial problems. Hidden costs such as equipment maintenance, database management, software licensing, and updates constitute a significant burden. Therefore, genebank managers must carefully evaluate the balance between investment planning, labour costs, and achieving objectives before starting projects. Furthermore, genebanks should focus on long-term studies that support well-planned, consistent phenotyping programmes in situ rather than short-term projects. Comprehensive planning and adequate resourcing should also be ensured to sustain research and breeding programmes in the long term to meet the future need for increased plant production. High technology allows large amounts of data to be generated quickly with standardised protocols. However, a substantial infrastructure investment in storage, backup, data management, and analysis is essential for effectively using these data. However, the need for an interdisciplinary approach should not be ignored. The accuracy of the data generated must also be confirmed and analysed before use. Since phenotypic data can vary over time, they have a more flexible structure compared to genomic data, which are fixed. Considering that only about 10% of genebank materials are used in plant breeding studies, the importance of high-quality phenotype and genotype data increases. Although accurate phenotypic data play a critical role in plant breeders' identification of potential parents with superior performance, phenotyping is one of the most costly stages of breeding programmes. Plant breeders are often reluctant to undertake this risk due to economic costs and uncertainties unless they can guarantee efficient results in their screening to identify existing gene resources or new alleles from a large amount of genebank material. However, supplementing genebank passport data with phenotypic and genomic information and integrating these data with accessions increases the value of both genebanks and accessions (Nguyen & Norton, 2020). Given the increasing flow of big data, improving the design and development methods of traditional computational and analytical data models is inevitable (Al-Jarraha et al., 2015). Advanced data analytics and machine learning algorithms can extract meaningful insights from large datasets by processing phenotypic data quickly and accurately (Gill et al., 2022). These techniques contribute to more adequately characterising genetic resources by revealing patterns, relationships, and correlations between genotypes and traits (Cobb et al., 2013). Phenomic technologies are often integrated with genetic information generated by sequencing technologies. Researchers can more efficiently identify genetic markers linked to desired traits in this context by correlating phenotypic traits with genetic variation. This approach facilitates the provision of phenotypic data for use with the increasing volume of genotype data, especially in crops such as wheat (Niu et al., 2023) and barley (Jayakodi et al., 2020).

Documentation of the material is of great importance in conserving plant genetic resources. This documentation includes basic information, origin or legal status, and data relevant to conservation management, such as storage quantities and germinability rates. Furthermore, the phenotypic and agronomic characterisation process plays a critical role in assessing the potential of this material for research and breeding. In recent years, significant efforts have been made to increase cooperation between genebanks to conserve plant biodiversity best. To this end, networks like the European Plant Genetic Resources Cooperation Programme have been established. Such initiatives have laid the foundation for developing international systems such as EURISCO and Genesys, which are centralised information platforms for research on plant genetic resources. Existing information systems are currently limited to passport data and phenotypic data. However, there is great potential for the sustainable utilisation of genebank collections by expanding the interactions between these systems and enriching the available information with data from other disciplines. Several pilot projects have recently been completed to evaluate genebank collections at the molecular level. These promising approaches should be continued in the future. In order to adapt to the ever-increasing diversity and significant data-driven nature of the modern research environment, genebanks' information systems must be integrated with other data sources and made accessible over networks. Such integrations can lead to new information on plant genetic resources, increasing the availability of these resources and strengthening the scientific reputation of genebanks (Ghamkar et al., 2024).

References

Al-Jarraha, O. Y., Yoo, P. D., Muhaidat, S., Karagiannidisa, G. K., & Taha, K. (2015). Efficient machine learning for big data: A review. *Big Data Research*. https://doi.org/10.1016/j.bdr.2015.04.001

Anonymous. (2025a). https://www.echocommunity.org/en/resources/6d00be7b-0ed6-46b9-8964-0ec761522658

Anonymous. (2025b). https://cropgenebank.sgrp.cgiar.org/index.php/wheat-mainmenu-362/conservation-mainmenu-381

Araus, J. L., & Kefauver, S. C. (2018). Breeding to adapt agriculture to climate change: Affordable phenotyping solutions. *Current Opinion in Plant Biology, 45*, 237–247. https://doi.org/10.1016/j.pbi.2018.05.003

Cobb, J. N., DeClerck, G., Greenberg, A., Clark, R., & McCouch, S. (2013). Next-generation phenotyping: Requirements and strategies for enhancing our understanding of genotype–phenotype relationships and its relevance to crop improvement. *Theoretical and Applied Genetics, 126*, 867–887. https://doi.org/10.1007/s00122-013-2066-0

Engels, J. M. M., & Ebert, A. W. (2021). A critical review of the current global ex situ conservation system for plant agrobiodiversity. II. Strengths and weaknesses of the current systemand recommendations for its improvement. *Plants, 2021*(10), 1904.

Engels, J. M. M., & Maggioni, L. (2018). Managing germplasm in a virtual European genebank (AEGIS) through networking. In N. C. Karafyllis (Ed.), *Theorien der Lebendsammlung: Pflanzen, Mikroben und Tiere als Biofakte in Genbanken* (Vol. 25, pp. 169–197). Verlag Karl Alber.

FAO. (2014). *Genebank Standards for Plant. Genetic Resources for Food and Agriculture; Commission on Genetic Resources for Food and Agriculture*. Food and Agriculture Organization of the United Nations.

Ghamkhar, K., Hay, F. R., Engbers, M., Dempewolf, H., & Schurr, U. (2024). Realizing the potential of plant genetic resources: the use of phenomics for genebanks. *Plants, People, Planet, 7*(1), 23–32.

Gill, T., Gill, S. K., Saini, D. K., Chopra, Y., de Koff, J. P., & Sandhu, K. S. (2022). A comprehensive review of high throughput phenotyping and machine learning for plant stress phenotyping. *Phenomics, 2*, 156–183. https://doi.org/10.1007/s43657-022-00048-z

Guo, X., Chen, F., Gao, F., Li, L., Liu, K., You, L., Hua, C., Yang, F., Liu, W., Peng, C., Wang, L., Yang, X., Zhou, F., Tong, J., Cai, J., Li, Z., Wan, B., Zhang, L., Yang, T., Zhang, M., Yang, L., Yang, Y., Zeng, W., Wang, B., Wei, X., & Xu, X. (2020). CNSA: a data repository for archiving omics data. *Database, 2020*, baaa055. https://doi.org/10.1093/database/baaa055

https://wheat.pw.usda.gov. Accessed 26 Dec 2024.

Jayakodi, M., Padmarasu, S., Haberer, G., Bonthala, V. S., Gundlach, H., Monat, C., Lux, T., Kamal, N., Lang, D., & Himmelbach, A. (2020). The barley pan-genome reveals the hidden legacy of mutation breeding. *Nature, 588*, 284–289. https://doi.org/10.1038/s41586-020-2947-8

Karagöz, A., Zencirci, N., Tan, A., Taşkın, T., Köksel, H., Sürek, M., Toker, M., & Özbek, K. (2010). Conservation and use of plant genetic resources. Turkish Agricultural Engineering VII. Technical Congress.

Nguyen, G. N., & Norton, S. L. (2020). Genebank phenomics: A strategic approach to enhance value and utilization of crop germplasm. *Plants, 9*, 817. https://doi.org/10.3390/plants9070817

Niu, J., Ma, S., Zheng, S., Zhang, C., Yaru, L., Si, Y., Tian, S., Shi, X., Liu, X., & Naeem, M. K. (2023). Whole-genome sequencing of diverse wheat accessions uncovers genetic changes during modern breeding in China and the United States. *The Plant Cell, 35*, 4199–4216. https://doi.org/10.1093/plcell/koad229

O'Driscoll, A., Daugelaite, J., & Sleator, R. D. (2013). 'Big data', Hadoop and cloud computing in genomics. *Journal of Biomedical Informatics., 46*, 774–781.

Özbek, K., Keskin, C. N., & Zencirci, N. (2024). Wheat genetic resources. In *Advances in wheat breeding: Towards climate resilience and nutrient security* (pp. 525–554). Springer.

Weise, S., Lohwasser, U., & Oppermann, M. (2020). Document or lose it: On the importance of information management for genetic resources conservation in genebanks. *Plants, 9*, 1050. https://doi.org/10.3390/plants9081050

Yao, E., Blake, V. C., Cooper, L., Wight, C. P., Michel, S., Cagirici, H. B., Lazo, G. R., Birkett, C. L., Waring, D. J., Jannink, J. L., & Holmes, I. (2022). GrainGenes: A data-rich repository for small grains genetics and genomics. *Database, 2022*, baac034. https://doi.org/10.1093/database/baac034

Zhao, Y., Jiang, B., Huo, Y., Yi, H., Tian, H., Wu, H., Wang, R., Zhao, J., & Wang, F. (2021). A high-performance database management system for managing and analyzing large-scale SNP data in plant genotyping and breeding applications. *Agriculture, 2021*(11), 1027. https://doi.org/10.3390/agriculture11111027

Chapter 5
High-Throughput Phenotyping in Wheat

Waqas Liaqat, Amjad Ali, Muhammed Tatar, Muhammad Tanveer Altaf, Parnaz Mortazvi, Mehmet Bedir, Muhammad Azhar Nadeem, and Faheem Shehzad Baloch

Abstract High-throughput phenotyping has become an important tool in modern wheat breeding, enabling rapid and precise measurement of plant traits. This chapter discusses the integration of advanced technologies, such as unmanned aerial vehicles, imaging sensors, and deep learning algorithms, in the phenotyping process. Key imaging techniques, including RGB cameras, multispectral, fluorescence, thermal, and 3D imaging, are explored for their ability to capture diverse phenotypic traits in wheat. The chapter also highlights the potential of satellite imagery and hyperspectral sensors to provide large-scale, high-resolution data. The incorporation of deep learning models further enhances the analysis of these complex datasets, facilitating improved prediction accuracy and trait selection. The convergence of

The original version of the chapter has been revised. A correction to this chapter can be found at https://doi.org/10.1007/978-3-031-99954-3_15

W. Liaqat · M. T. Altaf (✉)
Department of Field Crops, Faculty of Agriculture, Recep Tayyip Erdoğan University, Rize/Pazar, Türkiye

A. Ali
Department of Plant Protection, Faculty of Agricultural Sciences and Technologies, University of Science and Technology, Sivas, Türkiye

M. Tatar · P. Mortazvi · M. Bedir
Department of Plant Production and Technologies, Faculty of Agricultural Sciences and Technologies, Sivas University of Science and Technology, Sivas, Türkiye

M. A. Nadeem
Department of Biotechnology, Faculty of Science, Mersin University, Yenişehir, Mersin, Türkiye

F. S. Baloch (✉)
Department of Biotechnology, Faculty of Science, Mersin University, Yenişehir, Mersin, Türkiye

Department of Plant Resources and Environment, Jeju National University, Jeju City, Republic of Korea

N. Zencirci et al. (eds.), *Empowering Wheat Cultivation with GIS, Digital Approaches and Artificial Intelligence*, https://doi.org/10.1007/978-3-031-99954-3_5

these technologies offers promising advancements in wheat breeding by improving the speed and efficiency of phenotypic assessments.

Keywords Wheat breeding · Unmanned aerial vehicles · Thermal imaging · Hyperspectral sensors · Deep learning

Abbreviations

CNN	Convolutional neural network
CT	Computed tomography
DL	Deep learning
DSM	Digital surface model
DTM	Digital terrain model
HTP	High-throughput phenotyping
IR	Infrared
LiDAR	Light detection and ranging
ML	Machine learning
MRI	Magnetic resonance imaging
NIR	Near-infrared reflectance
PCA	Principal component analysis
PLSR	Partial least squares regression
QTL	Quantitative trait locus
RADAR	Radio detection and ranging
RF	Random forest
RGB	Red green blue (color model/camera)
RIL	Recombinant inbred line
ToF	Time-of-flight
UAV	Unmanned aerial vehicle

5.1 Introduction

Breeding crops for high yield and superior adaptability to new and variable climates is imperative to ensure continued food security, biomass production, and ecosystem services. Furthermore, new technologies are essential to enhance plant breeding to meet these challenges (Chawade et al., 2019; Fu & Jiang, 2022). Advancements in DNA marker assays and sequencing technologies have enabled high-throughput genotyping of many plants at relatively low costs, but phenotyping remains labor-intensive and expensive, especially in large-scale field trials (Bhat et al., 2020; Yalinkiliç et al., 2024). Thus, phenotyping has emerged as a bottleneck in modern plant breeding research.

High-throughput phenotyping (HTP), particularly through remote and proximal sensing, offers a nondestructive, rapid approach to evaluating plant traits. This new frontier in plant breeding allows for the efficient and large-scale assessment of phenotypes, overcoming traditional limitations (Cudjoe et al., 2023). Studies have shown that the grain yield of wheat genotypes can be reliably assessed through spectral sensing techniques (Cheng et al., 2024; Hassan et al., 2019). Additionally, ground-based HTP platforms not only save time and costs but also provide objective, digital data that can be stored for future reevaluation and selection (Hu et al., 2020). In recent years, unmanned aerial vehicles (UAVs) have further revolutionized high-throughput phenotyping. UAVs offer several advantages over ground-based platforms, such as the ability to generate real-time surface maps and measure plant parameters across many plots simultaneously (Xu et al., 2021; Gracia-Romero et al., 2019). UAVs, particularly those operating at low altitudes with high-resolution sensors, can overcome issues like uneven terrain, nonsimultaneous measurements, and plot-specific configurations that are inherent to ground-based systems (Sankaran et al., 2015). However, comparative data between aerial- and ground-based sensing for phenotyping wheat genotypes are still limited, highlighting an important area for future research (Hu et al., 2020).

The global demand for food is rising increasingly due to rapid population growth (Abbas et al., 2024; Ali et al., 2024). Therefore, breeding of wheat crop has become a major target to fulfill the demand of the growing population. Achieving these gains will require a combination of advances in wheat genetics, physiology, breeding, and agronomy. A critical component in this endeavor is the observation and quantification of plant phenotypes, or phenotyping (Sheikh et al., 2024). A plant's phenotype, comprising measurable traits, provides insights into its agronomic performance, underlying genetics, and environmental interactions.

Despite its importance, phenotypic data collection has lagged behind genetic data collection due to breakthroughs in genomics and molecular breeding. This disparity has led to the so-called phenotyping bottleneck, wherein progress in plant breeding is limited by current phenotyping capabilities (Fiorani & Schurr, 2013). Traditional phenotyping methods often involve manual measurements that are time-consuming and constrained by environmental variations, necessitating repeated, multilocation trials. To address this, various tools—mainly based on image analysis—have been developed to automate or expedite phenotyping processes (Gill et al., 2022).

For aboveground organs, high-throughput phenotyping has benefited from technological advancements in imaging from ground vehicles, UAVs, aircraft, and even satellites (Fig. 5.1). These tools allow for both high-resolution assessments of individual plants and large-scale population studies. Several excellent articles discuss the use of sensor technologies in HTP (Feng et al., 2021; Xie & Yang, 2020). This chapter focuses on imaging tools that enable detailed phenotypic resolution in wheat, particularly those that are cost-effective and user-friendly, making them accessible to a wide range of research projects.

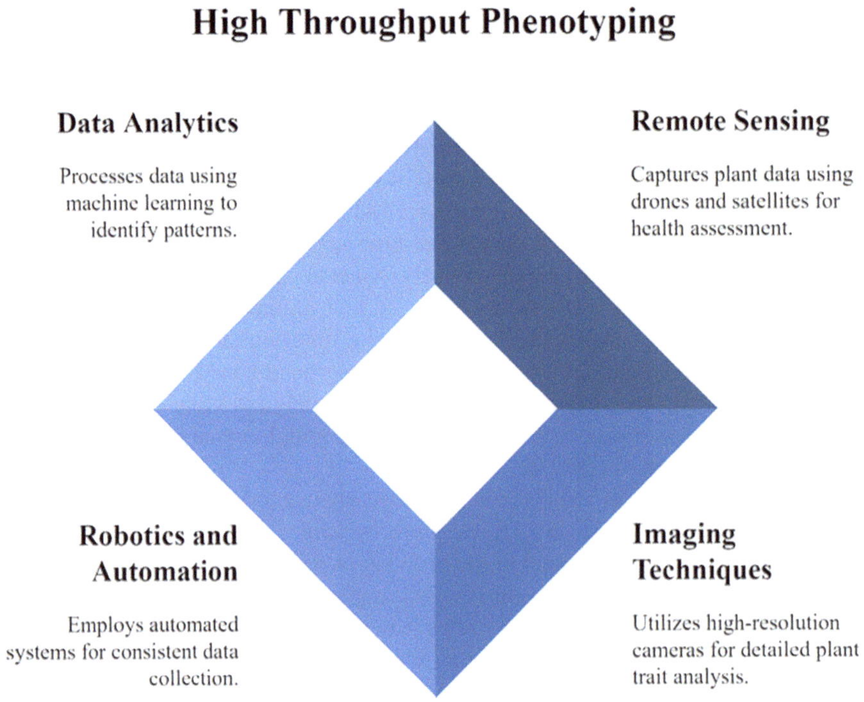

Fig. 5.1 An overview of a high-throughput phenotyping system

5.2 High-Throughput Phenotyping in Wheat Breeding

Recent advancements in HTP technologies have revolutionized the collection of extensive phenotypic data through automated systems and sophisticated imaging technologies (Mir et al., 2019). These innovations enable the rapid and precise measurement of several plant traits (Fig. 5.2). Traits such as plant height, leaf area, and growth patterns can be assessed with high resolution using technologies like drones, sensors, and multispectral and hyperspectral cameras (Yang et al., 2020). The integration of automation and robotics significantly streamlines data collection processes, allowing for the simultaneous analysis of a large number of plants or genetic lines, thereby accelerating research and breeding programs. Postcollection, the identification of correlations between phenotypic traits and genetic information necessitates the application of machine learning models and artificial intelligence. These capabilities facilitate the handling of large datasets and support informed decision-making. HTP is designed to ensure accurate and precise measurements, which are crucial for understanding genetic variations and their impacts on plant performance (Kuriakose et al., 2020). The regular deployment of instrumented platforms produces substantial volumes of time-series data, facilitating advanced data processing and analytics. Traits derived from HTP related to physiology and

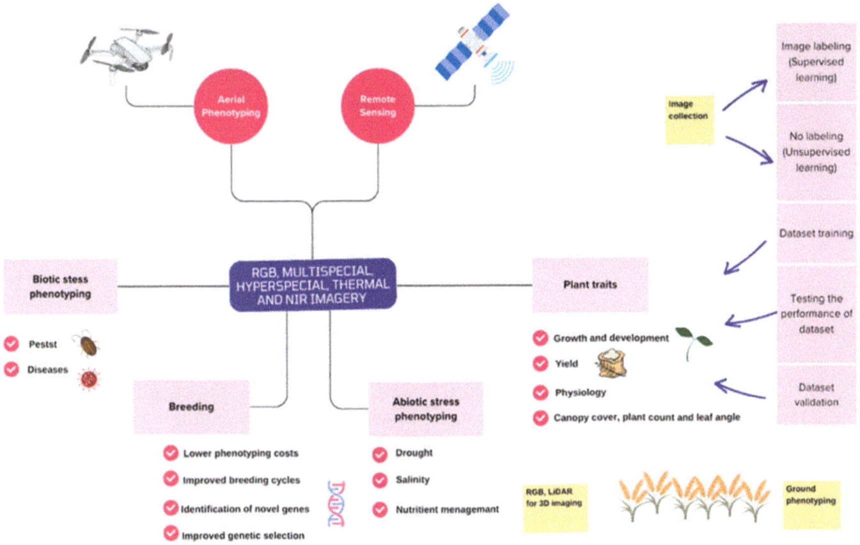

Fig. 5.2 An overview of high-throughput phenotyping for plant stress assessment

canopy architecture are highly informative and correlate with the observed variability in wheat yield under different levels of water stress. These traits offer valuable decision-making support for objective selection in wheat breeding research (Pérez-Ruiz et al., 2020). Integrating these technologies enhances the efficiency and effectiveness of plant breeding and research, thereby contributing to improvements in agricultural production and sustainability.

5.3 Key Technologies in High-Throughput Phenotyping

High-throughput phenotyping relies on several key technologies that collectively enhance the efficiency and accuracy of plant trait measurement (He et al., 2024). These technologies include:

5.3.1 Unmanned Aerial Vehicles (UAVs)

UAVs, commonly known as drones, are extensively employed in HTP for capturing aerial imagery and data from extensive crop fields (Fig. 5.3). Equipped with high-resolution cameras and sensors, drones provide valuable information on traits such as plant growth, canopy cover, and spatial distribution (Phade et al., 2023). They offer a comprehensive view of large-scale plant populations, facilitating the

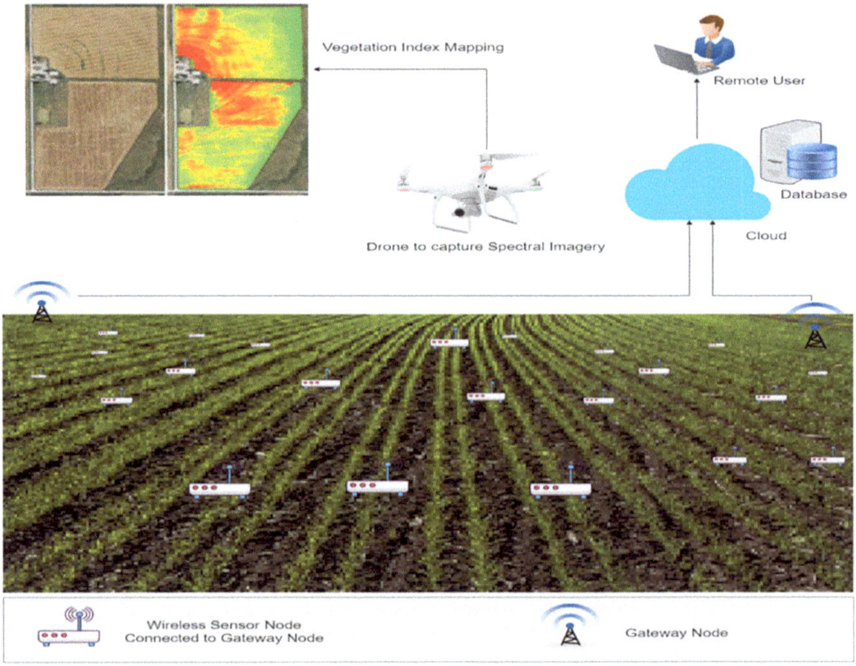

Fig. 5.3 Application of drone cameras in precision agriculture (Shafi et al., 2019)

assessment of plant health and development while alleviating monitoring challenges (Skobelev et al., 2019). Hu et al. (2020) explored the potential of UAVs for HTP in wheat breeding, particularly in early selection cycles where limited seeds and resources restrict breeders to small plots. The study aimed to assess the grain yield of 32 wheat genotypes grown in plots with varying row numbers (1, 2, 3, and 12 rows) using UAV- and ground-based spectral sensing. Results showed that near-infrared (NIR)-based spectral indices were significantly correlated with grain yield, regardless of plot size. UAV-based sensing demonstrated higher precision than ground-based methods, suggesting its potential to improve phenotyping efficiency in large-scale breeding programs. The study concludes that UAVs may enhance the selection of improved genotypes and aid in calibrating genomic data by capturing additional complex traits.

5.3.2 *Imaging Sensors*

Various imaging sensors play a crucial role in HTP by capturing detailed visual and nonvisual data. For assessing plant health, stress responses, and nutrient status, multispectral and hyperspectral sensors gather information across multiple light wavelengths (Bruning et al., 2020). Thermal imaging sensors measure plant temperature, indicating water stress or disease. These sensors are often integrated into UAVs or stationary platforms to generate extensive phenotypic data. The use of UAVs in precision agriculture has become increasingly prevalent in recent years (Radoglou-Grammatikis et al., 2020). Initially developed for military purposes, UAVs have since found diverse applications in agriculture, including supervision, spraying, and monitoring (Rejeb et al., 2022). They are employed for tasks such as chemical spraying, growth monitoring, livestock management, soil and field analysis, field mapping, plantation management, geo-fencing, and crop health monitoring (Kahveci, 2017; Kajal & Chauhan, 2024). The application of UAVs continues to be explored due to their vast potential in precision agriculture, including monitoring pests and diseases in field crops. This approach reduces labor and time requirements.

Research on UAVs for plant disease monitoring has been comprehensively reviewed in several studies (Feng et al., 2021). Spectral data from healthy and diseased plants vary due to changes in pigmentation, color, and water content. Studies have investigated using UAVs for monitoring and detecting wheat diseases (Heidarian Dehkordi et al., 2020; Guo et al., 2021a; Neupane & Baysal-Gurel, 2021). Additionally, changing environmental conditions have led to the diagnosis of wheat diseases in previously unaffected regions. UAVs provide an advanced and efficient approach for the swift identification and monitoring of wheat diseases by utilizing high-resolution aerial imaging, allowing for early detection of infections, even before the appearance of visible symptoms (Joshi et al., 2024). The choice of sensor in UAVs significantly impacts the quality of the data collected (Amarasingam et al., 2022). UAVs equipped with multispectral or hyperspectral cameras can detect plant diseases, such as those affecting wheat, at an early stage, even before visible symptoms appear. At the subfield level, UAV-based disease monitoring offers enhanced precision during detection (Kouadio et al., 2023). RGB and LiDAR sensors are primarily utilized for estimating crop traits such as canopy height and lodging, whereas multispectral cameras are particularly effective for detecting plant stress, pathogens, and nutrient deficiencies (Delavarpour et al., 2021).

5.3.3 RGB Cameras

RGB cameras are among the most commonly used sensors. They capture high-resolution images of plants and generate 3D models, including digital terrain models (DTMs) and digital surface models (DSMs), which are essential for assessing crop biomass (Sweet et al., 2022). However, the limitations in the spectral resolution of RGB cameras prevent them from capturing all the wavelengths necessary for accurate disease detection. Li et al. (2023) applied a combination of UAV-based RGB imaging and ground phenotyping to assess canopy senescence in wheat, using a RIL population from the cross Zhongmai 175/Lunxuan 987. Their findings revealed that UAV-based vegetation indices were as effective as ground-based traits for selecting slow-senescence genotypes, with significant variation observed among the genotypes. Five QTLs linked to senescence were mapped on chromosomes 1B,

2B, 3A, and 4B. A significant locus on chromosome 4B explained up to 23.07% of phenotypic variation for both UAV and ground traits at the late grain-filling stage. Additionally, QTL on chromosomes 1B, 3A, and 4B were validated in a natural population for slow senescence, chlorophyll levels, and canopy temperature. These results suggest that UAV-based RGB imagery offers an efficient and precise approach to evaluating senescence and its genetic regulation in wheat. The primary advantages of RGB sensors include their cost-effectiveness, lightweight design, and ease of use, making them practical for agricultural applications (Amarasingam et al., 2022).

5.3.4 *Multispectral Cameras*

Multispectral cameras, on the other hand, capture radiation across multiple bands, including visible and near-infrared spectra (Maes & Steppe, 2019). These cameras are capable of calculating indices such as the Normalized Difference Vegetation Index and are especially effective for disease screening in wheat (Skendžić et al., 2023). For identifying plant stress and pathogens, hyperspectral cameras are a valuable solution as they collect data in narrow spectral bands, providing more detailed information. Hyperspectral imaging has proven effective in detecting diseases such as yellow rust in wheat (Yao et al., 2019; Zhang et al., 2019), while UAS-based RGB and multispectral imaging have also shown significant success in identifying disease infections and severity in various wheat diseases, including leaf rust, stripe rust, and fusarium head blight (Heidarian Dehkordi et al., 2020; Herr et al., 2023; Guo et al., 2021a, b). Some commonly used open-source image-based tools for shoot and root phenotyping in wheat are presented in Table 5.1 (Bekkering et al., 2020).

Table 5.1 Commonly used open source image-based tools for shoot and root phenotyping in wheat (Bekkering et al., 2020)

Software	Operating system	Output examples
CoverageTool	Windows XP +	Leaf area, color profiles, leaf area index
LeafByte	iOS 9+	Leaf area, herbivory extent
Plant Screen Mobile	Android OS 4+	Leaf area and color profiles
Easy Leaf Area	Windows, Mac OS, Android	Total leaf area, leaf area index
RootNav	Windows XP+	Primary and lateral root count, angles, and lengths
DIRT	Platform independent (Web interface)	Root density and distribution
SmartRoot	Platform independent (ImageJ plugin)	Primary and lateral root length, root diameter, insertion angles, lateral root density
RootNav	Windows XP+	Primary and lateral root count, angles, and lengths
SegRoot	Windows 7+, Mac OS, Linux	Root length

5.3.5 *Visible Light Imaging*

Visible light imaging has gained widespread acceptance as a primary method for plant phenotyping, owing to its rapid imaging capabilities, cost-effectiveness, and ease of use. This method provides significant advantages, including the ability to directly capture many morphological traits of plants through visible light images (Garillos-Manliguez & Chiang, 2021). Moreover, physiological and pathological traits can be indirectly assessed through the analysis of these images. However, visible light imaging is not without its limitations. Its effectiveness is heavily reliant on optimal lighting conditions. In addition, accurately segmenting objects of interest, such as leaves and stems, from the background can be problematic, particularly when the background color closely matches that of the target objects. Shadows and occlusions within plant canopies can also introduce errors in image analysis (Ma et al., 2023). Despite these challenges, advancements in image processing algorithms have been reported to mitigate some of these issues. Recent literature highlights various improvements in this area, which contributes to enhancing the accuracy and reliability of visible light imaging in plant phenotyping (Guo et al., 2017).

5.3.6 *Fluorescence Imaging*

Fluorescence imaging offers a unique and valuable tool for measuring plant physiological and pathological features, but has limitations in sensitivity to water stress and scalability for larger plants. With further development, such as 3D reconstruction capabilities and improved robustness, fluorescence imaging could become an indispensable asset in detecting early signs of stress and taking timely preventative measures, but currently, its applications are primarily limited to controlled environments due to high-power requirements (Scholes & Rolfe, 2009; Zhang & Zhang, 2018). Arya et al. (2024) evaluate water deficit stress tolerance in 184 RILs of wheat using HTP and chlorophyll fluorescence imaging. Under water stress, RILs exhibited reduced Fv/Fm ratios, but stress-tolerant lines maintained higher values, minimizing yield losses. The study highlights the potential of fluorescence image-based HTP for identifying stress-tolerant wheat RILs and novel sensor-derived traits, advocating for using advanced data analytics like PCA, clustering, and ML to enhance crop improvement for stress tolerance and sustainable food production.

5.3.7 *Thermal Imaging*

Thermal imaging systems provide outstanding sensitivity, accuracy, and a broad measurement range, establishing them as dependable noncontact solutions for automated real-time monitoring and measurements. These features make thermal imaging a highly versatile and effective solution for a wide range of applications (Martínez et al., 2017). Thermal cameras provide a significant advantage over traditional temperature measurement tools by offering high-resolution temperature measurements over an area, rather than just a single point (Qin et al., 2021). This capability allows thermal imaging to rapidly measure multiple plots at the field scale and capture canopy temperatures under varying environmental conditions (Perich et al., 2020). Ashfaq et al. (2022) utilized thermal imaging as a key tool to assess drought and heat stress tolerance in 46 wheat genotypes. Infrared thermal imaging was employed to measure canopy temperature, a nondestructive and rapid method for evaluating water stress indices. Cooler canopy temperatures, as detected through thermal imaging, were positively associated with higher grain yields under both drought and heat stress conditions. This approach provided a reliable indicator of each genotype's ability to regulate temperature under stress, contributing to identifying genotypes with superior stress tolerance. Thermal imaging thus played a critical role in distinguishing high-performing genotypes based on their water-use efficiency and stress adaptation.

5.3.8 3D Imaging Technique

3D imaging technology provides a unique opportunity to measure intricate plant morphological features in both controlled and field environments, including branch structure, leaf angles, and root systems. However, despite its potential, the application of this technology is frequently hindered by factors such as scale, resolution, throughput, accessibility, and cost, limiting its use in many settings (Zhang & Zhang, 2018; Paulus, 2019). The effectiveness of stereovision in measuring plant phenotypes is highly dependent on the algorithm, with performance being greatly affected by lighting conditions, the plant's optical features (such as surface texture), and the need for plant rotations. The deployment of stereo systems in the field presents additional challenges, as canopy movement caused by wind can lead to significant difficulties. In root phenotyping, tomographic technologies, such as MRI, offer advantages in detecting and reconstructing root systems, as well as measuring water transport and quantifying photosynthesis. However, MRI's effectiveness is greatly influenced by the type of media used for plant growth. The limitations associated with high costs, low resolution, extended scanning times, and the lack of scalability for field applications have severely restricted the practicality of tomographic technologies (such as MRI, laser scanning, ground-penetrating radar, and X-ray CT) in plant research, as these methods are generally effective only for measuring large roots at relatively shallow depths and in specific soil types (Gibbs et al., 2016).

5.3.9 Satellite Imagery

Satellite imagery, commonly referred to as remote sensing, represents one of the earliest methods employed for HTP. It utilizes multiple sensors on satellites to capture ground information from large areas (Feng et al., 2021). These sensors encompass RGB, multispectral, hyperspectral, thermal, and time-of-flight (ToF) technologies. Multispectral sensors capture data across specific wavelength bands within the electromagnetic spectrum, typically covering 2–10 bands simultaneously. In addition to these bands, near-infrared (NIR) and infrared (IR) bands are also utilized, providing additional detail across the electromagnetic spectrum (Kim, 2020). Multispectral sensors can provide spectral resolutions up to 0.25 m, and this resolution continues to improve with ongoing technological advancements, delivering greater detail and accuracy in remote sensing. Satellite imaging technologies are crucial for gathering data applicable to various agricultural practices. Nonetheless, their high costs, low resolution for plot-level data collection, and extended revisit intervals present significant limitations that hinder their effectiveness in crop improvement efforts (Jangra et al., 2021).

5.3.10 Hyperspectral Sensors

Hyperspectral sensors target an extensive range of bands (up to thousands) within a narrower spectral range, providing more comprehensive and accurate data (Pettorelli, 2019). With their ability to target multiple bands, hyperspectral sensors offer higher spectral resolution than multispectral sensors, leading to investigations into their use for identifying plant diseases (Nagasubramanian et al., 2019). Multispectral and hyperspectral sensors predominantly depend on sunlight reflected from the Earth's surface to collect data and are typically categorized as passive sensors. In contrast, active sensors, such as radio detection and ranging (RADAR) and light detection and ranging (LiDAR), transmit radiation to obtain data, providing a distinct approach to gathering information about the Earth's surface (Teke et al., 2013). The data collected by multispectral and hyperspectral sensors across various spectral bands can be analyzed by applying a wide array of spectral indices to assess plant health and conditions by highlighting specific biochemical properties related to plant status (Sarić et al., 2022). Satellite imaging is often limited by its high construction and launch costs. Pettorelli (2019) provides a list of unique sensors and their spatial resolutions, including Resourcesat-2, Resourcesat-2A, Sentinel-2 A + B, EO-1 Hyperion, Spot-6, Spot-7, and KOMPSAT-3A, which have all been utilized in plant stress assessment. A brief summary of all possible features of HTP is given in Table 5.2.

Table 5.2 Key technologies and their applications in high-throughput phenotyping

Technology	Purpose	Key features	Applications	Limitations	References
Unmanned Aerial Vehicles	Capture aerial imagery and data over large fields.	High-resolution cameras, sensors (RGB, multispectral, hyperspectral).	Monitoring plant growth, canopy cover, spatial distribution, disease detection.	Wind interference, expensive, large-scale fields only.	Radoglou-Grammatikis et al. (2020)
RGB Cameras	Capture high-resolution images for 3D modeling.	Cost-effective, easy to use, 3D model generation.	Biomass assessment, disease detection (stripe rust, Fusarium head blight).	Limited spectral resolution for accurate disease detection.	Zhao et al. (2022)
Multispectral Cameras	Capture radiation across multiple bands.	Can detect plant stress, pathogens, nutrient deficiencies.	Disease screening (e.g., wheat diseases).	Limited detail in spectral resolution compared to hyperspectral.	Zhang et al. (2024)
Hyperspectral Cameras	Capture data in narrow spectral bands.	Provides more detailed spectral information for identifying stress/disease.	Early disease detection (e.g., yellow rust in wheat).	Expensive and complex to use in large-scale fields.	Zhang et al. (2019)
Thermal Imaging	Measure plant temperature for stress detection.	High precision, sensitivity, non-contact temperature measurement.	Water stress, disease detection.	Requires calibration for different environmental conditions.	Amarasingam et al. (2022)
Fluorescence Imaging	Measure plant physiological and pathological traits.	Early stress detection, photosynthetic performance.	Stress detection in controlled environments.	Not easily scalable, requires high power.	Singh et al. (2021)
Satellite Imagery	Remote sensing over large areas (e.g., counties).	Multispectral and hyperspectral sensors, high-resolution imagery.	Large-scale plant health and stress assessment.	High construction and launch costs, cloud interference.	Pettorelli (2019)
Machine Learning	Analyze large datasets for patterns and predictions.	Integrates data analysis, image processing, and decision-making.	Disease identification, stress prediction, phenotype-genotype correlations.	Requires large datasets and significant computing power.	Sandhu et al. (2021)

5.4 Application of Deep Learning in Plant Phenotyping

Deep learning (DL) encompasses a diverse range of network architectures, including deep neural networks, convolutional neural networks (CNNs), and recurrent neural networks. In recent years, CNNs have proven highly effective across various natural language processing and computer vision applications, such as segmentation and image classification, often surpassing traditional state-of-the-art methods (Zualkernan et al., 2023; Charisis & Argyropoulos, 2024). HTP relies significantly on nondestructive, image-based phenotyping approaches. Mathematically, a digital image is represented as a matrix of numerical values organized into rows and columns, resulting in large and complex datasets. Extracting meaningful insights from such extensive data remains a significant challenge (Gnädinger & Schmidhalter, 2017).

In recent years, DL-based solutions have emerged as a promising area within HTP, finding applications across numerous tasks and establishing itself as the state-of-the-art technology for multitask phenotyping (Arya et al., 2022; Hammers et al., 2023). Spike characteristics serve as proxy measures for grain yield; however, traditional methods face difficulties under field conditions due to feature overlap and varying visibility. The R-CNN model has demonstrated high accuracy (93.4%) and an F1 score of 0.95 in detecting and counting spikes under such conditions (Hasan et al., 2018). These deep learning techniques, particularly those applied to RGB images, are critical for the high-throughput quantitative assessment of numerous traits.

Field phenotyping presents unique challenges due to the complex and variable backgrounds encountered in outdoor environments (Ghosal et al., 2018). Sadeghi-Tehran et al. (2019) employed a U-Net architecture combined with a VGG model, using a superpixel-based SLIC approach and a binary cross-entropy loss function, to develop the DeepCount model for large-scale spike quantification under field conditions (Sadeghi-Tehran et al., 2019). This model demonstrated superior performance over traditional handcrafted image-based methods, achieving a correlation coefficient of 0.94 and a root mean square error of 25.1, largely attributed to the DL model's capability to manage the complexities of movement, shadow, and overlap in field environments.

For effective field phenotyping, integrating deep learning with robotics and advanced computer vision techniques is essential to develop robust phenotyping solutions. Such integration is expected to drive further advancements in automated HTP for large-scale breeding programs. The study by Moghimi et al. (2020) focuses on developing an automated framework for high-throughput yield phenotyping to enhance sustainable crop production. Using a hyperspectral camera mounted on an unmanned aerial vehicle, the framework collected high-resolution aerial imagery over two growing seasons from experimental wheat fields. The imagery enabled precise yield estimation through deep neural networks, achieving a coefficient of determination of 0.79 and 0.41 at subplot and plot scales, respectively. The framework demonstrated potential as a decision-support tool, streamlining the process of evaluating crop yield performance and optimizing plot size for more efficient variety selection.

The study by Kaur et al. (2024) investigates the use of UAVs equipped with hyperspectral sensors to enhance phenotyping in winter wheat breeding programs. Analyzed data from 2145 genotypes were used to predict traits like growth, biomass, leaf area, and canopy cover through hyperspectral measurements across wavelengths of 400–2500 nm. The study employed partial least squares regression (PLSR) and random forest (RF) regression techniques to evaluate plant attributes, revealing that integrating hyperspectral data and derived vegetation indices significantly improved prediction accuracy. Ultimately, the findings suggest that while both PLSR and RF provide similar performance, utilizing advanced machine learning algorithms is crucial for accurate trait predictions in winter wheat. Deep learning models should be included in a plant breeder's toolkit for large-scale breeding programs to enhance genetic gains for quantitative traits (Togninalli et al., 2023).

5.5 Conclusion and Future Outlooks

The adoption of high-throughput phenotyping technologies in wheat breeding represents a significant advancement in agricultural research. With the continued development of UAV-based platforms and advanced imaging systems, the precision and scalability of phenotyping can be further enhanced. Future research should focus on refining deep learning models to handle diverse environmental conditions and phenotypic variations. In addition, there is a need for standardizing data protocols and integrating multisource data to create more comprehensive phenotypic databases. As these advancements continue, high-throughput phenotyping is well positioned to play a critical role in accelerating genetic gain, optimizing resource use, and addressing the challenges posed by climate change in wheat production. These efforts will contribute to more resilient and higher-yielding wheat varieties, ensuring food security in a changing global landscape.

References

Abbas, O. K., Rahim, F. I., & Mohammed, M. K. (2024). Economic analysis of the Most important variables affecting the food security factors for the wheat crop in Iraq for the period 1995-2022. In *IOP conference series: Earth and environmental science* (Vol. 1371, No. 6, p. 062013). IOP Publishing.

Ali, A., Altaf, M. T., Bedir, M., Liaqat, W., Jamil, A., Nadeem, M. A., & Baloch, F. S. (2024). Biotechnological advancements toward sugarcane crop improvement. In N. Kumar (Ed.), *Industrial crop plants. Interdisciplinary biotechnological advances*. Springer. https://doi.org/10.1007/978-981-97-1003-4_13

Amarasingam, N., Gonzalez, F., Salgadoe, A. S. A., Sandino, J., & Powell, K. (2022). Detection of white leaf disease in sugarcane crops using UAV-derived RGB imagery with existing deep learning models. *Remote Sensing, 14*(23), 6137.

Arya, S., Sandhu, K. S., Singh, J., & Kumar, S. (2022). Deep learning: As the new frontier in high-throughput plant phenotyping. *Euphytica, 218*(4), 47.

Arya, S., Sahoo, R. N., Sehgal, V. K., Bandyopadhyay, K., Rejith, R. G., Chinnusamy, V., et al. (2024). High-throughput chlorophyll fluorescence image-based phenotyping for water deficit stress tolerance in wheat. *Plant Physiology Reports*, 1–16.

Ashfaq, W., Brodie, G., Fuentes, S., & Gupta, D. (2022). Infrared thermal imaging and morpho-physiological indices used for wheat genotypes screening under drought and heat stress. *Plants, 11*(23), 3269.

Bekkering, C. S., Huang, J., & Tian, L. (2020). Image-based, organ-level plant phenotyping for wheat improvement. *Agronomy, 10*(9), 1287.

Bhat, J. A., Deshmukh, R., Zhao, T., Patil, G., Deokar, A., Shinde, S., & Chaudhary, J. (2020). Harnessing high-throughput phenotyping and genotyping for enhanced drought tolerance in crop plants. *Journal of Biotechnology, 324*, 248–260.

Bruning, B., Berger, B., Lewis, M., Liu, H., & Garnett, T. (2020). Approaches, applications, and future directions for hyperspectral vegetation studies: An emphasis on yield-limiting factors in wheat. *The Plant Phenome Journal, 3*(1), e20007.

Charisis, C., & Argyropoulos, D. (2024). Deep learning-based instance segmentation architectures in agriculture: A review of the scopes and challenges. *Smart Agricultural Technology, 8*, 100448.

Chawade, A., van Ham, J., Blomquist, H., Bagge, O., Alexandersson, E., & Ortiz, R. (2019). High-throughput field-phenotyping tools for plant breeding and precision agriculture. *Agronomy, 9*(5), 258.

Cheng, T., Li, M., Quan, L., Song, Y., Lou, Z., Li, H., & Du, X. (2024). A multimodal and temporal network-based yield assessment method for different heat-tolerant genotypes of wheat. *Agronomy, 14*(8), 1694.

Cudjoe, D. K., Virlet, N., Castle, M., Riche, A. B., Mhada, M., Waine, T. W., et al. (2023). Field phenotyping for African crops: Overview and perspectives. *Frontiers in Plant Science, 14*, 1219673.

Delavarpour, N., Koparan, C., Nowatzki, J., Bajwa, S., & Sun, X. (2021). A technical study on UAV characteristics for precision agriculture applications and associated practical challenges. *Remote Sensing, 13*(6), 1204.

Feng, L., Chen, S., Zhang, C., Zhang, Y., & He, Y. (2021). A comprehensive review on recent applications of unmanned aerial vehicle remote sensing with various sensors for high-throughput plant phenotyping. *Computers and Electronics in Agriculture, 182*, 106033.

Fiorani, F., & Schurr, U. (2013). Future scenarios for plant phenotyping. *Annual Review of Plant Biology, 64*(1), 267–291.

Fu, X., & Jiang, D. (2022). High-throughput phenotyping: The latest research tool for sustainable crop production under global climate change scenarios. In *Sustainable crop productivity and quality under climate change* (pp. 313–381). Academic Press.

Garillos-Manliguez, C. A., & Chiang, J. Y. (2021). Multimodal deep learning and visible-light and hyperspectral imaging for fruit maturity estimation. *Sensors, 21*(4), 1288.

Ghosal, S., Blystone, D., Singh, A. K., et al. (2018). An explainable deep machine vision framework for plant stress phenotyping. *Proceedings of the National Academy of Sciences, 115*, 4613–4618.

Gibbs, J. A., Pound, M., French, A. P., Wells, D. M., Murchie, E., & Pridmore, T. (2016). Approaches to three-dimensional reconstruction of plant shoot topology and geometry. *Functional Plant Biology, 44*(1), 62–75.

Gill, T., Gill, S. K., Saini, D. K., Chopra, Y., de Koff, J. P., & Sandhu, K. S. (2022). A comprehensive review of high throughput phenotyping and machine learning for plant stress phenotyping. *Phenomics, 2*(3), 156–183.

Gnädinger, F., & Schmidhalter, U. (2017). Digital counts of maize plants by unmanned aerial vehicles (uavs). *Remote Sens (basel), 9*, 544.

Gracia-Romero, A., Kefauver, S. C., Fernandez-Gallego, J. A., Vergara-Díaz, O., Nieto-Taladriz, M. T., & Araus, J. L. (2019). UAV and ground image-based phenotyping: A proof of concept with durum wheat. *Remote Sensing, 11*(10), 1244.

Guo, P., Wu, F., Dai, J., Wang, H., Xu, L., & Zhang, G. (2017). Comparison of farmland crop classification methods based on visible light images of unmanned aerial vehicles. *Transactions of the Chinese Society of Agricultural Engineering, 33*(13), 112–119.

Guo, A., Huang, W., Dong, Y., Ye, H., Ma, H., Liu, B., et al. (2021a). Wheat yellow rust detection using UAV-based hyperspectral technology. *Remote Sensing, 13*(1), 123.

Guo, W., Carroll, M. E., Singh, A., Swetnam, T. L., Merchant, N., Sarkar, S., Singh, A. K., & Ganapathysubramanian, B. (2021b). UAS-based plant phenotyping for research and breeding applications. *Plant Phenomics, 2021*, 9840192.

Hammers, M., Winn, Z. J., Ben-Hur, A., Larkin, D., Murry, J., & Mason, R. E. (2023). Phenotyping and predicting wheat spike characteristics using image analysis and machine learning. *The Plant Phenome Journal, 6*(1), e20087.

Hasan, M. M., Chopin, J. P., Laga, H., & Miklavcic, S. J. (2018). Detection and analysis of wheat spikes using convolutional neural networks. *Plant Methods, 14*, 100.

Hassan, M. A., Yang, M., Rasheed, A., Yang, G., Reynolds, M., Xia, X., et al. (2019). A rapid monitoring of NDVI across the wheat growth cycle for grain yield prediction using a multispectral UAV platform. *Plant Science, 282*, 95–103.

He, S., Li, X., Chen, M., Xu, X., Tang, F., Gong, T., et al. (2024). Crop HTP technologies: Applications and prospects. *Agriculture, 14*(5), 723.

Heidarian Dehkordi, R., El Jarroudi, M., Kouadio, L., Meersmans, J., & Beyer, M. (2020). Monitoring wheat leaf rust and stripe rust in winter wheat using high-resolution UAV-based red-green-blue imagery. *Remote Sensing, 12*(22), 3696.

Herr, A. W., Adak, A., Carroll, M. E., Elango, D., Kar, S., Li, C., et al. (2023). Unoccupied aerial systems imagery for phenotyping in cotton, maize, soybean, and wheat breeding. *Crop Science, 63*(4), 1722–1749.

Hu, Y., Knapp, S., & Schmidhalter, U. (2020). Advancing high-throughput phenotyping of wheat in early selection cycles. *Remote Sensing, 12*(3), 574.

Jangra, S., Chaudhary, V., Yadav, R. C., & Yadav, N. R. (2021). High-throughput phenotyping: A platform to accelerate crop improvement. *Phenomics, 1*(2), 31–53.

Joshi, P., Sandhu, K. S., Dhillon, G. S., Chen, J., & Bohara, K. (2024). Detection and monitoring wheat diseases using unmanned aerial vehicles (UAVs). *Computers and Electronics in Agriculture, 224*, 109158.

Kahveci, M. (2017). Use of geographical information technologies in a precision agriculture management system for food traceability. In *Minimally processed refrigerated fruits and vegetables* (pp. 619–637).

Kajal, M., & Chauhan, A. (2024). Aerial sensing and imaging analysis for agriculture. In *Advances in aerial sensing and imaging* (pp. 383–410).

Kaur, S., Kakani, V. G., Carver, B., Jarquin, D., & Singh, A. (2024). Hyperspectral imaging combined with machine learning for high-throughput phenotyping in winter wheat. *The Plant Phenome Journal, 7*(1), e20111.

Kim, J. Y. (2020). Roadmap to high throughput phenotyping for plant breeding. *Journal of Biosystems Engineering, 45*, 43–55.

Kouadio, L., El Jarroudi, M., Belabess, Z., Laasli, S. E., Roni, M. Z. K., Amine, I. D. I., et al. (2023). A review on UAV-based applications for plant disease detection and monitoring. *Remote Sensing, 15*(17), 4273.

Kuriakose, S. V., Pushker, R., & Hyde, E. M. (2020). Data-driven decisions for accelerated plant breeding. *Accelerated Plant Breeding, 1*, 89–119.

Li, L., Hassan, M. A., Song, J., Xie, Y., Rasheed, A., Yang, S., et al. (2023). UAV-based RGB imagery and ground measurements for high-throughput phenotyping of senescence and QTL mapping in bread wheat. *Crop Science, 63*(6), 3292–3309.

Ma, W., Wang, K., Li, J., Yang, S. X., Li, J., Song, L., & Li, Q. (2023). Infrared and visible image fusion technology and application: A review. *Sensors, 23*(2), 599.

Maes, W. H., & Steppe, K. (2019). Perspectives for remote sensing with unmanned aerial vehicles in precision agriculture. *Trends in Plant Science, 24*(2), 152–164.

Martínez, J., Egea, G., Agüera, J., & Pérez-Ruiz, M. (2017). A cost-effective canopy temperature measurement system for precision agriculture: A case study on sugar beet. *Precision Agriculture, 18*, 95–110.

Mir, R. R., Reynolds, M., Pinto, F., Khan, M. A., & Bhat, M. A. (2019). High-throughput phenotyping for crop improvement in the genomics era. *Plant Science, 282*, 60–72.

Moghimi, A., Yang, C., & Anderson, J. A. (2020). Aerial hyperspectral imagery and deep neural networks for high-throughput yield phenotyping in wheat. *Computers and Electronics in Agriculture, 172*, 105299.

Nagasubramanian, K., Jones, S., Singh, A. K., Sarkar, S., Singh, A., & Ganapathysubramanian, B. (2019). Plant disease identification using explainable 3D deep learning on hyperspectral images. *Plant Methods, 15*, 1–10.

Neupane, K., & Baysal-Gurel, F. (2021). Automatic identification and monitoring of plant diseases using unmanned aerial vehicles: A review. *Remote Sensing, 13*(19), 3841.

Paulus, S. (2019). Measuring crops in 3D: Using geometry for plant phenotyping. *Plant Methods, 15*(1), 103.

Pérez-Ruiz, M., Prior, A., Martínez-Guanter, J., Apolo-Apolo, O. E., Andrade-Sanchez, P., & Egea, G. (2020). Development and evaluation of a self-propelled electric platform for high-throughput field phenotyping in wheat breeding trials. *Computers and Electronics in Agriculture, 169*, 105237.

Perich, G., Hund, A., Anderegg, J., Roth, L., Boer, M. P., Walter, A., et al. (2020). Assessment of multi-image unmanned aerial vehicle based high-throughput field phenotyping of canopy temperature. *Frontiers in Plant Science, 11*, 150.

Pettorelli, N. (2019). *Satellite remote sensing and the management of natural resources*. Oxford University Press.

Phade, G., Kishore, A. T., Omkar, S., & Suresh Kumar, M. (2023). IoT-enabled unmanned aerial vehicle: An emerging trend in precision farming. In *Drone technology: Future trends and practical applications* (pp. 301–324).

Qin, Q., Wu, Z., Zhang, T., Sagan, V., Zhang, Z., Zhang, Y., et al. (2021). Optical and thermal remote sensing for monitoring agricultural drought. *Remote Sensing, 13*(24), 5092.

Radoglou-Grammatikis, P., Sarigiannidis, P., Lagkas, T., & Moscholios, I. (2020). A compilation of UAV applications for precision agriculture. *Computer Networks, 172*, 107148.

Rejeb, A., Abdollahi, A., Rejeb, K., & Treiblmaier, H. (2022). Drones in agriculture: A review and bibliometric analysis. *Computers and Electronics in Agriculture, 198*, 107017.

Sadeghi-Tehran, P., Virlet, N., Ampe, E. M., et al. (2019). DeepCount: In-feld automatic quantifcation of wheat spikes using simple linear iterative clustering and deep convolutional neural networks. *Frontiers in Plant Science, 10*, 1176.

Sandhu, K. S., Lozada, D. N., Zhang, Z., Pumphrey, M. O., & Carter, A. H. (2021). Deep learning for predicting complex traits in spring wheat breeding program. *Frontiers in Plant Science, 11*, 613325.

Sankaran, S., Khot, L. R., Espinoza, C. Z., Jarolmasjed, S., Sathuvalli, V. R., Vandemark, G. J., et al. (2015). Low-altitude, high-resolution aerial imaging systems for row and field crop phenotyping: A review. *European Journal of Agronomy, 70*, 112–123.

Sarić, R., Nguyen, V. D., Burge, T., Berkowitz, O., Trtílek, M., Whelan, J., et al. (2022). Applications of hyperspectral imaging in plant phenotyping. *Trends in Plant Science, 27*(3), 301–315.

Scholes, J. D., & Rolfe, S. A. (2009). Chlorophyll fluorescence imaging as tool for understanding the impact of fungal diseases on plant performance: A phenomics perspective. *Functional Plant Biology, 36*(11), 880–892.

Shafi, U., Mumtaz, R., García-Nieto, J., Hassan, S. A., Zaidi, S. A. R., & Iqbal, N. (2019). Precision agriculture techniques and practices: From considerations to applications. *Sensors, 19*(17), 3796.

Sheikh, M., Iqra, F., Ambreen, H., Pravin, K. A., Ikra, M., & Chung, Y. S. (2024). Integrating artificial intelligence and high-throughput phenotyping for crop improvement. *Journal of Integrative Agriculture, 23*(6), 1787–1802.

Singh, A. K., Singh, A., Sarkar, S., Ganapathysubramanian, B., Schapaugh, W., Miguez, F. E., et al. (2021). High-throughput phenotyping in soybean. In *High-throughput crop phenotyping* (pp. 129–163). Springer International Publishing.

Skendžić, S., Zovko, M., Lešić, V., Pajač Živković, I., & Lemić, D. (2023). Detection and evaluation of environmental stress in winter wheat using remote and proximal sensing methods and vegetation indices—A review. *Diversity, 15*(4), 481.

Skobelev, P., Larukchin, V., Mayorov, I., Simonova, E., & Yalovenko, O. (2019). Smart farming-Open multi-agent platform and eco-system of smart services for precision farming. In *International Conference on Practical Applications of Agents and Multi-Agent Systems* (pp. 212–224). Springer International Publishing.

Sweet, D. D., Tirado, S. B., Springer, N. M., Hirsch, C. N., & Hirsch, C. D. (2022). Opportunities and challenges in phenotyping row crops using drone-based RGB imaging. *The Plant Phenome Journal, 5*(1), e20044.

Teke, M., Deveci, H. S., Haliloğlu, O., Gürbüz, S. Z., & Sakarya, U. (2013, June). A short survey of hyperspectral remote sensing applications in agriculture. In *2013 6th international conference on recent advances in space technologies (RAST)* (pp. 171–176). IEEE.

Togninalli, M., Wang, X., Kucera, T., Shrestha, S., Juliana, P., Mondal, S., et al. (2023). Multi-modal deep learning improves grain yield prediction in wheat breeding by fusing genomics and phenomics. *Bioinformatics, 39*(6), btad336.

Xie, C., & Yang, C. (2020). A review on plant high-throughput phenotyping traits using UAV-based sensors. *Computers and Electronics in Agriculture, 178*, 105731.

Xu, R., Li, C., & Bernardes, S. (2021). Development and testing of a UAV-based multi-sensor system for plant phenotyping and precision agriculture. *Remote Sensing, 13*(17), 3517.

Yalinkiliç, N. A., Başbağ, S., Altaf, M. T., Ali, A., Nadeem, M. A., & Baloch, F. S. (2024). Applicability of SCoT markers in unraveling genetic variation and population structure among sugar beet (Beta vulgaris L.) germplasm. *Molecular Biology Reports, 51*(1), 584. https://doi.org/10.1007/s11033-024-09526-1

Yang, W., Feng, H., Zhang, X., Zhang, J., Doonan, J. H., Batchelor, W. D., et al. (2020). Crop phenomics and high-throughput phenotyping: Past decades, current challenges, and future perspectives. *Molecular Plant, 13*(2), 187–214.

Yao, Z., Lei, Y., & He, D. (2019). Early visual detection of wheat stripe rust using visible/near-infrared hyperspectral imaging. *Sensors, 19*(4), 952.

Zhang, Y., & Zhang, N. (2018). Imaging technologies for plant high-throughput phenotyping: A review. *Frontiers of Agricultural Science and Engineering, 5*(4), 406–419.

Zhang, X., Han, L., Dong, Y., Shi, Y., Huang, W., Han, L., et al. (2019). A deep learning-based approach for automated yellow rust disease detection from high-resolution hyperspectral UAV images. *Remote Sensing, 11*(13), 1554.

Zhang, L., Wang, A., Zhang, H., Zhu, Q., Zhang, H., Sun, W., & Niu, Y. (2024). Estimating leaf chlorophyll content of winter wheat from UAV multispectral images using machine learning algorithms under different species, growth stages, and nitrogen stress conditions. *Agriculture, 14*(7), 1064.

Zhao, J., Kumar, A., Banoth, B. N., Marathi, B., Rajalakshmi, P., Rewald, B., et al. (2022). Deep-learning-based multispectral image reconstruction from single natural color RGB image-Enhancing UAV-based phenotyping. *Remote Sensing, 14*(5), 1272.

Zualkernan, I., Abuhani, D. A., Hussain, M. H., Khan, J., & ElMohandes, M. (2023). Machine learning for precision agriculture using imagery from unmanned aerial vehicles (uavs): A survey. *Drones, 7*(6), 382.

Chapter 6
Combination of High-Throughput Phenotyping and Genomics in Precise Wheat Breeding

Muhammad Tanveer Altaf, Amjad Ali, Parnaz Mortazvi, Waqas Liaqat, Muhammed Tatar, Rada Sucur, Mehmet Bedir, Jazib Javed, Jin-Ying Gou, Muhammad Azhar Nadeem, Ahmad Alsaleh, Hüsnü Aktas, and Faheem Shehzad Baloch

Abstract The advancement of wheat breeding is crucial to meeting the increasing global food demand, and integrating high-throughput phenotyping (HTP) with genomic approaches represents a transformative strategy for achieving precise wheat improvement. This chapter explores the multifaceted role of HTP technologies, including unmanned aerial vehicles (UAVs), imaging sensors, and satellite imagery, in enhancing phenotypic data acquisition. By utilizing various imaging techniques, such as RGB, multispectral, thermal, and hyperspectral sensors, breeders

The original version of the chapter has been revised. A correction to this chapter can be found at https://doi.org/10.1007/978-3-031-99954-3_15

M. T. Altaf (✉) · W. Liaqat
Department of Field Crops, Faculty of Agriculture, Recep Tayyip Erdoğan University, Rize/Pazar, Türkiye
e-mail: muhammadtanveer.altaf@erdogan.edu.tr

A. Ali
Department of Plant Protection, Faculty of Agricultural Sciences and Technologies, Sivas University of Science and Technology, Sivas, Türkiye

P. Mortazvi · M. Tatar · M. Bedir
Department of Plant Production and Technologies, Faculty of Agricultural Sciences and Technologies, Sivas University of Science and Technology, Sivas, Türkiye

R. Sucur
University of Novi Sad, Faculty of Agriculture, Novi Sad, Serbia

J. Javed · J.-Y. Gou
Frontiers Science Center for Molecular Design Breeding, China Agricultural University, Beijing, China

M. A. Nadeem
Department of Biotechnology, Faculty of Science, Mersin University, Yenişehir, Mersin, Türkiye

N. Zencirci et al. (eds.), *Empowering Wheat Cultivation with GIS, Digital Approaches and Artificial Intelligence*, https://doi.org/10.1007/978-3-031-99954-3_6

can capture detailed physiological traits, enabling more informed selection decisions. Coupled with genomic approaches, such as next-generation sequencing, quantitative trait loci mapping, and genome-wide association studies, the synergy between these methodologies significantly enhances prediction accuracy and optimizes complex traits, including drought resistance and yield potential. By harnessing big data and machine learning, breeders can analyze vast datasets to refine breeding strategies and improve varietal development efficiency. The comprehensive examination of these cutting-edge technologies highlights the critical need for collaborative efforts in wheat research aimed at developing resilient varieties capable of thriving under changing climate conditions. Overall, this chapter underscores the promise of combining HTP and genomics to revolutionize precision wheat breeding and contribute to sustainable food security.

Keywords Unmanned aerial vehicles · Thermal imaging · Next-generation sequencing · Genomic selection · Machine learning

Abbreviations

BSA	Bulk-segregant analysis
CIMMYT	International Maize and Wheat Improvement Center
DNA	Deoxyribonucleic acid
GBS	Genotyping by sequencing
GEBV	Genomic estimated breeding value
GS	Genomic selection
GWAS	Genome-Wide Association Studies
GY	Grain yield
HTP	High-throughput phenotyping
ICIM	Inclusive composite interval mapping
IWGSC	International Wheat Genome Sequencing Consortium
KASP	Kompetitive Allele-specific PCR
LD	Linkage disequilibrium
MAS	Marker-assisted selection
ML	Machine learning

A. Alsaleh
Department of Agriculture and Food, Hemp Research Institute, Yozgat Bozok University, Yozgat, Türkiye

H. Aktas
Kızıltepe Faculty of Agricultural Sciences and Technologies, Mardin, Türkiye

F. S. Baloch (✉)
Department of Biotechnology, Faculty of Science, Mersin University, Yenişehir, Mersin, Türkiye

Department of Plant Resources and Environment, Jeju National University, Jeju City, Republic of Korea

MSG	Multiplexed shotgun genotyping
MT-GS	Multi-trait genomic selection
NDVI	Normalized difference vegetation index
NGS	Next-generation sequencing
PCR	Polymerase chain reaction
QTL	Quantitative trait locus
R^2	Coefficient of determination
RAD-seq	Restriction-site associated DNA sequencing
RAINBOW	Haplotype-based GWAS method
RAPD	Random amplified polymorphic DNA
RFLP	Restriction fragment length polymorphism
RGB	Red green blue
RIL	Recombinant inbred line
RNA	Ribonucleic acid
RNA-seq	RNA sequencing
RRLs	Reduced-representation libraries
SBP	Sequence-based polymorphic markers
SeeD	Seeds of discovery
SNP	Single nucleotide polymorphism
SSR	Simple sequence repeat
UAV	Unmanned aerial vehicle

6.1 Introduction

In recent years, precise wheat breeding has emerged as a vital approach to tackling global agricultural challenges, particularly improving crop resilience, yield, and quality in response to changing environmental conditions (Lee et al., 2021). The combination of high-throughput phenotyping (HTP) and genomics has proven to be a game changer in wheat breeding, allowing for the precise measurement of complex traits and accelerating genetic improvements essential for sustainable agriculture. This integration enhances breeding efficiency and significantly shortens the time required to develop high-yielding, climate-resilient wheat varieties. HTP technologies have revolutionized the way phenotypic data is collected. With the help of automated systems and advanced imaging technologies, such as drones and multispectral and hyperspectral cameras, researchers can rapidly assess important plant traits like height, leaf area, growth patterns, and stress responses (Crain et al., 2018). These innovations, paired with machine learning (ML) models, allow for the analysis of large datasets, enabling a better understanding of the relationship between phenotypic traits and genetic markers. This data-driven approach enhances the effectiveness of breeding programs, leading to more robust crop performance and resilience. At the genetic level, next-generation sequencing (NGS) and molecular markers have transformed the landscape of wheat breeding (Baloch et al., 2024; Altaf et al., 2024a, b). Techniques like genome-wide association studies (GWAS), quantitative trait loci (QTL) mapping, and genomic selection (GS) provide breeders with powerful tools to identify genes associated with key traits such as disease resistance, drought tolerance, and yield optimization (Crain et al., 2022). By integrating these genomic technologies, breeding programs are able to make more precise and efficient selections of superior wheat varieties.

The synergy between HTP and genomics has further propelled wheat improvement by offering a multidimensional approach to trait analysis. This integration enhances the prediction accuracy of complex traits, such as yield and disease resistance, by allowing for a more precise selection of genotypes across varying environments (Crain et al., 2018). The alignment of phenotypic data with genomic information accelerates breeding cycles and improves overall genetic gain (Maccaferri et al., 2015). Despite these advancements, challenges remain, particularly in the management and analysis of the vast amounts of data generated by these technologies. To address this, more sophisticated ML models and big data analytics are needed to handle the complexities of phenotypic and genomic datasets (Tong & Nikoloski, 2021; Aasim et al., 2023). Future research should focus on refining the accuracy of trait measurements, especially under field conditions, and improving genomic prediction models to account for genotype-environment interactions. Looking ahead, the future of wheat breeding lies in the continued integration of cutting-edge technologies, such as remote sensing, artificial intelligence, and precision agriculture. These innovations will further enhance the speed and precision of breeding programs, enabling more rapid identification of high-performing wheat varieties tailored to diverse environmental conditions (Kaushal et al., 2024). Embracing emerging tools, such as multi-trait genomic selection, holds the potential to optimize combinations of traits for yield, quality, and stress tolerance, thereby meeting the growing demands of global agriculture and the challenges posed by climate change.

6.2 Big Data and Machine Learning

The growing use of HTP platforms in agriculture has resulted in a surge of vast and varied data collection at multiple time points, creating significant challenges in data analysis and comprehension (Gill et al., 2022). This highlights the need for advanced tools and methodologies to interpret and utilize this information to drive breakthrough discoveries in the field (Fig. 6.1). The McKinsey Global Institute predicts a staggering 40-fold increase in data since 2001, with data volume growing by approximately 50% per year. While storing such immense amounts of data was once deemed impossible, modern advancements have made it achievable. However, the reluctance to adopt HTP in large-scale agricultural operations stems from the difficulty in analyzing and interpreting the massive volumes of data generated. Several studies have successfully demonstrated the potential of ML approaches in image processing for the identification of a wide range of plant stresses, including powdery mildew in cucumber, aflatoxin levels in maize, leaf rust in wheat, salinity stress in chickpea, and others (Wani et al., 2022; Jackulin & Murugavalli, 2022; Rana & Bhushan, 2023). These examples showcase the promise of ML in

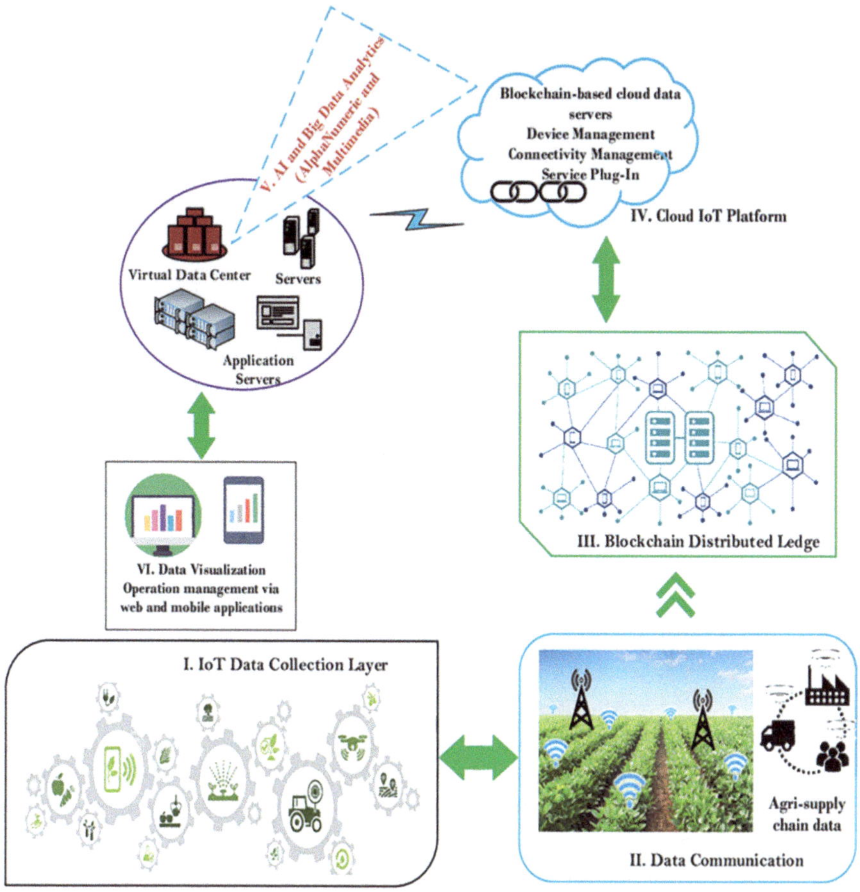

Fig. 6.1 Illustration of smart agriculture ecosystem (Bhat & Huang, 2021)

contributing to effective plant stress management and offer a promising alternative for plant researchers, as it enables the extraction of valuable information and the identification of patterns in data that were previously difficult to discern.

ML integrates techniques and tools from a wide array of disciplines, including probability, statistics, classification, regression, decision theory, data visualization, and neural networks. This interdisciplinary approach enables researchers to effectively link extracted data to plant phenotypic traits, offering a distinct advantage over traditional methods that often analyze individual traits in isolation. ML allows for the simultaneous analysis of multiple parameters and traits, enhancing the scope and depth of data interpretation. Moreover, ML has been successfully applied in various agricultural domains, including bioinformatics (Min et al., 2017), cell biology, epigenetics (Samantara et al., 2021), plant breeding, pathology (Mu et al., 2022), computer vision, image processing, voice recognition, and disease classification (Fuentes et al., 2017). The growing use of ML in agriculture is driven by both its adoption by commercial enterprises and the decreasing costs of sensors and imaging technologies. ML enables computers to learn the patterns present in a dataset without being explicitly instructed, thus allowing for reliable conclusions to be made. By leveraging experience to identify underlying structures, patterns, similarities, or dissimilarities in a dataset, ML can effectively address classification and prediction problems. ML models are typically calibrated using a large dataset known as a training set, with the model performance validated using an independent dataset, referred to as the testing set. The accuracy and precision of a calibrated model are critical factors influencing its applicability for future applications. Typically, two primary approaches are employed during the model training process: supervised and unsupervised ML. In supervised learning, data labels are provided during the training phase, allowing the model to learn associations between input features and corresponding outputs. In contrast, unsupervised learning enables the model to identify patterns and classify data without using predefined labels.

6.3 Genomic Approaches in Wheat Breeding

6.3.1 Next-Generation Sequencing and Molecular Markers in Wheat Breeding

Next-generation sequencing (NGS) platforms have revolutionized genetic and genomic research in crop plants, particularly in wheat, where the large and complex genome (~16 GB) presents significant challenges. NGS-based techniques have facilitated the identification of alleles and enabled genotype-by-sequencing (GBS) using thousands of markers to comprehensively analyze wheat genomes (Liu et al., 2014). Several approaches, including reduced-representation libraries (RRLs), restriction-site associated DNA sequencing (RAD-seq), sequence-based polymorphic markers (SBP), and multiplexed shotgun genotyping (MSG), have been developed to manage this complexity (Yang et al., 2012). Specifically, the GBS pipeline was designed to address the challenges of the wheat genome, targeting restriction sites after digestion to reduce redundancy. High-throughput genotyping platforms, such as DArT-Seq, SNPs, GBS markers, and population-specific targeted GBS (tGBS), have accelerated the accurate mapping of rust resistance-related genomic regions (Nsabiyera et al., 2020).

Despite these advancements, genomics-assisted selection has yet to make substantial contributions to improving drought resistance in wheat. Significant advancements in this field predominantly depend on conventional breeding techniques (Ali et al., 2022). Nonetheless, both private and public breeding programs are actively investigating drought tolerance through a combination of traditional methods and transgenic approaches (Saleem et al., 2016). Whole transcriptome analysis has further enhanced our understanding of gene expression changes in response to environmental stress, with transcriptomic studies in wheat and other non-model plants being increasingly utilized (Singh et al., 2016; Hong et al., 2016). Next-generation sequencing techniques, including RNA sequencing (RNA-seq), have become essential tools for crop breeding, particularly in plants lacking comprehensive genomic information. Proteome sequencing in wheat under drought and salinity stress has provided important insights into drought-responsive mechanisms (Guo et al., 2015), while RNA-seq data holds promise for identifying novel traits relevant to drought resistance.

6.3.2 Discovery of SNPs Using a Genotyping Array

The discovery of single nucleotide polymorphisms (SNPs) in wheat has greatly advanced with the development of genomics resources. Cavanagh et al. (2013) introduced a high-throughput SNP genotyping tool, the iSelect 9 K SNP Infinium BeadChip Assay, which was based on data from 2994 wheat accessions from around the world. This was followed by the development of a high-density SNP map, which incorporated the earlier SNP array and was expanded into the iSelect 90 K SNP Infinium array. This new array provided comprehensive coverage across the seven chromosome groups of wheat. These linked SNPs help address challenges associated with using DNA markers for germplasm characterization. The 90 K SNP array includes nearly 90,000 gene-linked SNPs that aid in evaluating genetic diversity in both tetraploid and hexaploid wheat (Wang et al., 2014).

Additionally, a GBS approach was employed to analyze over 40,000 wheat germplasm samples from CIMMYT as part of the Seeds of Discovery (SeeD) initiative (Li et al., 2015). A consensus map of GBS markers was developed, enabling the detection of new genes and QTLs related to leaf rust, stripe rust, and stem rust resistance. This map also facilitated the validation of known genes by aligning markers to the wheat genome reference (IWGSC RefSeq v1.0) (Appels et al., 2018). Trait-linked SNPs can be converted into Kompetitive allele-specific PCR (KASP) markers using PolyMarker, an automated bioinformatics tool (http://www.polymarker.info). Due to the evolutionary history of wheat and its diploidization, SNPs are found in different chromosomal locations. PolyMarker also designs KASP markers specific to genomes using the IWGSC RefSeq v1.0 (Ramirez-Gonzalez et al., 2015). However, before these markers can be effectively used in MAS, they need to be validated for Mendelian inheritance and their association with phenotypic traits within bi-parental mapping populations.

6.3.3 Quantitative Trait Loci Mapping in Wheat Breeding

Quantitative trait loci (QTL) mapping is an established methodology for elucidating the genetic foundations of critical traits in agricultural crops. Substantial progress has been accomplished in the identification of major QTLs and the isolation of genes correlated with essential agronomic traits, including seed size, yield, and kernel weight, across diverse cereal species. Particularly, various genes associated with kernel size and weight in rice have been identified, such as GW2 (Song et al., 2007), GS3 (Fan et al., 2006), qGL3 (Zhang et al., 2012), GS5 (Li et al., 2011), and GS8 (Wang et al., 2012). Additionally, investigations have emphasized the significance of the OsMKKK10-OsMKK4-OsMAPK6 signaling pathway in regulating seed size and weight (Xu et al., 2018).

In wheat, a high-density linkage map was developed by Su et al. (2018), which included 6312 SNP and SSR markers to identify QTLs associated with kernel traits, such as size and weight, using a recombinant inbred line (RIL) population. From five different environments, 78 QTLs related to kernel diameter, thousand seed weight, seed length, and seed width were identified through inclusive composite interval mapping (ICIM). Among them, six stable QTLs were found across four environments for kernel size and weight. Chromosome regions on 2D and 5B were key locations for these traits, making them ideal candidates for fine mapping and gene discovery (Su et al., 2018).

Similarly, Wu et al. (2015) identified QTLs for seed length and total seed weight in Chinese wheat. In another study, Kumar et al. (2016) pinpointed QTL-15, which contains an ortholog of the GS3 gene found in rice. Regarding heat stress in wheat, QTLs were identified using 251 RILs derived from HD2808 and HUW510, representing heat-tolerant and susceptible genotypes, respectively. These QTLs were mapped to chromosomes 2A, 2B, and 6D, and were linked to traits such as number of grains per spike, grain weight per spike, grain filling rate, thousand seed weight, grain yield, and duration of grain filling. These consistent QTLs serve as promising targets for enhancing heat tolerance in wheat (Bhusal et al., 2017).

A recent study by Ilyas et al. (2020) employed a linkage map comprising 21 groups covering the A, B, and D genomes, identifying a total of 60 QTLs with both major and minor effects. Notably, the B and D genomes were found to contain genes that confer salinity tolerance, which could be instrumental for MAS in developing high-yielding wheat varieties adapted to saline environments.

6.4 Genome-Wide Association Studies (GWAS) in Wheat Breeding

Genome-wide association studies and linkage mapping have become essential in crop genetics for identifying loci and genes associated with various traits, especially in the era of large-scale genotyping and whole-genome resequencing (Karaköy et al., 2024). Techniques such as QTL mapping and bulk-segregant analysis (BSA) are frequently utilized to pinpoint genomic regions linked to traits of interest. High-quality genetic linkage maps provide a foundation for accurately locating these regions. In comparison, GWAS, which assesses broader genetic diversity, enhances the precision of identifying associated loci. Before conducting association mapping, it is important to assess the population structure and level of linkage disequilibrium (LD). However, traits influenced by multiple genes, QTLs, or haplotypes such as yield are often governed by genes with small effects that interact with environmental factors, making them difficult to detect (Bernardo & Yu, 2007). This explains why many QTLs identified in bi-parental populations have limited utility in breeding programs due to their instability or prevalence across current varieties. Environmental interactions further complicate the identification of consistent QTLs across different genetic backgrounds. Consequently, finding rare QTLs and accurately quantifying those with minor effects remain challenging for both linkage and association mapping, as these are highly influenced by environmental conditions (Crossa et al., 2017; Altaf et al., 2024c).

Haplotypes, defined as multi-allelic combinations of two or more SNPs in strong LD, offer more comprehensive insights than individual SNP markers. The emerging concept of "haplotype assembly" presents a promising approach for enhancing crop traits through the strategic combination of favorable haplotypes (Bevan et al., 2017). Since QTL alleles are distributed differently than SNP markers, utilizing haplotype data can reveal associations that might otherwise be missed when using single SNPs (Lorenz et al., 2010). The RAINBOW method, a cutting-edge haplotype-based GWAS approach, has been developed to more effectively detect rare variants and complex traits by minimizing false positives and improving the identification of causal variants without dependence on LD (Hamazaki & Iwata, 2020). This haplotype-based GWAS methodology has demonstrated efficacy in various crops, including wheat (Voss-Fels et al., 2019) and barley (Lorenz et al., 2010), among others (Qian et al., 2017).

In the context of wheat, both SNP-based and haplotype-based GWAS have successfully identified previously unknown QTLs, including those associated with flag leaf area on chromosome 6B (Chen et al., 2021). Additionally, using haplotypes instead of SNPs has significantly increased the predictive power for traits like leaf rust resistance in wheat, with haplotype-based GWAS improving predictability from 11.72% to 22.86% (Liu et al., 2020). Haplotypes found in wild relatives and crop progenitors often encompass extensive genetic diversity, and selecting haplotypes with fewer deleterious alleles can drive genetic improvement in crops (Bevan et al., 2017).

6.5 Marker-Assisted Selection for Precision Wheat Breeding

Marker-assisted selection (MAS) represents a critical advancement in plant breeding, facilitating the selection of individuals based on molecular markers associated with specific traits (Cobb et al., 2019). Since the 1990s, various molecular marker systems have been established, resulting in a substantial increase in the availability of polymorphic markers for wheat breeding (Rasheed et al., 2017). MAS is particularly advantageous for selecting traits governed by genes with substantial effects, including adaptability, grain yield, disease resistance, end-use quality, and tolerance to abiotic stresses. For these traits, functional or diagnostic markers derived from major genes are particularly effective and have been instrumental in the progression of molecular breeding strategies (Rasheed & Xia, 2019).

Molecular markers provide distinct advantages over traditional morphological and biochemical methods, as they are not affected by environmental variables and developmental stages and allow for detecting variations at the DNA level (Ashraf et al., 2014). The utilization of these markers began in the 1990s with restriction fragment length polymorphism (RFLP) markers, which were employed for mapping genetic diversity and identifying chromosomes in wheat (Gupta et al., 1999). However, due to the labor-intensive nature, high costs, and low frequency of RFLP, the scientific community shifted toward PCR-based markers, including randomly amplified polymorphic DNA (RAPD) and simple sequence repeats (SSR). The latter has proven more reproducible and is widely applied in genome-specific studies (Alotaibi et al., 2021).

SSR markers have been instrumental in advancing wheat breeding by enhancing the identification of novel loci and improving the resolution of yield traits. The initial documentation of SSR markers in wheat in 1998 marked a significant breakthrough in trait enhancement (Alotaibi et al., 2021). Furthermore, functional markers, derived from specific gene sequences, have been extensively utilized in wheat breeding programs. To date, approximately 157 functional markers associated with over 100 loci have been reported for traits related to disease resistance, grain yield, quality, and abiotic stress tolerance (Liu et al., 2012).

MAS has proven effective in improving quality traits such as kernel texture and grain protein content (Kumar et al., 2018). For example, MAS was used to develop backcross-derived lines and release 45 MAS-derived wheat lines, including the hard red winter wheat variety 'Farnum,' which carried the Gpc-B1 gene for grain protein content. These examples highlight MAS as a promising method for wheat quality improvement. Despite its many advantages, MAS has limitations when predicting complex traits influenced by multiple genes and environmental factors (Xu & Crouch, 2008). To address these challenges, genomic selection (GS) has been integrated with MAS, allowing breeders to account for gene interactions and environmental effects, thus improving selection accuracy (Xu & Crouch, 2008).

6.6 Genomic Selection for Accelerating Wheat Improvement

Genomic selection (GS) marks a transformative shift in wheat breeding, providing enhanced prediction accuracy compared to traditional methods like phenotypic selection and MAS. While MAS focuses on specific markers associated with important traits, GS employs genome-wide markers to create predictive models that assess the genomic estimated breeding value (GEBV) of individuals. These models are developed using data from a training population that has both genotype and phenotype information. As the training population grows and the models are fine-tuned, their predictive accuracy increases, making GS especially effective for traits influenced by multiple genes (Lebedev et al., 2020).

One of the advantages of GS is that it does not depend on prior identification of specific QTLs, allowing for efficient assessment of traits governed by numerous genes. Research by Battenfield et al. (2016) indicates that expanding the size of training populations and increasing the marker count can significantly boost prediction accuracy for traits such as grain yield and end-use quality. Additionally, GS facilitates quicker selection processes, often shortening the breeding cycle by 2–3 years compared to conventional techniques. Recent developments in GS have concentrated on optimizing various aspects, including marker density and predictive modeling, to further improve accuracy. For instance, models that consider genotype × environment interactions have shown promise in predicting performance across diverse environmental conditions. There is also ongoing work on multi-trait GS models, which aim to enhance the prediction of primary traits by utilizing data from correlated secondary traits (Moeinizade et al., 2020).

The integration of HTP has significantly advanced GS, offering comprehensive data on traits like plant height, disease resistance, and growth rate. Incorporating these detailed phenotypic measurements into GS models has led to notable increases in prediction accuracy. For example, research conducted by Rutkoski et al. (2016) and Crain et al. (2018) demonstrated that combining phenotypic data with GS could enhance prediction accuracy by as much as 67% for grain yield. GS is revolutionizing the field of wheat breeding by providing faster and more precise selection processes. The incorporation of emerging technologies such as HTP and the development of advanced statistical models suggest that GS has substantial potential to enhance wheat yield, quality, and adaptability in response to changing environmental conditions.

6.7 Synergy Between High-Throughput Phenotyping and Genomics

Combining HTP and genomics offers a promising approach for multidimensional trait analysis in wheat improvement (Fig. 6.2). This integration enhances the prediction accuracy of complex traits, facilitating the development of high-yielding and resilient wheat varieties.

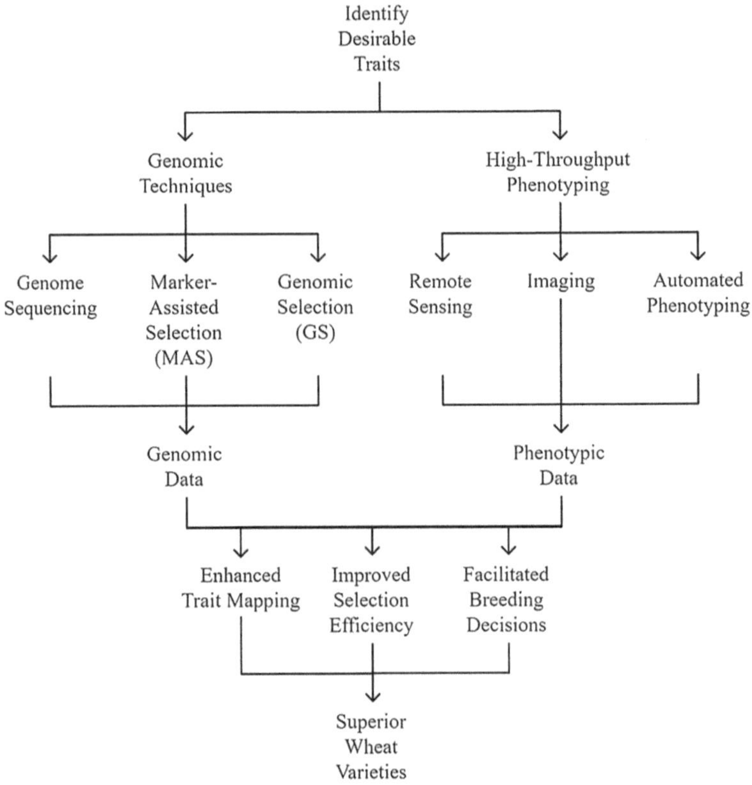

Fig. 6.2 Genomic selection and high-throughput phenotyping in wheat for trait improvement

6.7.1 *Enhanced Prediction Accuracy*

HTP, particularly through UAV-assisted methods, significantly improves the prediction of traits like grain yield and grain protein content using deep learning models, achieving R^2 values of 0.76 and 0.75, respectively (Kaushal et al., 2024). Multi-trait genomic selection (MT-GS) models have shown up to a twofold increase in prediction accuracy for end-use quality traits, such as bake absorption and loaf volume, by incorporating correlated secondary traits (Gill et al., 2023).

6.7.2 *Complex Trait Optimization*

Multi-trait ensemble genomic prediction models have been effective in optimizing genetic gains by improving prediction accuracy for nearly 90% of traits, including GY, by 3–52% (Fradgley et al. 2023b). Integrating nonparametric models like Random Forest has outperformed traditional models, enhancing the understanding of trait relationships and facilitating better selection strategies (Fradgley et al., 2023a; Silva et al., 2023).

6.8 Case Studies: Successes in Integrating High-Throughput Phenotyping and Genomics for Wheat Improvement

The integration of HTP and genomic selection has resulted in significant advancements in wheat improvement, particularly in enhancing complex traits such as drought tolerance and yield. These combined approaches have accelerated genetic gain by enabling a more precise selection of genotypes with desirable traits across diverse environments.

6.8.1 Drought Resistance and Yield Prediction

One notable case study comes from the wheat breeding program at CIMMYT (International Maize and Wheat Improvement Center), where HTP was used in conjunction with GS to improve drought tolerance. Rutkoski et al. (2016) employed canopy temperature and normalized difference vegetation index (NDVI) measurements through HTP platforms to identify genotypes with superior drought tolerance. These phenotypic data were then integrated with genomic data, leading to an improvement in prediction accuracy for grain yield under drought conditions by 67%. The study demonstrated that incorporating HTP data into GS models accelerates the selection process for drought-tolerant lines, significantly enhancing the efficiency of a breeding program. A study utilized an automated HTP platform to analyze 28,392 image-based traits, identifying 17,073 drought-related traits and 5320 significant SNPs linked to drought tolerance (Zhang et al., 2024).

HTP and GS have also been effectively combined to increase grain yield in wheat. Crain et al. (2018) conducted a study using over 1.1 million phenotypic data points from 1170 CIMMYT wheat lines, collected via HTP under heat and drought stress conditions. By integrating these phenotypic data with genome-wide markers, they improved prediction accuracy for grain yield by up to 33% compared to standard models. This combination of HTP and genomic data allowed for faster identification of high-yielding genotypes, even under challenging environmental conditions. Machine learning models developed from these traits provided robust predictions for yield and drought resistance, facilitating the rapid breeding of drought-resistant wheat varieties (Zhang et al., 2024). In the study by Sun et al. (2019), GS models were tested for their ability to predict grain yield in wheat across multiple growing cycles, addressing the challenge of low predictive accuracy due to genetic and environmental variability. Conducted at the International Wheat and

Maize Improvement Center, the research focused on integrating HTP traits (canopy temperature and green normalized difference vegetation index) to enhance grain yield prediction. Results demonstrated that incorporating these secondary traits increased prediction accuracy by 146% across three cycles, especially when measured during wheat heading and early grain filling stages, offering promising avenues for improving wheat breeding strategies.

6.9 Multi-Trait Genomic Selection

Research has demonstrated the effectiveness of UAV-assisted HTP combined with deep learning, achieving high prediction accuracies for grain yield and other traits across multiple locations (Kaushal et al., 2024). Incorporating HTP traits into multi-trait genomic selection models significantly improved predictive abilities, showcasing the potential for enhanced breeding efficiency. Kumar et al. (2021) emphasized that, while recent advances in plant breeding and agronomy have contributed to modest genetic gains, these gains are insufficient to meet future food demands. GS, a promising modern breeding technology, is now being incorporated into plant breeding programs to accelerate genetic improvement. The authors highlight the use of GS in wheat for improving traits like yield, stress tolerance, and disease resistance. However, challenges remain in predicting nonadditive traits and genotype-environment interactions, necessitating multi-trait and multi-environment modeling to improve prediction accuracy.

6.10 Potential for Accelerating Genetic Gain

The combined use of HTP and GS offers the potential to accelerate genetic gain in wheat by enabling earlier and more accurate selection of desirable traits. Michel et al. (2018) demonstrated that genomic selection alone can reduce the breeding cycle by 2–3 years compared to traditional methods. When combined with HTP, which allows for the rapid collection of large-scale phenotypic data, this process can be further accelerated. HTP systems can monitor traits like plant height, biomass, and photosynthetic efficiency in real time, providing breeders with critical data to improve selection accuracy.

Furthermore, advances in HTP technologies, such as drone-based imagery and multispectral sensors, enable phenotypic data collection across large field plots and diverse environments, capturing genotype × environment interactions more effectively. This capacity to gather detailed, high-resolution phenotypic data strengthens genomic prediction models, thereby improving the selection process for complex traits like yield and drought tolerance (Reynolds et al., 2020). Integrating HTP with genomic tools has led to significant improvements in wheat breeding by enhancing selection accuracy and accelerating genetic gain. Case studies have shown that traits like drought tolerance and grain yield have benefited from these combined approaches, offering a promising path forward for addressing the challenges of food security and climate change. The continued development of HTP technologies and genomic models will likely result in further advancements in wheat improvement.

6.11 Conclusion

The future of precise wheat breeding lies in the integration of high-throughput phenotyping and genomics. With advancements in unmanned aerial vehicles, advanced imaging sensors, and AI-driven models, we can achieve more accurate trait measurements and deeper insights into plant responses to environmental challenges. Ongoing research and development are crucial for improving predictive accuracy within breeding programs, facilitating the rapid release of high-performing wheat varieties tailored to diverse climatic conditions. This synergy not only enhances wheat yield and resilience but also addresses the pressing challenges of global food security amid climate change.

References

Aasim, M., Ali, S. A., Altaf, M. T., Ali, A., Nadeem, M. A., & Baloch, F. S. (2023). Artificial neural network and decision tree facilitated prediction and validation of cytokinin-auxin induced in vitro organogenesis of sorghum (Sorghum bicolor L.). *Plant Cell, Tissue and Organ Culture (PCTOC), 153*(3), 611–624. https://doi.org/10.1007/s11240-023-02498-3

Ali, A., Altaf, M. T., Nadeem, M. A., Karaköy, T., Shah, A. N., Azeem, H., & Chung, Y. S. (2022). Recent advancement in OMICS approaches to enhance abiotic stress tolerance in legumes. *Frontiers in Plant Science, 13*, 952759. https://doi.org/10.3389/fpls.2022.952759

Alotaibi, F., Alharbi, S., Alotaibi, M., Al Mosallam, M., Motawei, M., & Alrajhi, A. (2021). Wheat omics: Classical breeding to new breeding technologies. *Saudi Journal of Biological Sciences, 28*(2), 1433–1444. https://doi.org/10.1016/j.sjbs.2020.11.083

Altaf, M. T., Nadeem, M. A., Ali, A., Liaqat, W., Bedir, M., Baran, N., & Baloch, F. S. (2024a). Applicability of Start Codon Targeted (SCoT) markers for the assessment of genetic diversity in bread wheat germplasm. *Genetic Resources and Crop Evolution*, 1–14. https://doi.org/10.1007/s10722-024-02016-0

Altaf, M. T., Liaqat, W., Bedir, M., Ali, A., Nadeem, M. A., & Baloch, F. S. (2024b). Wheat biofortification: A promising approach to improve public health. In N. Zencirci, F. Altay, F. S. Baloch, M. A. Nadeem, & N. Ludidi (Eds.), *Advances in wheat breeding*. Springer. https://doi.org/10.1007/978-981-99-9478-6_16

Altaf, M. T., Liaqat, W., Ali, A., Jamil, A., Bedir, M., Nadeem, M. A., & Baloch, F. S. (2024c). Conventional and biotechnological approaches for the improvement of industrial crops. In N. Kumar (Ed.), *Industrial crop plants. Interdisciplinary biotechnological advances*. Springer. https://doi.org/10.1007/978-981-97-1003-4_1

Appels, R., Eversole, K., Stein, N., Feuillet, C., Keller, B., Rogers, J., et al. (2018). Shifting the limits in wheat research and breeding using a fully annotated reference genome. *Science, 361*(6403), eaar7191. https://doi.org/10.1126/science.aar7191

Ashraf, K., Ahmad, A., Chaudhary, A., Mujeeb, M., Ahmad, S., Amir, M., & Mallick, N. (2014). Genetic diversity analysis of Zingiber officinale Roscoe by RAPD collected from subcontinent of India. *Saudi Journal of Biological Sciences, 21*(2), 159–165. https://doi.org/10.1016/j.sjbs.2013.09.005

Baloch, F. S., Ali, A., Tajibayev, D., Nadeem, M. A., Ölmez, F., Aktaş, H., et al. (2024). Stripe rust resistance gene Yr15 in Turkish and Kazakhstan wheat germplasms and the potential of Turkish wild emmer for stripe rust breeding. *Genetic Resources and Crop Evolution, 71*(6), 2699–2719. https://doi.org/10.1007/s10722-023-01804-4

Battenfield, S. D., Guzmán, C., Gaynor, R. C., Singh, R. P., Peña, R. J., Dreisigacker, S., Allan, K. F., et al. (2016). Genomic selection for processing and end-use quality traits in the CIMMYT spring bread wheat breeding program. *Plant Genome, 9*, 1–12. https://doi.org/10.3835/plantgenome2016.01.0005

Bernardo, R., & Yu, J. (2007). Prospects for genome-wide selection for quantitative traits in maize. *Crop Science, 47*(3), 1082–1090. https://doi.org/10.2135/cropsci2006.11.0690

Bevan, M. W., Uauy, C., Wulff, B. B., Zhou, J., Krasileva, K., & Clark, M. D. (2017). Genomic innovation for crop improvement. *Nature, 543*(7645), 346–354. https://doi.org/10.1038/nature22011

Bhat, S. A., & Huang, N. F. (2021). Big data and ai revolution in precision agriculture: Survey and challenges. *IEEE Access, 9*, 110209–110222. https://doi.org/10.1109/ACCESS.2021.3102227

Bhusal, N., Sarial, A. K., Sharma, P., & Sareen, S. (2017). Mapping QTLs for grain yield components in wheat under heat stress. *PLoS One, 12*(12), e0189594. https://doi.org/10.1371/journal.pone.0189594

Cavanagh, C. R., Chao, S., Wang, S., Huang, B. E., Stephen, S., Kiani, S., Forrest, K., et al. (2013). Genome-wide comparative diversity uncovers multiple targets of selection for improvement in hexaploid wheat landraces and cultivars. *Proceedings. National Academy of Sciences. United States of America, 110*(20), 8057–8062. https://doi.org/10.1073/pnas.1217133110

Chen, S., Liu, F., Wu, W., Jiang, Y., & Zhan, K. (2021). A SNP-based GWAS and functional haplotype-based GWAS of flag leaf-related traits and their influence on the yield of bread wheat (*Triticum aestivum* L.). *Theoretical and Applied Genetics, 134*(12), 3895–3909. https://doi.org/10.1007/s00122-021-03935-7

Cobb, J. N., Biswas, P. S., & Platten, J. D. (2019). Back to the future: Revisiting MAS as a tool for modern plant breeding. *Theoretical and Applied Genetics, 132*, 647–667. https://doi.org/10.1007/s00122-018-3266-4

Crain, J., Mondal, S., Rutkoski, J., Singh, R. P., & Poland, J. (2018). Combining high-throughput phenotyping and genomic information to increase prediction and selection accuracy in wheat breeding. *The Plant Genome, 11*(1), 170043. https://doi.org/10.3835/plantgenome2017.05.0043

Crain, J., Wang, X., Evers, B., & Poland, J. (2022). Evaluation of field-based single plant phenotyping for wheat breeding. *The Plant Phenome Journal, 5*(1), e20045. https://doi.org/10.1002/ppj2.20045

Crossa, J., Pérez-Rodríguez, P., Cuevas, J., Montesinos-López, O., Jarquín, D., de los Campos, G., Burgueño, et al. (2017). Genomic selection in plant breeding: Methods, models, and perspectives. *Trends in Plant Science, 22*(11), 961–975. https://doi.org/10.1016/j.tplants.2017.08.011

Fan, C., Xing, Y., Mao, H., Lu, T., Han, B., Xu, C., Li, X., & Zhang, Q. (2006). GS3, a major QTL for grain length and weight and minor QTL for grain width and thickness in rice, encodes a putative transmembrane protein. *Theoretical and Applied Genetics, 112*, 1164–1171. https://doi.org/10.1007/s00122-006-0218-1

Fradgley, N., Gardner, K. A., Bentley, A. R., Howell, P., Mackay, I. J., Scott, M. F., Mott, R., et al. (2023a). Multi-trait ensemble genomic prediction and simulations of recurrent selection highlight importance of complex trait genetic architecture for long-term genetic gains in wheat. *In Silico Plants, 5*(1), diad002. https://doi.org/10.1093/insilicoplants/diad002

Fradgley NS, Bacon J, Bentley AR, Costa-Neto G, Cottrell A, Crossa J et al. & Gardner KA (2023b) Prediction of near-term climate change impacts on UK wheat quality and the potential for adaptation through plantbreeding. *Global Change Biology, 29*(5), 1296–1313

Fuentes, A., Yoon, S., Kim, S. C., & Park, D. S. (2017). A robust deep-learning-based detector for real-time tomato plant diseases and pests recognition. *Sensors, 17*(9), 2022. https://doi.org/10.3390/s17092022

Gill, T., Gill, S. K., Saini, D. K., Chopra, Y., de Koff, J. P., & Sandhu, K. S. (2022). A comprehensive review of high throughput phenotyping and machine learning for plant stress phenotyping. *Phenomics, 2*(3), 156–183. https://doi.org/10.1007/s43657-022-00048-z

Gill, H. S., Brar, N., Halder, J., Hall, C., Seabourn, B. W., Chen, Y. R., Amand, P. S., et al. (2023). Multi-trait genomic selection improves the prediction accuracy of end-use quality traits in hard winter wheat. *The Plant Genome, 16*(4), e20331. https://doi.org/10.1002/tpg2.20331

Guo, R., Yang, Z., Li, F., Yan, C., Zhong, X., Liu, Q., Xia, X., et al. (2015). Comparative metabolic responses and adaptive strategies of wheat (*Triticum aestivum*) to salt and alkali stress. *BMC Plant Biology, 15*, 1–13. https://doi.org/10.1186/s12870-015-0546-x

Gupta, P. K., Varshney, R. K., Sharma, P. C., & Ramesh, B. J. P. B. (1999). Molecular markers and their applications in wheat breeding. *Plant Breeding, 118*(5), 369–390. https://doi.org/10.1046/j.1439-0523.1999.00401.x

Hamazaki, K., & Iwata, H. (2020). RAINBOW: Haplotype-based genome-wide association study using a novel SNP-set method. *PLoS Computational Biology, 16*(2), e1007663. https://doi.org/10.1371/journal.pcbi.1007663

Hong, Z., Yan, J., & Wenshan, H. (2016). Review and research on health and safety issues for magneticallycoupled resonant wireless power transfer systems. *Transactions of China Electrotechnical Society, 31*(2), 1–12.

Ilyas, N., Amjid, M. W., Saleem, M. A., Khan, W., Wattoo, F. M., Rana, R. M., Maqsood, et al. (2020). Quantitative trait loci (QTL) mapping for physiological and biochemical attributes in a Pasban90/Frontana recombinant inbred lines (RILs) population of wheat (*Triticum aestivum*) under salt stress condition. *Saudi Journal of Biological Sciences, 27*(1), 341–351. https://doi.org/10.1016/j.sjbs.2019.10.003

Jackulin, C., & Murugavalli, S. (2022). A comprehensive review on detection of plant disease using machine learning and deep learning approaches. *Measurement: Sensors, 24*, 100441. https://doi.org/10.1016/j.measen.2022.100441

Karaköy, T., Toklu, F., Karagöl, E. T., Uncuer, D., Çilesiz, Y., Ali, A., et al. (2024). Genome-wide association studies revealed DArTseq loci associated with agronomic traits in Turkish faba bean germplasm. *Genetic Resources and Crop Evolution, 71*(1), 181–198. https://doi.org/10.1007/s10722-023-01615-7

Kaushal, S., Gill, H. S., Billah, M. M., Khan, S. N., Halder, J., Bernardo, A., Amand, P. S., et al. (2024). Enhancing the potential of phenomic and genomic prediction in winter wheat breeding using high-throughput phenotyping and deep learning. *Frontiers in Plant Science, 15*, 1410249. https://doi.org/10.3389/fpls.2024.1410249

Kumar, A., Mantovani, E. E., Seetan, R., Soltani, A., Echeverry-Solarte, M., Jain, S., Simsek, S., et al. (2016). Dissection of genetic factors underlying wheat kernel shape and size in an elite× nonadapted cross using a high density SNP linkage map. *The Plant Genome, 9*(1), plantgenome2015-09. https://doi.org/10.3835/plantgenome2015.09.0081

Kumar, A., Jain, S., Elias, E. M., Ibrahim, M., & Sharma, L. K. (2018). An overview of QTL identification and marker-assisted selection for grain protein content in wheat. In *Eco-friendly agro-biological techniques for enhancing crop productivity* (pp. 245–274). https://doi.org/10.1007/978-981-10-6934-5_11

Kumar, N., Rana, M., Kumar, B., Chand, S., Shiv, A., Wani, S. H., & Kumar, S. (2021). Genomic selection for wheat improvement. In *Physiological, molecular, and genetic perspectives of wheat improvement* (pp. 175–207). https://doi.org/10.1007/978-981-10-6934-5_11

Lebedev, V. G., Lebedeva, T. N., Chernodubov, A. I., & Shestibratov, K. A. (2020). Genomic selection for forest tree improvement: Methods, achievements and perspectives. *Forests, 11*(11), 1190. https://doi.org/10.3390/f11111190

Lee, G. H., Choi, C. H., & Kim, J. Y. (2021). Current status of precision breeding technology and plant transformation for development of wheat breeding material. *Korean Society of Breeding Science, 53*(3), 250–265. https://doi.org/10.9787/KJBS.2021.53.3.250

Li, Y., Fan, C., Xing, Y., Jiang, Y., Luo, L., Sun, L., Shao, D., et al. (2011). Natural variation in GS5 plays an important role in regulating grain size and yield in rice. *Nature Genetics, 43*(12), 1266–1269. https://doi.org/10.1038/ng.977

Li, H., Vikram, P., Singh, R. P., Kilian, A., Carling, J., Song, J., et al. (2015). A high density GBS map of bread wheat and its application for dissecting complex disease resistance traits. *BMC Genomics, 16*(1), 216. https://doi.org/10.1186/s12864-015-1424-5

Liu, Y., He, Z., Appels, R., & Xia, X. (2012). Functional markers in wheat: Current status and future prospects. *Theoretical and Applied Genetics, 125*, 1–10.

Liu, H., Bayer, M., Druka, A., Russell, J. R., Hackett, C. A., Poland, J., et al. (2014). An evaluation of genotyping by sequencing (GBS) to map the Breviaristatum-e (ari-e) locus in cultivated barley. *BMC Genomics, 15*(1), 1–11. https://doi.org/10.1186/1471-2164-15-104

Liu, F., Jiang, Y., Zhao, Y., Schulthess, A. W., & Reif, J. C. (2020). Haplotype-based genome-wide association increases the predictability of leaf rust (*Puccinia triticina*) resistance in wheat. *Journal of Experimental Botany, 71*(22), 6958–6968. https://doi.org/10.1093/jxb/eraa387

Lorenz, A. J., Hamblin, M. T., & Jannink, J. L. (2010). Performance of single nucleotide polymorphisms versus haplotypes for genome-wide association analysis in barley. *PLoS One, 5*(11), e14079. https://doi.org/10.1371/journal.pone.0014079

Maccaferri, M., Ricci, A., Salvi, S., Milner, S. G., Noli, E., Martelli, P. L., et al. (2015). A high-density, SNP-based consensus map of tetraploid wheat as a bridge to integrate durum and bread wheat genomics and breeding. *Plant Biotechnology Journal, 13*(5), 648–663. https://doi.org/10.1111/pbi.12288

Michel, S., Kummer, C., Gallee, M., Hellinger, J., Ametz, C., Akgöl, B., Epure, et al. (2018). Improving the baking quality of bread wheat by genomic selection in early generations. *Theoretical and Applied Genetics, 131*, 477–493. https://doi.org/10.1007/s00122-017-2998-x

Min, S., Lee, B., & Yoon, S. (2017). Deep learning in bioinformatics. *Briefings in Bioinformatics, 18*(5), 851–869. https://doi.org/10.1093/bib/bbw068

Moeinizade, S., Kusmec, A., Hu, G., Wang, L., & Schnable, P. S. (2020). Multi-trait genomic selection methods for crop improvement. *Genetics, 215*(4), 931–945. https://doi.org/10.1534/genetics.120.303305

Mu, H., Wang, B., & Yuan, F. (2022). Bioinformatics in plant breeding and research on disease resistance. *Plants, 11*(22), 3118. https://doi.org/10.3390/plants11223118

Nsabiyera, V., Baranwal, D., Qureshi, N., Kay, P., Forrest, K., Valárik, M., Doležel, J., et al. (2020). Fine mapping of Lr49 using 90K SNP chip array and flow sorted chromosome sequencing in wheat. *Frontiers in Plant Science, 10*, 1787. https://doi.org/10.3389/fpls.2019.01787

Qian, L., Hickey, L. T., Stahl, A., Werner, C. R., Hayes, B., Snowdon, R. J., & Voss-Fels, K. P. (2017). Exploring and harnessing haplotype diversity to improve yield stability in crops. *Frontiers in Plant Science, 8*, 1534. https://doi.org/10.3389/fpls.2017.01534

Ramirez-Gonzalez, R. H., Uauy, C., & Caccamo, M. (2015). PolyMarker: A fast polyploid primer design pipeline. *Bioinformatics, 31*(12), 2038–2039. https://doi.org/10.1093/bioinformatics/btv069

Rana, M., & Bhushan, M. (2023). Machine learning and deep learning approach for medical image analysis: Diagnosis to detection. *Multimedia Tools and Applications, 82*(17), 26731–26769.

Rasheed, A., & Xia, X. (2019). From markers to genome-based breeding in wheat. *Theoretical and Applied Genetics, 132*(3), 767–784. https://doi.org/10.1007/s00122-019-03286-4

Rasheed, A., Hao, Y., Xia, X., Khan, A., Xu, Y., Varshney, R. K., & He, Z. (2017). Crop breeding chips and genotyping platforms: Progress, challenges, and perspectives. *Molecular Plant, 10*(8), 1047–1064. https://doi.org/10.1016/j.molp.2017.06.008

Reynolds, M., Chapman, S., Crespo-Herrera, L., Molero, G., Mondal, S., Pequeno, D. N., Pinto, F., et al. (2020). Breeder friendly phenotyping. *Plant Science, 295*, 110396. https://doi.org/10.1016/j.plantsci.2019.110396

Rutkoski, J., Poland, J., Mondal, S., Autrique, E., Pérez, L. G., Crossa, J., Reynolds, M., & Singh, R. (2016). Canopy temperature and vegetation indices from high-throughput phenotyping improve accuracy of pedigree and genomic selection for grain yield in wheat. *G3: Genes, Genomes Genetics, 6*(9), 2799–2808. https://doi.org/10.1534/g3.116.032888

Saleem, B., Khan, A. S., Shahzad, M. T., & Ijaz, F. (2016). Estimation of heritability and genetic advance for various metric traits in seven F2 populations of bread wheat (*Triticum aestivum* L.). *Journal of Agricultural Sciences, Belgrade, 61*(1), 1–9. https://doi.org/10.2298/JAS1601001S

Samantara, K., Shiv, A., de Sousa, L. L., Sandhu, K. S., Priyadarshini, P., & Mohapatra, S. R. (2021). A comprehensive review on epigenetic mechanisms and application of epigenetic modifications for crop improvement. *Environmental and Experimental Botany, 188*, 104479. https://doi.org/10.1016/j.envexpbot.2021.104479

Silva, J. V., van Heerwaarden, J., Reidsma, P., Laborte, A. G., Tesfaye, K., & van Ittersum, M. K. (2023). Big data, small explanatory and predictive power: Lessons from random forest modeling of on-farm yield variability and implications for data-driven agronomy. *Field Crops Research, 302*, 109063. https://doi.org/10.1016/j.fcr.2023.109063

Singh, S., Parihar, P., Singh, R., Singh, V. P., & Prasad, S. M. (2016). Heavy metal tolerance in plants: Role of transcriptomics, proteomics, metabolomics, and ionomics. *Frontiers in Plant Science, 6*, 1143. https://doi.org/10.3389/fpls.2015.01143

Song, X. J., Huang, W., Shi, M., Zhu, M. Z., & Lin, H. X. (2007). A QTL for rice grain width and weight encodes a previously unknown RING-type E3 ubiquitin ligase. *Nature Genetics, 39*(5), 623–630. https://doi.org/10.1038/ng2014

Su, Q., Zhang, X., Zhang, W., Zhang, N., Song, L., Liu, L., Xue, X., et al. (2018). QTL detection for kernel size and weight in bread wheat (*Triticum aestivum* L.) using a high-density SNP and SSR-based linkage map. *Frontiers in Plant Science, 9*, 1484. https://doi.org/10.3389/fpls.2018.01484

Sun, J., Poland, J. A., Mondal, S., Crossa, J., Juliana, P., Singh, R. P., Rutkoski, R. J., et al. (2019). High-throughput phenotyping platforms enhance genomic selection for wheat grain yield across populations and cycles in early stage. *Theoretical and Applied Genetics, 132*, 1705–1720. https://doi.org/10.1007/s00122-019-03309-0

Tong, H., & Nikoloski, Z. (2021). Machine learning approaches for crop improvement: Leveraging phenotypic and genotypic big data. *Journal of Plant Physiology, 257*, 153354. https://doi.org/10.1016/j.jplph.2020.153354

Voss-Fels, K. P., Stahl, A., Wittkop, B., Lichthardt, C., Nagler, S., Rose, T., Chen, T.-W., et al. (2019). Breeding improves wheat productivity under contrasting agrochemical input levels. *Nature Plants, 5*(7), 706–714. https://doi.org/10.1038/s41477-019-0445-5

Wang, S., Wu, K., Yuan, Q., Liu, X., Liu, Z., Lin, X., Zeng, R., et al. (2012). Control of grain size, shape and quality by OsSPL16 in rice. *Nature Genetics, 44*(8), 950–954. https://doi.org/10.1038/ng.2327

Wang, S., Wong, D., Forrest, K., Allen, A., Chao, S., Huang, B. E., Maccaferri, F., et al. (2014). Characterization of polyploid wheat genomic diversity using a high-density 90000 single nucleotide polymorphism array. *Plant Biotechnology Journal, 12*(6), 787–796. https://doi.org/10.1111/pbi.12183

Wani, J. A., Sharma, S., Muzamil, M., Ahmed, S., Sharma, S., & Singh, S. (2022). Machine learning and deep learning based computational techniques in automatic agricultural diseases detection: Methodologies, applications, and challenges. *Archives of Computational Methods in Engineering, 29*(1), 641–677. https://doi.org/10.1007/s11831-021-09588-5

Wu, Q. H., Chen, Y. X., Zhou, S. H., Fu, L., Chen, J. J., Xiao, Y., & Zhang, D. (2015). High-density genetic linkage map construction and QTL mapping of grain shape and size in the wheat population Yanda1817× Beinong6. *PLoS One, 10*(2), e0118144. https://doi.org/10.1371/journal.pone.0118144

Xu, Y., & Crouch, J. H. (2008). Marker-assisted selection in plant breeding: From publications to practice. *Crop Science, 48*(2), 391–407. https://doi.org/10.2135/cropsci2007.04.0191

Xu, R., Duan, P., Yu, H., Zhou, Z., Zhang, B., Wang, R., et al. (2018). Control of grain size and weight by the OsMKKK10-OsMKK4-OsMAPK6 signaling pathway in rice. *Molecular Plant, 11*(6), 860–873. https://doi.org/10.1016/j.molp.2018.04.004

Yang, H., Tao, Y., Zheng, Z., Li, C., Sweetingham, M. W., & Howieson, J. G. (2012). Application of next-generation sequencing for rapid marker development in molecular plant breeding: A case study on anthracnose disease resistance in *Lupinus angustifolius* L. *BMC Genomics, 13*, 1–12. https://doi.org/10.1186/1471-2164-13-318

Zhang, L., Zhao, Y. L., Gao, L. F., Zhao, G. Y., Zhou, R. H., Zhang, B. S., & Jia, J. Z. (2012). TaCKX6-D1, the ortholog of rice OsCKX2, is associated with grain weight in hexaploid wheat. *New Phytologist, 195*(3), 574–584. https://doi.org/10.1111/j.1469-8137.2012.04194.x

Zhang, L., Wang, A., Zhang, H., Zhu, Q., Zhang, H., Sun, W., & Niu, Y. (2024). Estimating leaf chlorophyll content of winter wheat from UAV multispectral images using machine learning algorithms under different species, growth stages, and nitrogen stress conditions. *Agriculture, 14*(7), 1064. https://doi.org/10.3390/agriculture14071064

Chapter 7
Predictive Modeling of Wheat Production in Turkiye: A NARX Network Approach Incorporating Climatic and Economic Factors

Aslı Akıllı and Mustafa Kan

Abstract The agricultural production process in the contemporary era is significantly influenced by environmental factors, rapidly changing market conditions, and macroeconomic indicators. Evaluating the wheat production process—a strategic component of agricultural activities—and the factors affecting this process through statistical analysis methods provides valuable insights into production quantity forecasts. In forecasting studies of agricultural data structures in a time series format, both traditional statistical methods and artificial intelligence-based models are widely used. In this study, nonlinear autoregressive with exogenous inputs (NARX) networks, characterized by their dynamic and flexible structure, have been utilized to predict wheat production in Turkiye. Meteorological data (temperature, precipitation), macroeconomic indicators (wheat price, barley price), and other parameters (year, population, wheat cultivation area, wheat yield) associated with past (1950–2021) wheat production quantities have been identified as system inputs. The output variable has been designated as the wheat production quantity for specific periods (1950–2021). The NARX model has been trained using the time series data of these variables, optimized through various training algorithms, and its success in production forecasts has been evaluated using statistical performance metrics. According to the analysis results, they highlight the dominance of temporal, demographic, and economic variables in predicting wheat production, while environmental factors and specific agricultural practices appear to play a minor role in this particular model configuration. This study demonstrates that LSTM-based NARX networks have achieved high accuracy in predicting wheat production levels. The use of such AI-based methods can enable more precise and reliable management of agricultural production processes and offer valuable contributions to policymakers and practitioners in making strategic decisions. Future research could extend these

A. Akıllı (✉) · M. Kan (✉)
Kırşehir Ahi Evran University, Agriculture Faculty, Department of Agricultural Economics, Kırşehir, Turkiye
e-mail: asliakilli@ahievran.edu.tr; mustafa.kan@ahievran.edu.tr

N. Zencirci et al. (eds.), *Empowering Wheat Cultivation with GIS, Digital Approaches and Artificial Intelligence*, https://doi.org/10.1007/978-3-031-99954-3_7

findings by applying similar models to different agricultural products and geographical regions.

Keywords Artificial intelligence · NARX · Time series · Wheat production

Abbreviations

ADAM	Adaptive moment estimation
ADAMAX	Variant of adam optimizer
AI	Artificial intelligence
ANN	Artificial neural network
ARIMA	Autoregressive integrated moving average
C3S	Copernicus climate change service
CNN	Convolutional neural network
ECMWF	European Centre for Medium-Range Weather Forecasts
EU	European Union
EUR	Euro
FAO	Food and Agriculture Organization
FAOSTAT	FAO Statistical Database
LSTM	Long short-term memory
MAPE	Mean absolute percentage error
MD	Moderate drought
MSE	Mean squared error
NADAM	Nesterov-accelerated adaptive moment estimation
NARX	Nonlinear autoregressive with exogenous inputs
PV	Photovoltaic
R^2	Coefficient of determination
R^2_{Adj}	Adjusted coefficient of determination
RMSE	Root mean squared error
RNN	Recurrent neural network
SGD	Stochastic gradient descent
SLD	Yield in slight drought
SVD	Severe drought regions
TGB	Turkish Grain Board
TMO	Turkish Grain Board
TUIK	Turkish Statistical Institute
USA	The United States
USD	United States Dollar
USDA	United States Department of Agriculture
ÇKS	Farmer Registration System

7.1 Introduction

The agricultural sector, as one of the fundamental pillars of economic and social development, not only provides an indispensable source of food for humanity but also constitutes the livelihood of millions of people. This makes the agricultural sector one of the most strategic industries. In addition to supplying raw materials to the food industry, agriculture also plays a crucial role in providing essential inputs to numerous other important sectors such as textiles, animal feed, tobacco, paper, and energy. As highlighted in Maslow's hierarchy of needs theory, the agricultural sector's role in fulfilling the essential need for food underscores its irreplaceability by any other sector. Particularly, the production of staple foods, which vary across countries but include items such as wheat, corn, rice, and potatoes, is vital for ensuring food security.

Wheat ranks among the most strategic crops both globally and in Turkiye. It is cultivated in nearly every region of the world, including Turkiye; it is significant not only due to its large base of producers but also because it is the primary raw material for bread, a staple food for humans. In terms of cultivated area, wheat is grown on approximately 13.93% of the world's total agricultural land, according to FAO's 2022 Crop Land data. From a production perspective, global wheat production amounts to 808.4 million tons (FAOSTAT, 2024a, b). An analysis of global wheat production for the 2023–2024 season reveals that China is the largest producer, followed by the European Union (EU), India, Russia, and the United States, all of which are key economic and political entities. Turkiye is also one of the world's top 10 wheat-producing countries (Fig. 7.1).

When analyzing global wheat production trends, it becomes clear that factors such as climatic conditions, as well as economic and political circumstances, can lead to fluctuations in the production on a country-by-country basis. In particular, the recent developments, including the Ukraine-Russia war and the effects of climate change have caused shifts in the rankings of the top 10 wheat-producing countries.

Top Producing Countries

Market	% of Global Production	Total Production (2023/2024, Metric Tons)
China	17%	136.59 Million
European Union	17%	134.15 Million
India	14%	110.55 Million
Russia	12%	91.5 Million
United States	6%	49.31 Million
Canada	4%	31.95 Million
Pakistan	4%	28.16 Million
Australia	3%	26 Million
Ukraine	3%	23 Million
Turkey	3%	21 Million

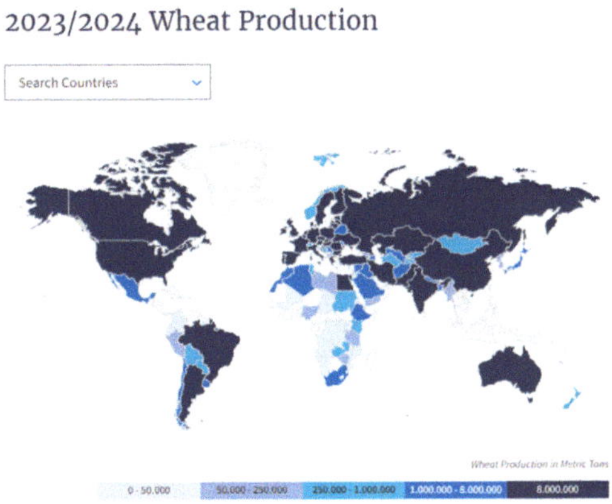

Fig. 7.1 Distribution of global wheat production statistics (FAS-USDA, 2024)

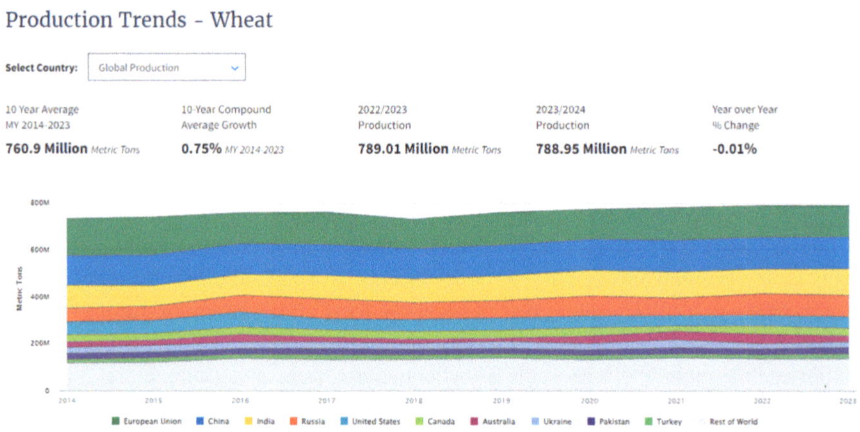

Fig. 7.2 Global wheat production trends (FAS-USDA, 2024)

However, despite these changes, a review of the overall averages and trends reveals that there are no significant alterations among the leading producers. It can be observed that there is a slight upward trend in total production, albeit modest. Between 2014 and 2023, the increase in production volume was approximately 0.75% (Fig. 7.2).

Wheat, beyond being merely an agricultural product, holds significant economic, social, and cultural value for Turkiye. As one of the centers of origin for wheat, Turkiye is particularly important from the perspective of plant genetic resources. Wheat, with a history spanning 10,000 years in Anatolia (Harlan, 1995), is not only a strategic crop but also a cultural heritage (Kan et al., 2017). Turkiye is recognized as one of the key regions where wheat was first domesticated (Zohary & Hopf, 1988). The historical significance of wheat within the borders of Turkiye dates back to even earlier than any known civilizations. Wheat has been at the heart of revolutionary changes that have shaped the modern way of life, particularly in the very geography that we inhabit today. Archaeological findings, especially in Southeastern Anatolia and the Fertile Crescent, indicate that this region is the homeland of wheat (Nesbitt & Samuel, 1996; Tanno & Willcox, 2006).

From a production standpoint, wheat holds strategic importance in Turkiye. According to the Farmer Registration System (ÇKS), approximately 964,608 agricultural enterprises (MoAF, 2024) and their families are involved in wheat production, and almost the entire population is affected by its consumption. Given its broad impact, wheat production is a critical issue in Turkiye. The effective management of production processes for essential food products like wheat is crucial for global food security and sustainable agricultural practices (Shiferaw et al., 2013; Rehman et al., 2022). However, this process is continuously influenced by dynamic factors such as climate change, population growth, market fluctuations, and political uncertainties (Raynold & Brown, 2022; Farooq et al., 2023).

Every country strives to achieve a minimum level of self-sufficiency in essential food products. Ensuring self-sufficiency requires that the domestic supply meets demand without surplus production (Eraktan, 2001). Thus, production planning is a leading policy for achieving self-sufficiency. This is important not only for ensuring food security through current production but also for creating a strategic stock plan for critical food products. Therefore, it is essential to conduct production planning for such products and to understand the factors that influence their production. Predicting production quantities in agriculture plays a vital role in coping with these dynamic factors, shaping agricultural policies, and ensuring the efficient use of resources. Time series analyses, in particular, allow for a better understanding of trends and seasonal variations in the production process. Although traditional statistical methods have long been used for such data structures, they may be insufficient in fully capturing the complexity of today's dynamic agricultural systems (Tao et al., 2023; Mahmoud et al., 2024). At this juncture, artificial intelligence-based methods, particularly artificial neural networks, emerge as powerful tools for agricultural production forecasting.

The flexible and customizable nature of artificial neural networks, combined with their capacity to model complex data relationships, allows for more accurate and reliable predictions (Khaki & Wang, 2019; Sadenova et al., 2023). In this context, nonlinear autoregressive networks with exogenous inputs (NARX) are highlighted in the literature as a prominent model in applied sciences. The fundamental feature of NARX networks is their ability to forecast future conditions by considering past system data and external factors (Buevich et al., 2021). This feature is particularly advantageous for agricultural production forecasting, as crop production depends not only on historical production data but also on external factors such as meteorological conditions (Shen et al., 2024), economic policies, and market demands (Wang et al., 2021).

The development and implementation of NARX networks have been greatly influenced by a number of pivotal studies across various fields. A foundational contribution to this area was made by Leontaritis and Billings in 1985, who introduced essential concepts and methods for modeling nonlinear systems using autoregressive structures with external inputs (Leontaritis & Billings, 1985; Alsumaiei, 2020). This groundbreaking work established the theoretical basis for later advancements in NARX network architectures and their practical applications. NARX networks, initially conceived in the late 1980s and early 1990s, were designed to address the challenges of modeling complex dynamic systems. Early investigations primarily targeted applications within engineering and physical sciences. Over subsequent years, the deployment of these models has expanded into agricultural domains. Currently, the application of NARX models in predicting agricultural yields is advancing rapidly, reflecting their growing importance and effectiveness in this field.

Research in the scientific literature indicates that NARX networks hold substantial promise for predicting agricultural yields, particularly in wheat yield applications. These models enhance forecasting precision by integrating diverse data sources and accurately capturing complex variable interactions. As the agricultural sector continues to face challenges such as climate change and resource limitations, employing advanced modeling techniques like NARX networks significantly contributes to sustainable and efficient food production. The use of NARX networks in agricultural yield prediction has garnered significant attention due to their ability to model complex nonlinear relationships. These models are particularly useful for forecasting crop yields, where multiple variables and external factors influence outcomes. The literature includes numerous studies utilizing neural networks for yield prediction in various crops, such as soybeans and maize (Kaul et al., 2005), sugarcane (Kumar et al., 2015), wheat (Bhojani & Bhatt, 2020; Bali & Singla, 2021; Latifi & Shabanali Fami, 2022; Sun et al., 2022), and rice (Pradhan et al., 2012; Jabjone & Jiamrum, 2013; Jang & Choe, 2014; Thimmegowda et al., 2023). Additionally, research has focused on maize (Singh & Prajneshu, 2008) and various other agricultural products (Kaleeswaran et al., 2020). Furthermore, there are comparative studies on neural networks and ARIMA methods for forecasting rubber production (Kosanan & Kantanantha, 2014), oilseed production (Mithiya et al., 2019), and sugarcane yield (Paswan et al., 2022). Additionally, LSTM-recurrent neural networks have been employed in the prediction of palm oil yields (Sugiyarto & Abadi, 2019).

The incorporation of neural network models with NARX networks into agricultural crop yield predictions has garnered considerable attention in recent years. The increasing interest in these data-driven methodologies is reflected in the volume of studies, and a diverse range of research has highlighted the utility of NARX networks across various agricultural contexts, particularly in predicting wheat yields. The study by Abraham et al. (2020) aims to predict soybean harvest area, yield, and production using artificial neural networks (ANN) and compares these predictions with classical time series analysis methods. The results indicate that ANN is the most effective approach for predicting soybean harvest area and production, while classical linear functions are more effective for predicting soybean yield. The study conducted by Boussaada et al. (2018) examines the use of a nonlinear autoregressive exogenous (NARX) neural network model for predicting daily solar radiation levels. The NARX model addresses the complexities of forecasting solar radiation for photovoltaic (PV) energy systems by considering variables such as cloud cover and the dynamic position of the Sun. Furthermore, Guo and Xue (2014) conducted a comparative analysis of spatial and temporal neural network models, including NARX networks, for crop yield forecasting. Their study demonstrated that NARX models excel in capturing temporal dependencies, significantly enhancing prediction accuracy. The results of the analysis demonstrate the usefulness of NARX networks as a valuable tool in agricultural forecasting, especially in optimizing crop yield prediction by effectively modeling dynamic temporal relationships. In their study, Sivaranjani and Vimal (2023) utilized predictive maintenance techniques with artificial neural networks (ANNs) to assist decision-makers in selecting the best crop. For crop yield prediction based on crop selection criteria, this study presents a nonlinear autoregressive with external input (NARX) ANN. The research findings suggest that by choosing crops suitable for specific soil conditions and climate, yields can be increased while reducing irrigation water use.

In this study, NARX networks were employed to forecast wheat production in Turkiye by evaluating various inputs such as meteorological data, macroeconomic indicators, and historical production quantities. Thus, the impact of variables likely to influence wheat production will be identified, and an important forecasting tool for production planning will be evaluated.

7.2 Material and Methods

Data Description In this research, nonlinear autoregressive with exogenous inputs (NARX) networks are utilized to predict wheat production in Turkiye. Meteorological data and historical wheat production figures have been selected as input variables, while the output variable is defined as the wheat production quantity for specific periods. The NARX model was trained using the time series data of these variables, and its effectiveness in forecasting production was assessed using statistical performance metrics. During the process of training and optimizing the model, time series data and various training algorithms were used, and the model's performance was evaluated using statistical criteria. This process aims to demonstrate the potential of NARX networks in agricultural production forecasting and to positively contribute to minimizing uncertainties in agricultural production.

In our study, in addition to agricultural production and price data, the distribution of climatic parameters is presented in detail, along with numerical findings concerning the potential dynamics of these variables over time. The analysis of these data, which form the foundation of our modeling process using NARX networks, aims to provide a strong reference point for in-depth analyses in the agricultural field. In this study, to incorporate the changes and trends observed in average air temperature in Turkiye into the model, the ERA5-Land reanalysis dataset was used. This dataset, created according to the land parameters of ERA5 (ECMWF Reanalysis v5), is produced by the Copernicus Climate Change Service (C3S) of the European Centre for Medium-Range Weather Forecasts (ECMWF) (Muñoz-Sabater et al., 2021; Copernicus Climate Change Service, 2021; Güler & Erlat, 2023; Yeşilköy & Demir, 2024). The dataset covers the period from 1950 to 2021 to ensure compatibility with the other parameters. ERA5-Land monthly mean temperature at 2 m above the ground and total precipitation data were used. Population data were obtained from FAO (FAOSTAT, 2024c). Wheat price ($/kg); barley price ($/kg); and wheat and barley cultivation, production (yield), and sales between 1950 and 2021 were provided by the Turkish Grain Board (TMO, 2024) and Turkish Statistical Institute (TUIK, 2024). Detailed information on the variables in the model is provided in Table 7.1. The methodology was developed using the Python programming language (Python 3.12) in combination with libraries compatible with the Jupyter Notebook platform (Randles et al., 2017). Specifically, the libraries Pandas (McKinney, 2011), Matplotlib (Bisong, 2019), NumPy (Oliphant, 2006), Seaborn

Table 7.1 Agricultural and climatic parameters: descriptions and definitions

Parameters	Descriptions
Population (000 people)	The total number of individuals residing in the studied area, which may influence agricultural practices and food demand.
Wheat price ($/kg)	The market price of wheat per kilogram, reflecting the economic value of wheat in the given time period.
Barley price ($/kg)	The market price of barley per kilogram, indicating its economic significance in the agricultural sector.
Season 1 temperature (September–March) (Celcius)	2 m temperature parameter refers to the temperature of the air at 2 m above the surface of land, sea, or inland waters.
Season 2 emperature (April–July) (Celcius)	
Season 1 total precipitation (September--March) (mm/day)	This parameter represents the accumulated liquid and frozen water, including rain and snow, that falls to the Earth's surface. It is the sum of large-scale precipitation and convective precipitation
Season 2 total precipitation (April–July) (mm/day)	
Wheat cultivation (Ha)	The total area of land, measured in hectares, dedicated to wheat cultivation, representing the scale of agricultural activity.
Wheat yield (kg/Da)	The average amount of wheat produced per decare, indicating the productivity of the wheat cultivation.
Wheat production (ton)	The total wheat production, measured in tons, reflecting the output of wheat farming activities.

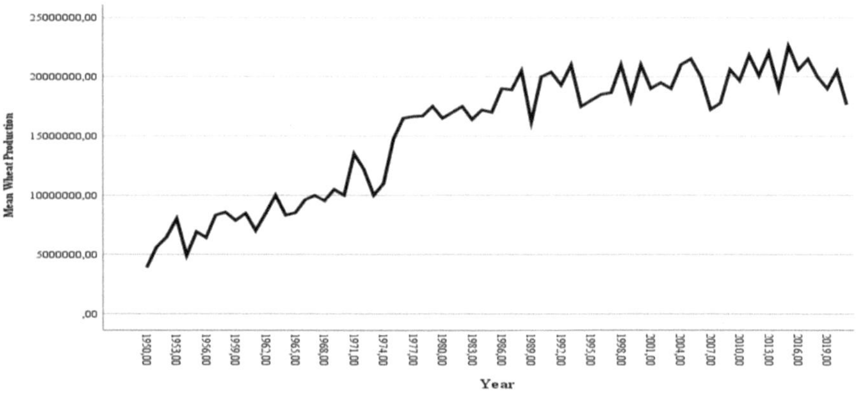

Fig. 7.3 Wheat production (ton) mean of Turkiye over time

(Bisong, 2019), and Scikit-learn (Pedregosa et al., 2011) were utilized to create the wheat production forecasting methodology.

Figure 7.3 presents the annual average wheat production quantities (in tons) from 1950 to 2021. The horizontal axis represents the years, while the vertical axis indicates the corresponding wheat production quantities. Upon examination of the graph, it is evident that, although there were fluctuations in wheat production

Table 7.2 Descriptive statistics of variables

Parameters	Mean ± Std. Error of Mean.	Skewness	Kurtosis	Minimum	Maximum
Population	50721.998 ± 2290.3428	0.142	−1.228	20977.7	84775.4
Wheat price (t − 1)	0.175 ± 0.0092	1.047	0.443	0.08	0.41
Barley price (t − 1)	0.140 ± 0.0084	1.157	0.851	0.06	0.37
Season 1 TP	147.4104 ± 2.5942	0.264	0.809	96.14	219.17
Season 2 TP	69.5352 ± 1.6160	0.747	1.589	44.27	117.7
Season 1 temp	6.2004 ± 0.1164	0.145	−0.401	4.32	8.4
Season 2 temp	16.4671 ± 0.0926	0.34	−0.092	14.92	18.52
Wheat cultivation	8338097.888 ± 136334.698	−1.182	1.552	4,477,191	9,800,000
Wheat yield	182.5139 ± 7.0511	0.043	−1.047	77	296
Wheat production	15355429.527 ± 632776.044	−0.593	−1.122	3,871,926	22,600,000

between 1950 and the early 1970s, a general upward trend is discernible. During this period, production experienced occasional declines, yet the overall tendency was toward an increase. From the mid-1970s onward, a marked upward trend in wheat production became apparent. This increase persisted until the early 1980s, after which production levels stabilized. A renewed upward trend emerged in the mid-1990s and continued into the early 2000s. However, from the late 2000s onward, production exhibited notable fluctuations, leading to a somewhat volatile pattern from that period until 2021. Overall, it can be concluded that there was a long-term upward trend in wheat production between 1950 and 2021. Nevertheless, within this overarching trend, there were distinct periods of fluctuation and stability. By 2021, it is noteworthy that production quantities exhibited a slight decline in comparison to previous years.

Table 7.2 presents the descriptive statistics for our model parameters, including mean values, standard errors of the mean, skewness, kurtosis, and the observed range (minimum and maximum values) for variables relevant to agricultural production, pricing, and climatic conditions. The population parameter, with a mean of 50,721.998 and a standard error of 2290.343, exhibits a nearly symmetric distribution, as indicated by a skewness of 0.142, and a relatively flat distribution, with a kurtosis of −1.228. This suggests that the population data are fairly evenly distributed, with no significant outliers or extreme values. Pricing data for wheat and barley reveal distinct statistical characteristics. The mean wheat price is 0.175$/kg, accompanied by a positive skewness of 1.047 and a kurtosis of 0.443, indicating the presence of occasional higher prices that significantly influence the overall distribution. Similarly, the barley price, averaging 0.140$/kg, also displays positive skewness (1.157) and a kurtosis of 0.851, suggesting that extreme values, though infrequent, play a crucial role in shaping the price distribution. Climatic parameters, particularly seasonal precipitation and temperatures, exhibit stability with low standard errors and relatively symmetric distributions. The mean Season 1 temperature (September–March) is 6.2004 °C, with a slight positive skewness of 0.145 and a kurtosis of −0.401, indicating a minor tendency toward lower values but a generally stable temperature range during this period. Meanwhile, the mean Season 2 temperature (April–July) is 16.4671 °C, with a skewness of 0.340 and a kurtosis of −0.092, reflecting a similarly stable distribution, albeit with a slight skew toward higher temperatures. For precipitation, the mean Season 1 total precipitation (September–March) is 147.4104 mm, with a skewness of 0.264 and a kurtosis of 0.809, suggesting a relatively symmetric distribution with a minor inclination toward higher precipitation values. The mean Season 2 total precipitation (April–July) is 69.5352 mm, exhibiting a skewness of 0.747 and a kurtosis of 1.589, indicating a greater tendency toward higher precipitation values during this period, though still within a relatively stable range. Metrics related to wheat and barley cultivation, yield, and production exhibit notable variability. The wheat cultivation parameter, with a mean of 8,338,097.888 ha and a standard error of 136,334.698, shows negative skewness (−1.182) and high kurtosis (1.552). This suggests that there are occasional large outliers that significantly affect the distribution. The data for wheat yield and production, while generally symmetrical, also reflect a wide distribution around the mean, as indicated by flatter kurtosis values. The statistical analysis of the model parameters suggests that there are no significant asymmetric structures, outliers, or abnormal variability across the dataset.

Table 7.3 illustrates the statistical relationships among the variables using the Spearman rank correlation coefficient, which was chosen due to the non-normal distribution of most variables. Upon examining the correlations between the wheat production variable and other factors, significant relationships were identified. The most notable is the strong positive correlation between wheat production and wheat yield ($\rho = 0.939^{**}, p < 0.01$). This indicates that increases in wheat yield are closely associated with increases in wheat production, underscoring the critical role of productivity improvements in driving overall production levels. Additionally, wheat production shows a highly significant positive correlation with population ($\rho = 0.899^{**}, p < 0.01$). This relationship suggests that as the population grows, the demand for wheat also increases, leading to higher production levels. The positive correlation with population highlights the demand-driven nature of wheat production, where agricultural output scales with demographic changes. The analysis also reveals a strong positive correlation between wheat production and time (year) ($\rho = 0.899^{**}$, $p < 0.01$), indicating a consistent upward trend in production over the years. This finding suggests that wheat production has benefited from sustained growth, likely driven by advancements in agricultural technology, improvements in farming practices, and the expansion of cultivated areas. A significant positive correlation is also observed between wheat production and wheat price ($\rho = 0.674^{**}$, $p < 0.01$). This relationship suggests that higher wheat prices incentivize greater production, as farmers are more likely to increase output in response to favorable market conditions. The price-production dynamic reflects how economic incentives play a crucial role in shaping production decisions.

Table 7.3 Correlation coefficients of variables

Variables	V1	V2	V3	V4	V5	V6	V7	V8	V9	V10	V11
Year	.	0.999**	0.748**	0.788**	−0.161	−0.052	0.273*	0.477**	0.238*	0.952**	0.899**
Population	.	.	0.748**	0.788**	−0.161	−0.052	0.273*	0.477**	0.238*	0.952**	0.899**
Wheat price (t − 1)	.	.	.	0.978**	−0.156	−0.053	0.172	0.369**	−0.002	0.750**	0.674**
Barley price (t − 1)	.	.	.	.	−0.180	−0.057	0.180	0.379**	0.030	0.786**	0.715**
Season 1 TP	.	.	.	.	.	0.119	−0.057	−0.135	0.067	−0.144	−0.170
Season 2 TP	.	.	.	.	.	.	−0.128	−0.114	0.218	−0.074	−0.034
Season 1 temp	.	.	.	.	.	.	.	0.432**	−0.283*	0.284*	0.182
Season 2 temp	.	.	.	.	.	.	.	.	−0.248*	0.403**	0.279*
Wheat cultivation	.	.	.	.	.	.	.	.	.	0.214	0.431**
Wheat yield	.	.	.	.	.	.	.	.	.	.	0.939**
Wheat production	.	.	.	.	.	.	.	.	.	.	.

** *Statistically significant at 0.01 ($p < 0.01$)*, * *Statistically significant at 0.05 ($p < 0.05$)*

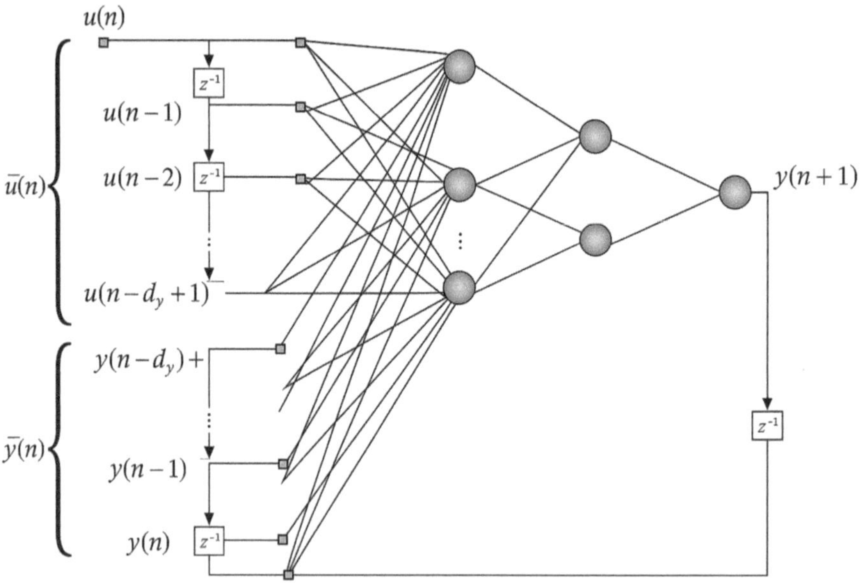

Fig. 7.4 NARX neural network architecture (Menezes & Barreto, 2008)

NARX Network Nonlinear autoregressive with exogenous inputs (NARX) networks are a specialized class of recurrent neural networks designed for modeling complex nonlinear systems. These networks are characterized by their ability to combine both past outputs and external inputs to predict future outputs, making them highly effective for tasks involving time series forecasting and modeling nonlinear dynamics. By reintegrating network outputs into inputs, NARX networks can capture complex dependencies and feedback loops in data, allowing for accurate representation and prediction of various real-world phenomena across different applied sciences, particularly in agricultural sciences.

The nonlinear autoregressive with exogenous inputs (NARX) network is an important model used for representing discrete-time nonlinear systems. NARX network's architecture is shown in Fig. 7.4. Its primary goal is to forecast future outputs by considering both past outputs and inputs. Mathematically, the NARX model is represented in Eq. (7.1) as follows:

$$y(n+1) = f\left[\, y(n), \ldots, y(n-dy+1); u(n-k), u(n-k+1), \ldots, u(n-du-k+1) \,\right] \tag{7.1}$$

where $y(n) \in \mathbb{R}$ ve $u(n) \in \mathbb{R}$ represent the output and input values of the model at time n, respectively. d_u ve d_y satisfying the condition $d_u \geq 1$, $d_y \geq 1$, and $d_u \leq d_y$ represent the input memory layout and output memory layout, respectively.

Satisfying the condition $k \geq 0$, the $'k'$ parameter is known as a delay term and is also referred to as the process dead time. Assuming $k = 0$, as reported by Menezes and Barretto (2008), the NARX model takes the mathematical form shown in Eq. (7.2) and expressed in vector form as in Eq. (7.3), where the vectors $y(n)$ and $u(n)$ represent the output and input regressors, respectively. The function $f(\cdot)$ denotes the nonlinear function. According to Cocianu et al. (2022), LSTM neural networks can be utilized to determine $f(\cdot)$.

$$y(n+1) = f\left[\, y(n), \ldots, y(n-dy+1); u(n), u(n-1), \ldots, u(n-d_u+1) \,\right] \tag{7.2}$$

$$y(n+1) = f\left[\, y(n), u(n) \,\right] \tag{7.3}$$

The training of the NARX network can be performed in two different modes: series-parallel (SP) and parallel (P). The primary difference between these two modes is that in the series-parallel mode, the output regressor consists only of the actual output values of the system. In contrast, in parallel mode, the outputs predicted by the NARX networks are included in the output regressor through feedback. Equations (7.4) and (7.5) provide the mathematical representations of the model structures of NARX networks in series-parallel and parallel modes, respectively (Menezes & Barretto, 2008; Boussaada et al., 2018; Massaoudi et al., 2021; Moursi et al., 2021), where (^) denotes the estimated values or functions.

$$\begin{aligned} \widehat{y}\,(n+1) &= \widehat{f}\left[\, y_{sp}(n), u(n) \,\right] \\ &= \widehat{f}\left[\, y(n), \ldots, y(n-d_y+1); u(n), u(n-1), \ldots, u(n-d_u+1) \,\right] \end{aligned} \tag{7.4}$$

$$\begin{aligned} \widehat{y}\,(n+1) &= \widehat{f}\left[\, y_p(n), u(n) \,\right] \\ &= \widehat{f}\left[\, \widehat{y}(n), \ldots, \widehat{y}(n-d_y+1); u(n), u(n-1), \ldots, u(n-d_u+1) \,\right] \end{aligned} \tag{7.5}$$

The NARX neural network comprises two layers of feedforward networks, with a linear transfer function in the output layer and a sigmoid function, $\sigma(x)$, in the hidden layer, calculated in Eq. (7.6) (Massaoudi et al., 2021; Di Nunno et al., 2022).

$$\sigma(x) = \frac{1}{1 + \exp(-\sigma)} \tag{7.6}$$

The NARX network is a powerful tool for identifying nonlinear systems and has been effectively utilized in numerous real-world input-output modeling applications. In the literature, analysis methods that enhance forecasting power and assist in selecting appropriate parameter values by combining LSTM and NARX models are available. Cocianu and Avramescu (2020) integrated LSTM networks into the

NARX model to forecast the evolution of EUR-USD exchange rates in their study. Massaoudi et al. (2021) used this combination to predict the power output of photovoltaic stations. Moursi et al. (2021) proposed a new IoT-enabled and edge computing-based system for air quality monitoring and prediction using long short-term memory (LSTM), random forest, extra trees, gradient boost, extreme gradient boost, and random forests in XGBoost with the proposed NARX architecture. Moursi et al. (2022) proposed a combination of convolutional neural networks (CNNs) and LSTMs with the NARX model to predict air pollution containing particulate matter with a diameter of less than 2.5 μm. Xu and Zhang (2021) used the LSTM method for short-term prediction and then integrated the results into the NARX model to achieve better performance.

LSTM cells were introduced by Hochreiter and Schmidhuber (1997) to address the problem of vanishing gradients in RNN architectures (Massaoudi et al., 2019) in order to improve training performance (Wei et al., 2020). This problem occurs when gradients become too small, hindering the model's ability to learn from long sequences of data. LSTMs utilize a unique gate control system and a specialized memory mechanism to exploit long dependencies in data. The entire process is managed by a series of LSTM cells. An LSTM architecture is designed to reduce gradient effects by using four types of gates: these are an input gate, a forget gate, an output gate, and a self-repeating unit that regulates the interactions between memory units. Gate operations are based on functions such as hyperbolic tangent (tanh), sigmoid function, and matrix multiplication.

The input gate (I_t) assesses the impact of the input signal on modifying the state of the memory cell, as described in Eq. (7.7), where W_t and U_t are weight matrices, x_t is the input at time t, h_{t-1} is the hidden state from the previous step, and b_I is the bias. Forget gate (f_t) determines which information to discard from the memory cell state and ensures that irrelevant data that could mislead the forecasting process is discarded. Values range between 0 and 1, where a higher value indicates more important information as described in Eq. (7.8). As can be seen in Eq. (7.9), output gate (o_t) controls the influence of the current cell state on other memory cells and evaluates the effect of the cell state on the overall output (Thapa et al., 2020; Massaoudi et al., 2021; Moursi et al., 2021). The cell state is given in Eq. (7.10). The function denoted by "⊙" represents the multiplication of the corresponding elements. Hidden state (h_t) is given in Eq. (7.11) (Massaoudi et al., 2021; Gao et al., 2023) and tanh(·) is the tanh function, specified in Eq. (7.12).

$$I_t = \sigma(W_I x_t + U_i h_{t-1} + b_I) \tag{7.7}$$

$$f_t = \sigma\left(W_f x_t + U_f h_{t-1} + b_I\right) \tag{7.8}$$

$$o_t = \sigma(W_o x_t + U_o h_{t-1} + b_o) \tag{7.9}$$

$$\tilde{c} = f_t \odot \mathrm{c}_{t-1} \tag{7.10}$$

$$h_t = o_t \odot \tanh(\mathrm{c_t}) \tag{7.11}$$

$$\tanh(x) = \frac{e^x - e^{-x}}{e^x + e^{-x}} \tag{7.12}$$

The input sequence is described as $x = [x_1, \ldots, x_t]$ for LSTM cells, $x_t \in \mathbb{R}$.

Initially, the hidden state h_{t-1}, cell state c_{t-1}, and current input x_t are processed to determine the new state values. At this stage, x_t and h_{t-1} are passed to the forget gate to measure the importance of the information, which is scaled between 0 and 1. The forget gate decides whether to keep the information or not, and the cell state is updated based on the information from the previous and current cell states. Thus, calculations are performed by the final hidden layer of the LSTM to obtain the updated state value (Massaoudi et al., 2021). LSTM networks are effective in handling long-term dependencies and sequence prediction tasks due to their ability to selectively retain information over extended periods, addressing the limitations of traditional RNNs. In this study, the ADAM optimizer, known for its computational efficiency and suitability for nonstationary complex targets, was used during the LSTM training process (Cocianu et al., 2022; Gao et al., 2023).

Data Preprocessing The prediction accuracy of machine learning algorithms is heavily dependent on the quality and reliability of the input data. Thus, in scientific research on machine learning methods, normalization techniques play a crucial role in data structures with a large number of observations and variables, particularly when there are dimensional differences between input and observation values (Akıllı & Atıl, 2020). The data preprocessing stage encompasses the processes of cleaning the data and making it suitable for analysis and model training. This study focuses on the normalization process. By normalizing time series data before the training stage, the aim is to ensure that all features contribute equally to distance calculations in algorithms. Normalization is a critical step in data preprocessing, especially in time series analysis, as it helps mitigate issues arising from differences in scale among various features. This step ensures that the model's performance is not biased toward features with larger scales, thereby improving the accuracy and reliability of machine learning models. Min-Max normalization is a widely used method in machine learning research. This technique applies a linear transformation to the original data, mapping values to a range between "0" and "1" or "−1" and "1." A key advantage of Min-Max normalization is its ability to preserve the relationships between the original data values. However, this method is sensitive to extreme values or outliers, as large values can disproportionately influence the results (Jain et al., 2005). The mathematical representation of Min-Max normalization technique is given in Eq. (7.13), where $x_i^{'}$ is normalized value of x_i and $x_{\min}$ and $x_{\max}$ are minimum and maximum value of x_i, respectively

Table 7.4 Statistical Error Criteria and Equations

Statistical Error Criteria	Equations
Root mean square error (RMSE)	$\text{RMSE} = \sqrt{\frac{1}{n}\sum_{i=1}^{n}(y_i - \widehat{y_i})^2}$
Mean absolute percentage error (MAPE)	$\text{MAPE} = \left(\frac{100}{n}\right)\sum_{i=1}^{n}\left\lvert\frac{(y_i - \widehat{y_i})}{y_i}\right\rvert$
Adj. coefficient of determination (R^2_{Adj})	$R^2_{\text{Adj}} = 1 - \left(1 - R^2\right)\frac{(n-1)}{(n-p-1)}$

Where for the ith record, $\widehat{y_i}$: predicted value, y_i: actual value, n: number of records

$$x_i' = \frac{x_i - x_{i\,\min}}{x_{\max} - x_{\min}} \tag{7.13}$$

Evaluation Metrics In scientific studies involving time series data, the success of the forecasting method is evaluated by various statistical performance metrics. Performance can often be described by error metrics or forecast trend indicators. To evaluate the effectiveness of NARX-based models, various metrics were employed within the framework of statistical error criteria. The first metric, the adjusted coefficient of determination (Adj. R^2), was used to enhance the reliability of model comparison by accounting for the number of predictors (a higher adj. R^2 value shows a better fit). The second metric, mean absolute percentage error (MAPE), calculates the average absolute percentage difference between observed and predicted values, providing a scale-independent measure of accuracy. The third metric is root mean square error (RMSE), which measures the square root of the average of the squared differences between the model's predicted values and the actual values, indicating the standard deviation of the residuals. Lower RMSE and MAPE value shows a better fit. The evaluation metrics were calculated separately for the training and validation phases, allowing for an assessment of the neural network's accuracy in both short- and long-term forecasting periods. The statistical error metrics used in predictive modeling are detailed in Table 7.4.

7.3 Results and Discussion

In this study, a long short-term memory (LSTM)-based nonlinear autoregressive with exogenous inputs (NARX) neural network analysis was conducted to model wheat production. Initially, the dataset was extracted from an Excel file as part of the data preprocessing phase and transformed into a format suitable for neural network analysis. The Min-Max scaling technique was employed to normalize the data. The training set comprised 80% of the data. The model training phase involved the creation of an NARX model incorporating LSTM layers to effectively capture the temporal dependencies in the dataset and enhance model accuracy. The model architecture consists of two LSTM layers, with different hidden node numbers. The

LSTM analysis utilized the Adam optimizer, known for its computational efficiency and suitability for nonfixed complex targets. Experiments were additionally conducted using various optimizers. Hyperparameters, including the optimization algorithm, number of LSTM layers, hidden units, loss function, learning rate, batch size, and dropout, are meticulously chosen according to performance. To mitigate the risk of overfitting, an early stopping mechanism was implemented, halting training when improvements in model performance plateaued. During the performance evaluation phase, the model's effectiveness was assessed using root mean square error (RMSE), mean absolute percentage error (MAPE), and the adjusted coefficient of determination (R^2) as performance metrics. Additionally, permutation importance analysis was applied, which involves randomly shuffling the input variable values and observing the resultant changes in model performance to evaluate the impact of input variables on prediction performance. This technique enables the determination of the relative contribution of each input variable to the accuracy of the LSTM-based NARX model's predictions.

The batch size value of 32 has been suggested in various scientific studies (Bengio, 2012; Thapa et al., 2020). In our study, mean square error (MSE) was used as the loss function. The number of LSTM layers and hidden units was carefully determined based on a detailed literature review and various experiments. Both layers incorporate dropout. The dropout technique, which is used to prevent overfitting in the context of LSTM analysis, is quite effective. In this study, different dropout values (0, 0.1, 0.2, 0.3) were tested, and 0.2 was found to be the most appropriate value (Thapa et al., 2020; Massaoudi et al., 2021; Gao et al., 2023). Analysis of NARX networks was performed using different algorithms. The optimal number of hidden nodes was determined through trial and error. Feedback delay was assessed based on the autocorrelation of the target dataset, while input delay was evaluated according to performance. The optimizer's learning rate was set to 0.002, and the model was trained for 300 epochs. To prevent overfitting, the EarlyStopping method was used. The patience parameter was set to 30, meaning that if there was no improvement in the validation loss for 30 consecutive epochs, the training would be halted, and the best weights would be restored.

Figure 7.5 presents a comparative analysis of the Adam optimizer's performance with varying numbers of hidden nodes (10, 20, 50, 100, and 150). The figure demonstrates that the model's predictions align closely with the actual values across these different configurations of hidden nodes. The analysis reveals that, regardless of the number of hidden nodes, the model's predictions consistently track the actual values, indicating that the Adam optimizer maintains high accuracy across all tested configurations. Notably, as the number of hidden nodes increases, the prediction lines converge more closely around the actual values, particularly in configurations with 50, 100, and 150 hidden nodes. This suggests that increasing the number of hidden nodes may marginally enhance the model's predictive consistency and accuracy. Figure 7.6 presents a comparative analysis of various optimizers (ADAM, Nadam, RMSprop, SGD, Adadelta, ADAMAX, and Adagrad) in relation to actual and predicted values. Figure 7.6 illustrates how closely each optimizer's predictions align with the actual data points across the examined configurations. The

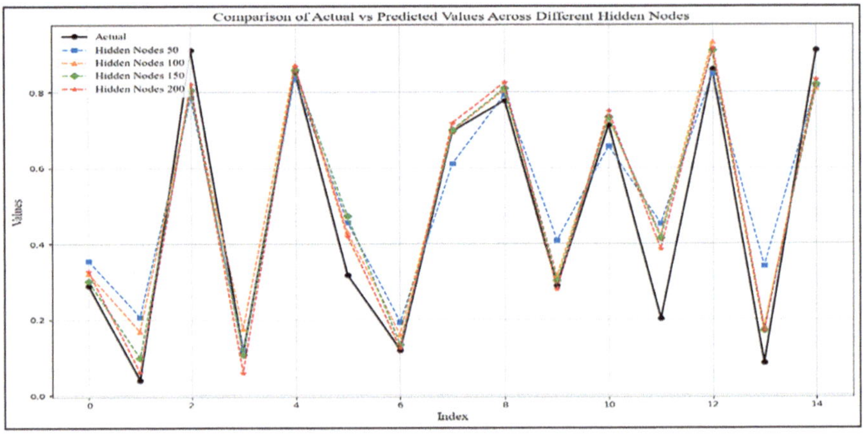

Fig. 7.5 Comparison of Adam optimizer performance with different numbers of hidden nodes

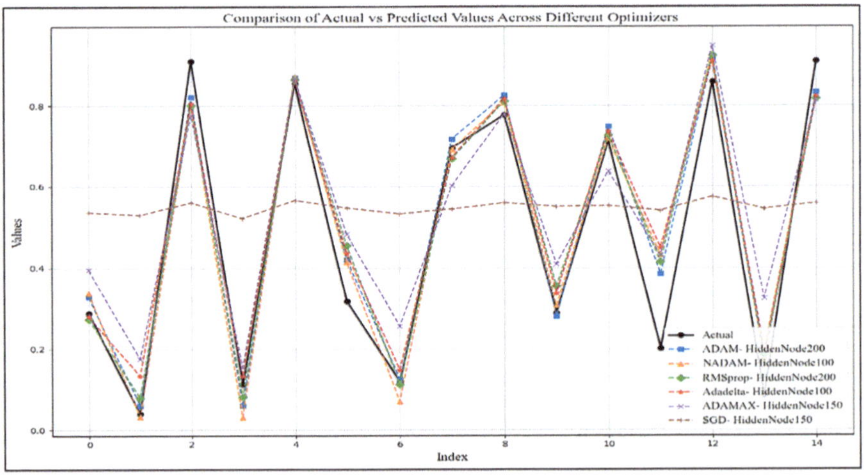

Fig. 7.6 Comparison of optimizers for actual and predicted values

figure illustrates a comparison between actual values and the predictions made by various neural network models, each optimized using different algorithms—ADAM, NADAM, RMSprop, Adadelta, ADAMAX, and SGD—across specific hidden node configurations. The majority of the optimizers, particularly ADAM, NADAM, and RMSprop, exhibit a strong alignment with the actual values. ADAM and RMSprop, in particular, demonstrate remarkable consistency in capturing both general trends and specific fluctuations within the data, especially at indices 2, 5, 8, and 13. These results suggest that these optimizers are highly effective for neural network optimization. NADAM also performs well, closely mirroring the actual values, although there are slight deviations at certain points, such as around index 6. Adadelta, while generally aligned with the actual values, shows potential sensitivity to model

Table 7.5 Actual and predicted values of algorithms

Actual	ADAM Hidden node 200	Nadam Hidden node 100	RMSprop Hidden node 200	Adadelta Hidden node 100	ADAMAX Hidden node 150	SGD Hidden node 150	Adagrad Hidden node 200
0.288	0.327	0.338	0.273	0.28	0.395	0.536	0.524
0.039	0.06	0.032	0.079	0.133	0.175	0.529	0.509
0.909	0.821	0.802	0.8	0.805	0.776	0.56	0.582
0.112	0.059	0.031	0.081	0.132	0.154	0.521	0.492
0.853	0.868	0.871	0.863	0.862	0.867	0.566	0.597
0.316	0.42	0.414	0.455	0.436	0.483	0.547	0.551
0.118	0.124	0.07	0.112	0.148	0.255	0.532	0.516
0.694	0.717	0.691	0.667	0.671	0.601	0.544	0.545
0.776	0.825	0.808	0.811	0.816	0.779	0.559	0.581
0.288	0.279	0.306	0.353	0.337	0.409	0.55	0.556
0.711	0.749	0.718	0.727	0.737	0.637	0.552	0.564
0.2	0.384	0.431	0.414	0.449	0.43	0.54	0.537
0.858	0.914	0.925	0.923	0.91	0.947	0.574	0.618
0.084	0.169	0.174	0.184	0.194	0.324	0.544	0.542
0.909	0.832	0.82	0.816	0.822	0.812	0.558	0.580

complexity or data noise, which could account for its occasional divergence. In contrast, ADAMAX and SGD display the most significant deviations from the actual values. The variability in prediction accuracy across different indices highlights the importance of selecting an appropriate optimizer. Models optimized with ADAM and RMSprop consistently capture both minor fluctuations and significant shifts in the data, whereas those optimized with ADAMAX and SGD may require further tuning or alternative configurations to improve accuracy.

Table 7.5 displays the prediction outcomes generated by different optimization algorithms applied to nonlinear autoregressive with exogenous inputs (NARX) neural networks, alongside the actual observed values. Table 7.5 provides a comparative analysis of the predictive accuracy of various optimization algorithms—namely ADAM, Nadam, RMSprop, SGD, Adadelta, ADAMAX, and Adagrad—each configured with different numbers of hidden nodes. The table reveals that ADAM and Nadam, particularly with 150 hidden nodes, consistently produce predictions that are closely aligned with the actual values across a range of data points, demonstrating their robustness and effectiveness in generalizing from training data. In contrast, SGD and Adadelta show more significant discrepancies between predicted and actual values, especially at lower data points, indicating potential limitations in these algorithms' optimization processes. The analysis further emphasizes the critical role of hidden node configuration in predictive accuracy, with a higher number of hidden nodes generally enhancing model performance across most algorithms. However, the degree of improvement varies, highlighting the optimizer-dependent nature of these enhancements. Table 7.5 illustrates the importance of selecting both the appropriate optimizer and model complexity to achieve optimal

Table 7.6 Statistical performance values of algorithms

Optimizer	Hidden nodes	RMSE	MAPE	R^2
ADAM	50	0.125118	71.63687	0.856307
	100	0.090663	48.88169	0.924551
	150	0.083323	31.12711	0.936272
	200	0.077209	29.32363	0.945281
ADAMAX	50	0.13746	68.22799	0.82656
	100	0.169387	101.5108	0.736634
	150	0.129705	72.35221	0.845578
	200	0.228602	150.7513	0.520313
ADADELTA	50	0.09564	34.90083	0.916039
	100	0.091237	42.02901	0.923592
	150	0.09946	37.15548	0.909198
	200	0.094807	37.56443	0.917495
ADAGRAD	50	0.314654	206.678	0.091207
	100	0.320558	209.0746	0.056786
	150	0.317712	204.2073	0.07346
	200	0.311049	200.955	0.111913
NADAM	50	0.114522	52.91897	0.879614
	100	0.084883	29.50504	0.933864
	150	0.090375	32.09843	0.925028
	200	0.104895	42.34359	0.899003
RMSprop	50	0.091993	37.77092	0.92232
	100	0.08738	31.32145	0.929915
	150	0.091191	32.84352	0.923669
	200	0.085322	31.49376	0.933178
SGD	50	0.338864	222.5409	−0.05402
	100	0.332258	216.1964	−0.01332
	150	0.326029	208.6289	0.024317
	200	0.32646	208.9215	0.021731

accuracy in artificial neural networks, with ADAM, Nadam, and RMSprop emerging as particularly consistent choices.

Table 7.6 presents a comprehensive evaluation of the statistical performance of various optimization algorithms (Adadelta, Adagrad, ADAM, ADAMAX, Nadam, RMSprop, and SGD) across different configurations of hidden nodes. The performance metrics include root mean square error (RMSE), mean absolute percentage error (MAPE), and the coefficient of determination (R^2), each providing insights into the accuracy and predictive power of the models. The ADAM optimizer emerges as the most effective across the configurations, particularly with 200 hidden nodes, where it achieves the lowest RMSE (0.0772) and the highest R^2 (0.945281). This performance indicates a strong capacity to minimize error and deliver highly accurate predictions, further evidenced by a relatively low MAPE (29.32363). The observed trend with ADAM suggests that it not only stabilizes effectively with

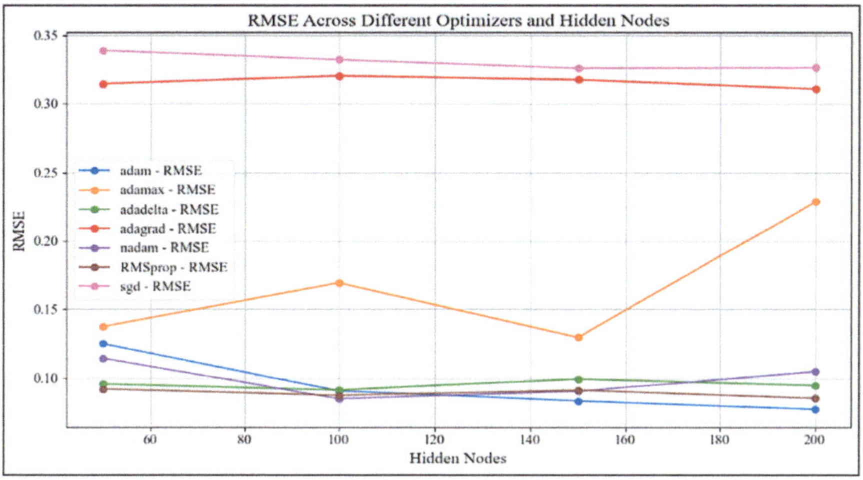

Fig. 7.7 RMSE distribution of different algorithms

increasing model complexity but also optimizes the learning process efficiently, making it the most reliable choice among the tested optimizers. Similarly, RMSprop and Nadam demonstrate good performance, particularly in the 100–150 hidden node configurations, with RMSprop achieving an RMSE of 0.085322 and an R^2 of 0.933178 with 200 hidden nodes, and Nadam showing competitive results with an RMSE of 0.084883 and an R^2 of 0.933864 at 100 hidden nodes. Results indicated that RMSprop and Nadam are robust alternatives to ADAM, especially when slight variations in model architecture are considered. In contrast, Adagrad and SGD display significantly weaker performance metrics across all hidden node configurations, with consistently high RMSE values (e.g., Adagrad at 50 nodes: 0.314654; SGD at 50 nodes: 0.338864) and correspondingly low or even negative R^2 scores (e.g., SGD at 50 nodes: −0.05402). This indicates substantial difficulty in learning from the data, likely due to their inherent characteristics—SGD's lack of momentum in capturing gradients effectively and Adagrad's diminishing learning rates over time. Meanwhile, ADAMAX and Adadelta fall into an intermediate category; while they outperform Adagrad and SGD, their performance does not match the effectiveness of ADAM, Nadam, or RMSprop. This is exemplified by Adadelta's RMSE of 0.091237 and R^2 of 0.923592 at 100 hidden nodes, and ADAMAX's fluctuating performance, particularly with higher node configurations, where it shows a substantial decline in accuracy (e.g., 200 nodes: RMSE 0.228602, R^2 0.520313).

The Fig. 7.7 illustrates RMSE values across different optimization algorithms—ADAM, ADAMAX, Adadelta, Adagrad, Nadam, RMSprop, and SGD—evaluated with varying hidden node configurations. ADAM, RMSprop, and Nadam consistently achieve low RMSE values across all configurations, indicating their robustness and effectiveness in minimizing prediction errors, regardless of model complexity. In contrast, ADAMAX and Adagrad exhibit higher RMSE values, particularly as the number of hidden nodes increases, suggesting difficulties in

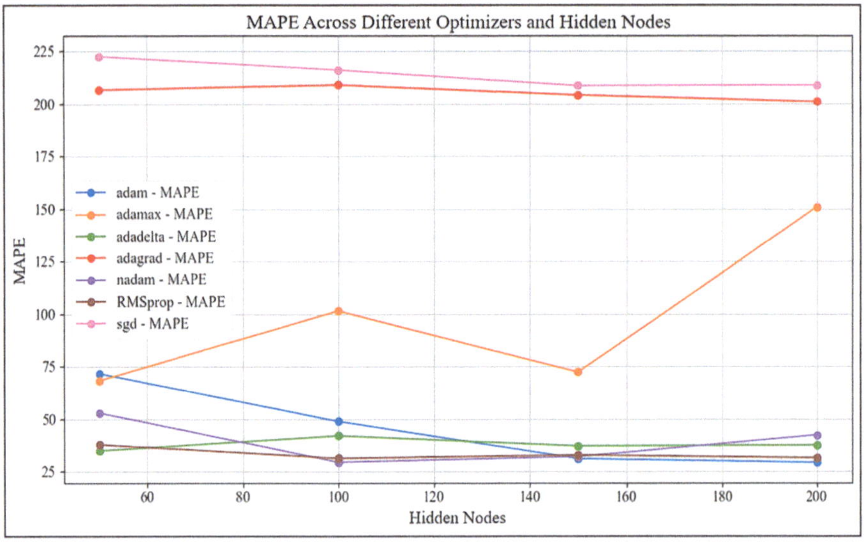

Fig. 7.8 MAPE distribution of different algorithms

optimizing more complex models. SGD, with the highest RMSE values across all configurations, is the least effective optimizer in this context, likely due to its simple gradient descent approach. These findings underscore the critical role of optimizer selection in neural network training, particularly as model complexity increases, with ADAM, RMSprop, and Nadam emerging as the most reliable choices for minimizing prediction errors. Figure 7.8 presents the MAPE values for various optimization algorithms—ADAM, ADAMAX, Adadelta, Adagrad, Nadam, RMSprop, and SGD—across different hidden node configurations. ADAM, Nadam, and RMSprop consistently demonstrate low MAPE values, indicating their robustness and accuracy across varying model complexities. Adadelta performs reliably, albeit slightly less effectively, maintaining moderate MAPE values. In contrast, ADAMAX and Adagrad exhibit higher MAPE values, particularly as the number of hidden nodes increases, with Adagrad consistently yielding the poorest performance. SGD shows the highest MAPE values across all configurations, highlighting its limitations in achieving accurate predictions. These findings underscore the critical importance of optimizer selection in neural network training, with ADAM, Nadam, and RMSprop emerging as the most reliable choices for maintaining prediction accuracy. Figure 7.9 illustrates the distribution of R^2 values across various optimization algorithms—ADAM, ADAMAX, Adadelta, Adagrad, Nadam, RMSprop, and SGD. R^2, the coefficient of determination, measures how well the predicted values align with the actual data, with higher values indicating better model performance. ADAM, Nadam, and RMSprop exhibit the highest R^2 values, consistently close to 1, demonstrating their strong predictive capabilities and robustness across different configurations. ADAMAX also shows relatively high R^2 values, although with greater variability, suggesting some sensitivity to model

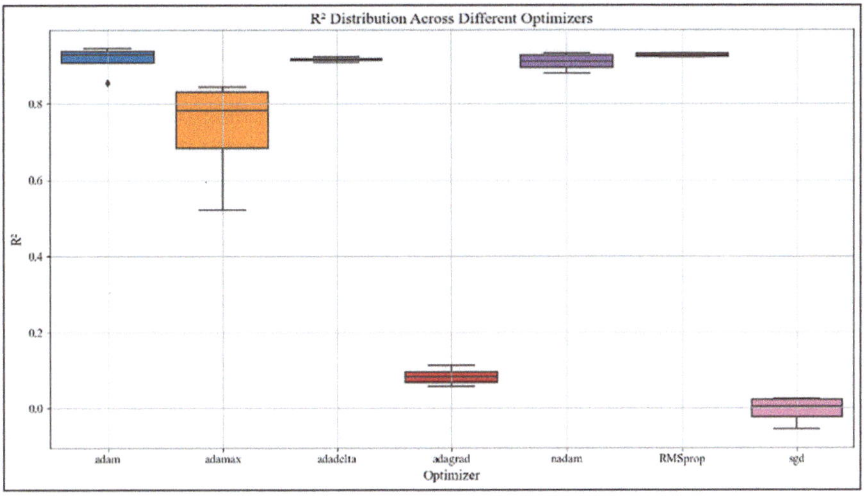

Fig. 7.9 Coefficient of determination distribution of different algorithms

Table 7.7 Results of permutation importance scores

Variable	Permutation importance score
Year	0.0636
Population	0.0301
Wheat price (t − 1)	0.0067
Barley price (t − 1)	0.0026
Season 1temp (September–March)	0.0012
Season 2 temp (April–July)	−0.0002
Season 1 TP(September–March)	−0.00019
Season 2 TP (April–July)	0.00047
Wheat cultivation	0.00024
Wheat yield	0.00015

complexity. In contrast, Adadelta presents moderate R^2 values, indicating reasonable but less optimal performance. Adagrad and SGD perform the poorest, with consistently low R^2 values, reflecting their limited effectiveness in capturing the underlying data patterns. Figure 7.9 shows that ADAM, Nadam, and RMSprop are highly successful and reliable in neural network training, while Adagrad and SGD have low prediction accuracy.

The permutation importance scores derived from the LSTM-based NARX model reveal the relative significance of each input feature in predicting wheat production (Table 7.7). Among the features, year emerges as the most influential, with an importance score of 0.0636, underscoring the critical role of temporal factors in the model. Especially between 1950 and 2021, both technological changes and changes in the inputs used play an important role here. Especially in the newly bred wheat varieties, a significant improvement has been achieved both in terms of yield and quality. In the model, the positive effect of wheat yield on the amount of wheat production was determined. Increasing yield and ensuring its stability are important parameters in both bread and durum wheat breeding studies in Turkiye (Özberk et al., 2010, 2016). Zencirci and Baran (1992) reported that a total genetic improvement of 74% was achieved with the registered varieties in the period 1932–91. In different studies conducted in Turkiye, it is stated that genetic progress has been achieved especially in yield and yield stability (Gummadov et al., 2015; Akın et al., 2017). Population also holds considerable significance, with a score of 0.0301, indicating that demographic variables play a substantial role, likely through their impact on demand and resource allocation. It is thought that an increase in population will lead to an increase in wheat demand, which in turn will lead to an increase in wheat production. When the model is analyzed, it is seen that the increase in population has a positive effect on wheat production. This is due to the fact that wheat and its products are considered as a staple food in Turkiye. One-year lagged "wheat price" and 1-year lagged "barley price" are included as economic variables in the model. Wheat price (0.0067) and barley price (0.0026) contribute to the model's predictions, although to a lesser extent, reflecting the influence of market conditions on agricultural decisions. Barley is included in the model as a competitor of wheat. One of the most important variables affecting supply is the price of the product. In the agricultural products market, the previous year's prices are generally taken as a reference for the farmer's production decision. The situation is similar in the wheat sector. In studies conducted on this subject, Özçelik et al. (2010) examined the effect of the Turkish grain board (TGB) wheat purchase price, farmer's price, TGB barley purchase price, Turkiye's wheat export price, and Turkiye's wheat import price on the amount of wheat production in Turkiye between 1980 and 2008 using the variance decomposition method obtained from VAR analysis. According to the variance decomposition results, they found that the contribution of the price received by the farmer and the TGB purchase price on the amount of wheat production was 40% and 12%, respectively, in the second period. Dörtok and Aksoy (2018) reported that the value of wheat price in the previous year has a positive effect on wheat production. Kızıloğlu et al. (2003) found the effect of wheat producer price on the amount of production to be positive and statistically significant in their study conducted in Erzurum province. It is also possible to come across the opposite cases in the literature. Çetinkaya (2012) analyzed the relationship between the amount of wheat production in Turkiye and TMO purchase prices by using the Koyck-Almon technique for the period between 2002 and 2011 and concluded that TMO purchase prices have no effect on the amount of production. It is theoretically expected that the change in the prices of competing products will have a negative effect on the production of the product under consideration. In this model, barley is selected as a rival crop for the period 1950–2021. When the literature is examined, it is seen that barley, chickpea, lentil, maize, and sunflower are considered as competitors to wheat, depending on irrigated and dry conditions (Konyalı, 2008; Bayramoğlu, 2011; Dörtok & Aksoy, 2018). Environmental variables, including T1 (temperature) and S1 (precipitation), exhibit minimal influence, with scores as low as 0.0005 and even a slightly negative value for T1 (−0.0002), suggesting that these factors either add noise or have their effects captured by other, more dominant features. Although the effect of climatic factors on the estimation of wheat production amount is low, there are studies in the literature on the effect of temperature and precipitation data, which we consider in the model, on wheat yield. Doğan and Kan (2019) reported that temperature has a negative effect on yield and precipitation has a positive effect on yield in slight drought (SLD), moderate drought (MD), and severe drought regions (SVD) in Turkiye. Kayam et al. (2002) reported that a 10% decrease in rainfall in April and May decreased yield by 22.9 kg/ha in their study conducted in the Aegean region of Turkiye. Şimşek and Çakmak (2012) predicted that temperature increases and precipitation decreases will decrease wheat yield in their prediction with the AgrometShell program. Overall, the analysis highlights the dominance of temporal, demographic, and economic variables in predicting wheat production, while environmental factors and specific agricultural practices appear to play a minor role in this particular model configuration.

7.4 Conclusion

Considering both its importance in human nutrition and its indispensability in animal production with its grains and stems, wheat is one of the strategic products. With the development of modern technology and science, we can say that today there is a quantitative increase in wheat, thanks to plant breeding methods. With this development, it is seen that modern wheat varieties have entered our lives more and the industry based on this also demands these varieties. However, the crises experienced led people to become more conscious about healthy consumption and to produce higher quality products by using more environmentally friendly techniques. In this regard, the increasing perception turns into more attitudes and behaviors, that is, there is a shift in preference toward the products that create the quality perception of the consumers. In this context, concepts such as local product, niche product, geographically indicated product, traditional product, and organic product have become more audible in the market, and these concepts have begun to be brought together with the perception of healthy products. In this context, local seeds and local varieties/populations have become preferred in more production and consumption.

In this new market created, the concept of healthy products has gained an important place in an increasing trend today. The COVID-19 pandemic we are currently experiencing has made important contribution to the development of this new trend. This situation also provides a positive contribution to the conservation and sustainable use of local varieties (landraces) which are important in terms of genetic resources. Spelled wheat is one of these genetic resources (Einkorn, Emmer, Spelt). The results of the research on these wheats showed that these wheat types have quantitative advantages when compared with the existing wheat varieties, and this new trend led to the inclusion of spelt. In addition to the studies carried out in Turkiye in the Fertile Crescent, which is known as the homeland of wheat, findings such as the positive effects of these wheat species on human health in Europe, their resistance to climate change, their resistance to biotic and abiotic stress conditions have made these wheats more popular.

This study demonstrates that LSTM-based NARX networks have achieved high accuracy in predicting wheat production levels. The analysis reveals that the variables *Year* and *Population* exert the most significant influence on wheat production. The NARX model effectively forecasts future wheat production by considering not only historical production data but also external factors. This underscores the flexibility and robustness of LSTM-based neural networks, particularly in complex agricultural systems. Although economic variables such as *Wheat Price* and *Barley Price* also have notable effects, their influence is found to be less decisive compared to *Year* and *Population*. The findings suggest that the impact of meteorological data on agricultural production forecasts may be less critical than other macroeconomic and environmental factors.

The results establish that LSTM-based NARX networks are powerful tools capable of predicting wheat production with high accuracy. The use of such AI-based methods can enable more precise and reliable management of agricultural production processes and offer valuable contributions to policymakers and practitioners in making strategic decisions. Future research could extend these findings by applying similar models to different agricultural products and geographical regions. Furthermore, a more detailed examination of the effects of external factors such as climate change, population growth, market fluctuations, and political uncertainties could further enhance the accuracy and reliability of agricultural production forecasts. The findings on LSTM-based NARX networks highlight the significant potential of these techniques to provide innovative and effective solutions for agricultural production forecasting in future research.

References

Abraham, E. R., Mendes dos Reis, J. G., Vendrametto, O., Oliveira Costa Neto, P. L. D., Carlo Toloi, R., Souza, A. E. D., & Oliveira Morais, M. D. (2020). Time series prediction with artificial neural networks: An analysis using Brazilian soybean production. *Agriculture, 10*(10), 475.

Akıllı, A., & Atıl, H. (2020). Evaluation of normalization techniques on neural networks for the prediction of 305-day milk yield. *Turkish Journal of Agricultural Engineering Research, 1*(2), 354–367.

Akın, B., Sohail, Q., Ünsal, R., Dinçer, M. N., Demir, L., Geren, H., Sevim, İ., Orhan, Ş., Yaktubay, Ş., & Morgunov, A. (2017). Genetic gains in grain yield in spring wheat in Turkey. *Turkish Journal of Agriculture and Forestry, 41*(2), 103–112. https://doi.org/10.3906/tar-1611-50

Alsumaiei, A. A. (2020). A nonlinear autoregressive modeling approach for forecasting groundwater level fluctuation in urban aquifers. *Water, 12*(3), 820.

Bali, N., & Singla, A. (2021). Deep learning based wheat crop yield prediction model in Punjab region of North India. *Applied Artificial Intelligence, 35*(15), 1304–1328.

Bayramoğlu, Z. (2011). Buğday üretim faaliyetini etkileyen başlıca faktörler. *Bitkisel Araştırma Dergisi, 1*, 28–34.

Bengio, Y. (2012). Practical recommendations for gradient-based training of deep architectures. In G. Montavon, G. B. Orr, & K.-R. Müller (Eds.), *Neural networks: Tricks of the trade* (pp. 1–33). Springer.

Bhojani, S. H., & Bhatt, N. (2020). Wheat crop yield prediction using new activation functions in neural network. *Neural Computing and Applications, 32*(17), 13941–13951.

Bisong, E. (2019). *Building machine learning and deep learning models on Google cloud platform*. Springer.

Boussaada, Z., Curea, O., Remaci, A., Camblong, H., & Mrabet, B. N. (2018). A nonlinear autoregressive exogenous (NARX) neural network model for the prediction of the daily direct solar radiation. *Energies, 11*(3), 620.

Buevich, A., Sergeev, A., Shichkin, A., & Baglaeva, E. (2021). A two-step combined algorithm based on NARX neural network and the subsequent prediction of the residues improves prediction accuracy of the greenhouse gases concentrations. *Neural Computing and Applications, 33*, 1547–1557.

Çetinkaya, Ş. (2012). Türkiye'nin Son On Yıllık Buğday Üretimindeki Miktar-Fiyat İlişkisinin Ekonometrik Analizi: Kocky-Almon Tekniği. *İnsan ve Toplum Bilimleri Araştırmaları Dergisi, 1*(4), 52–66.

Cocianu, C. L., & Avramescu, M. Ş. (2020). The use of LSTM neural networks to implement the NARX model: A case study of EUR-USD exchange rates. *Informatica Economica, 24*(1), 5.

Cocianu, C. L., Uscatu, C. R., & Avramescu, M. (2022). Improvement of LSTM-based forecasting with NARX model through use of an evolutionary algorithm. *Electronics, 11*(18), 2935.

Copernicus Climate Change Service, Climate Data Store. (2021). *ERA5-Land monthly averaged data from 1950 to present*. Copernicus Climate Change Service (C3S) Climate Data Store (CDS). https://climate.copernicus.eu/

Di Nunno, F., Race, M., & Granata, F. (2022). A nonlinear autoregressive exogenous (NARX) model to predict nitrate concentration in rivers. *Environmental Science and Pollution Research, 29*(27), 40623–40642.

Doğan, H. G., & Kan, A. (2019). The effect of precipitation and temperature on wheat yield in Turkey: A panel FMOLS and panel VECM approach. *Environment, Development and Sustainability, 21*, 447–460. https://doi.org/10.1007/s10668-018-0298-5

Dörtok, A., & Aksoy, A. (2018). Türkiye Buğday Sektörünün Eşanlı Model Yöntemiyle Tahmini. *KSÜ Tarım ve Doğa Derg, 21*(4), 580–586.

Eraktan, G. (2001). *Tarım Politikası Temelleri ve Türkiye'de Tarımsal Destekleme Politikası*. Uzel Yayınları.

FAOSTAT. (2024a). *Crops and livestock products*. https://www.fao.org/faostat/en/#data/QCL

FAOSTAT. (2024b). *Land Use*. https://www.fao.org/faostat/en/#data/RL

FAOSTAT. (2024c). *Population and Employment*. https://www.fao.org/faostat/en/#data/OA

Farooq, A., Farooq, N., Akbar, H., Hassan, Z. U., & Gheewala, S. H. (2023). A critical review of climate change impact at a global scale on cereal crop production. *Agronomy, 13*(1), 162.

FAS-USDA. (2024). *Production-Wheat*. https://fas.usda.gov/data/production/commodity/0410000

Gao, H., Qiu, S., Fang, J., Ma, N., Wang, J., Cheng, K., Wang, H., Zhu, Y., Hu, D., Liu, H., & Wang, J. (2023). Short-term prediction of PV power based on combined modal decomposition and NARX-LSTM-LightGBM. *Sustainability, 15*(10), 8266.

Gummadov, N., Keser, M., Akin, B., Cakmak, M., Mert, M., Taner, S., Ozturk, I., Topal, A., Yazar, S., & Morgounov, A. I. (2015). Genetic gains in wheat in Turkey: Winter wheat for irrigated conditions. *Crop Journal, 3*, 507–516.

Guo, W. W., & Xue, H. (2014). (2014) crop yield forecasting using artificial neural networks: A comparison between spatial and temporal models. *Mathematical Problems in Engineering, 1*, 857865.

Güler, H., & Erlat, E. (2023). Türkiye'de 1950–2022 döneminde ortalama hava sıcaklıklarında gözlenen değişim ve eğilimler. *Ege Coğrafya Dergisi, 32*(1), 1–17.

Harlan, J. R. (1995). *The living fields: Our agricultural heritage*. Cambridge Univ. Press.

Hochreiter, S., & Schmidhuber, J. (1997). Long short-term memory. *Neural Computation, 9*(8), 1735–1780.

Jabjone, S., & Jiamrum, C. (2013). Artificial neural networks for predicting the rice yield in Phimai District of Thailand. *International Journal of Electrical Energy, 1*(3), 177–181.

Jain, A., Nandakumar, K., & Ross, A. (2005). Score normalization in multimodal biometric systems. *Pattern Recognition, 38*(12), 2270–2285.

Jang, I. H., & Choe, Y. C. (2014). Forecasting rice productivity using a neural network method: A global warming scenario. *Advances in Science, Technology and Letters, 49*, 222–228.

Kaleeswaran, V., Dhamodharavadhani, S., & Rathipriya, R. (2020). A comparative study of activation functions and training algorithm of NAR neural network for crop prediction. In *2020 4th international conference on electronics, communication and aerospace technology (ICECA)* (pp. 1073–1077). IEEE.

Kan, M., Küçükçongar, M., Morgounov, A., Keser, M., Özdemir, F., Muminjanov, H., & ve Qualset CO. (2017). Türkiye'de Yerel Buğday Popülasyonlarının Durumu ve Yerel Buğday Üreten Üreticilerin Üretim Kararlarında Etkili Olan Faktörlerin Belirlenmesi. *Journal of Agricultural Faculty of Gaziosmanpaşa University (JAFAG), 34*(2), 54–64. https://doi.org/10.13002/jafag4225

Kaul, M., Hill, R. L., & Walthall, C. (2005). Artificial neural networks for corn and soybean yield prediction. *Agricultural Systems, 85*(1), 1–18.

Kayam, Y., Ozsoy, U., Lomas, J., Oden, O., Mandel, M., & Gurbuz, M. (2002). *The impact of climate change on wheat production of Aegean region in Turkey: The effect of a reduction of rainfall and the increase of temperature on wheat yields*. http://www.fao.org/sd/climagrimed/pdf/ws01_26.pdf.

Khaki, S., & Wang, L. (2019). Crop yield prediction using deep neural networks. *Frontiers in Plant Science, 10*, 621.

Kızıloğlu, S., Dağdemir, V., & Erem, T. (2003). Erzurum İlinde Bitkisel Ürünlerin Üretim Ekonomileri ve Arz Fonksiyonlarının Ekonometrik Analizi. In *VI. Ulusal Ekonometri ve İstatistik Sempozyumu, Mayıs 2003*.

Konyalı, S. (2008). *Türkiye'de Buğdayda Uygulanan Tarım Politikalarının Üreticiler Ve Tüketiciler Üzerindeki Etkileri: Trakya Bölgesi Örneği*. Namık Kemal Üniversitesi, Fen Bilimleri Enstitüsü, Tarım Ekonomisi A.B.D. Doktora Tezi, Tekirdağ, Türkiye.

Kosanan, O., & Kantanantha, N. (2014). Thailand's Para rubber production forecasting comparison. *Proceedings of the International Multi Conference of Engineers and Computer Scientists, 2*.

Kumar, S., Kumar, V., & Sharma, R. K. (2015). Sugarcane yield forecasting using artificial neural network models. *International Journal of Artificial Intelligence & Applications (IJAIA), 6*(5), 51–68.

Latifi, Z., & Shabanali Fami, H. (2022). Forecasting wheat production in Iran using time series technique and artificial neural network. *Journal of Agricultural Science and Technology, 24*(2), 261–273.

Leontaritis, I. J., & Billings, S. A. (1985). Input-output parametric models for non-linear systems part I: Deterministic non-linear systems. *International Journal of Control, 41*(2), 303–328.

Mahmoud, A., Mohammed, A., & Khalil, A. A. (2024). Time series forecasting of wheat crop productivity in Egypt using deep learning techniques. *International Journal of Data Science and Analytics*, 1–16.

Massaoudi, M., Chihi, I., Sidhom, L., Trabelsi, M., Refaat, S. S., & Oueslati, F. S. (2019). A novel approach based deep RNN using hybrid NARX-LSTM model for solar power forecasting. arXiv preprint arXiv:1910.10064.

Massaoudi, M., Chihi, I., Sidhom, L., Trabelsi, M., Refaat, S. S., Abu-Rub, H., & Oueslati, F. S. (2021). An effective hybrid NARX-LSTM model for point and interval PV power forecasting. *IEEE Access, 9*, 36571–36588.

McKinney. (2011). A foundational python library for data analysis and statistics. *Python for High Performance and Scientific Computing, 14*, 1–9.

Menezes, J. M. P., Jr., & Barreto, G. A. (2008). Long-term time series prediction with the NARX network: An empirical evaluation. *Neurocomputing, 71*(16–18), 3335–3343.

Mithiya, D., Datta, L., & Mandal, K. (2019). Time series analysis and forecasting of oilseeds production in India: Using autoregressive integrated moving average and group method of data handling–neural network. *Asian Journal of Agricultural Extension, Economics & Sociology, 30*, 1–14.

MoAF (Ministry of Agriculture and Forestry-Turkiye). (2024). *Farmer Records System Statistics.* (Unpublished data).

Moursi, A. S. A., El-Fishawy, N., Djahel, S., & Shouman, M. A. (2021). An IoT enabled system for enhanced air quality monitoring and prediction on the edge. *Complex & Intelligent Systems, 7*(6), 2923–2947.

Moursi, A. S. A., El-Fishawy, N., Djahel, S., & Shouman, M. A. (2022). Enhancing PM2.5 prediction using NARX-based combined CNN and LSTM hybrid model. *Sensors, 22*(12), 4418.

Muñoz-Sabater, J., Dutra, E., Agustí-Panareda, A., Albergel, C., Arduini, G., Balsamo, G., Thépaut, J. N. (2021). ERA5-land: A state-of-the-art global reanalysis dataset for land applications. *Earth Syst Sci Data Discuss.*

Nesbitt, M., & Samuel, D. (1996). From stable crop to extinction? The archaeology and history of the hulled wheats. In S. Padulosi, K. Hammer, & J. Heller (Eds.), *Hulled wheats* (pp. 41–100). International Plant Genetic Resources Institute.

Oliphant, T. E. (2006). *A Guide to NumPy* (Vol. 1, p. 371p). The MIT Press.

Özberk, İ., Zencirci, N., Özkan, H., Özberk, F., & Eser, V. (2010). *Dünden bugüne makarnalık buğday ıslahı ve geleceğe bakış*. Makarnalık Buğday ve Mamulleri Konferansı, 17–18 Mayıs, 2010, pp. 43–66.

Özberk, F., Karagöz, A., Özberk, İ., & Atlı, A. (2016). Buğday Genetik Kaynaklarından Yerel ve Kültür Çeşitlerine; Türkiye'de Buğday ve Ekmek. *Tarla Bitkileri Merkez Araştırma Enstitüsü Dergisi, 25*(2), 218–233.

Özçelik, A., Kayalak, S., & ve Özer, O. O. (2010). *Türkiye'nin Buğday Üretimi için Bir Öngörü Modeli: VAR Yaklaşımı. Türkiye IX* (pp. 25–32). Tarım Ekonomisi Kongresi, Şanlıurfa.

Paswan, S., Paul, A., Paul, A., & Noel, A. S. (2022). Time series prediction for sugarcane production in Bihar using ARIMA & ANN model. *The Pharma Innovation Journal, 11*(4), 1947–1956.

Pedregosa, F., Varoquaux, G., Gramfort, A., Michel, V., Thirion, B., Grisel, O., Blondel, M., Prettenhofer, P., Weiss, R., Dubourg, V., Vanderplas, J., Passos, A., & Cournapeau, D. (2011). Scikit-learn: Machine learning in python. *Journal of Machine Learning Research, 12*, 2825–2830.

Pradhan, K. K., Hota, S. K., & Satpathy, B. (2012). Artificial neural network methodology for modeling and resource use optimization in rice yield. *International Journal of Creative Mathematical Sciences & Technology (IJCMST), 2*(1), 35–41.

Randles, B. M., Pasquetto, I. V., Golshan, M. S., & Borgman, C. L. (2017). Using the Jupyter notebook as a tool for open science: An empirical study. In *2017 ACM/IEEE joint conference on digital libraries (JCDL)* (pp. 1–2). IEEE.

Rehman, A., Farooq, M., Lee, D. J., & Siddique, K. H. (2022). Sustainable agricultural practices for food security and ecosystem services. *Environmental Science and Pollution Research, 29*(56), 84076–84095.

Reynolds, M. P., & Braun, H. J. (2022). *Wheat improvement: Food security in a changing climate.* Springer Nature.

Sadenova, M., Beisekenov, N., Varbanov, P. S., & Pan, T. (2023). Application of machine learning and neural networks to predict the yield of cereals, legumes, oilseeds and forage crops in Kazakhstan. *Agriculture, 13*(6), 1195.

Shen, D., Zhao, X., Chai, L., Guo, Z., & Leng, C. (2024). Analysis of the agricultural economic value of a weather forecasting service based on a survey of peasant households in Chinese provinces. *Humanities and Social Sciences Communications, 11*(1), 1–7.

Shiferaw, B., Smale, M., Braun, H. J., Duveiller, E., Reynolds, M., & Muricho, G. (2013). Crops that feed the world 10. Past successes and future challenges to the role played by wheat in global food security. *Food Security, 5*, 291–317.

Singh, R. K. (2008). Artificial neural network methodology for modelling and forecasting maize crop yield. *Agricultural Economics Research Review, 21*(1), 5–10.

Şimşek, O., & Çakmak, B. (2012). Future-proof scenarios and risk analysis for wheat by agrometshell model. *Journal of Agricultural Sciences, 18*(2012), 187–196.

Sivaranjani, T., & Vimal, S. P. (2023). Application of NARX neural network for predicting suitable crop for cultivation: A comparative analysis. In *2023 5th international conference on smart systems and inventive technology (ICSSIT)* (pp. 1333–1336). IEEE.

Sugiyarto, A. W., & Abadi, A. M. (2019). Prediction of Indonesian palm oil production using long short-term memory recurrent neural network (LSTM-RNN). In *2019 1st international conference on artificial intelligence and data sciences (AiDAS)* (pp. 53–57). IEEE.

Sun, Z., Li, Q., Jin, S., Song, Y., Xu, S., Wang, X., Cai, J., Zhou, Q., Ge, Y., Zhang, R., Zang, J., & Jiang, D. (2022). Simultaneous prediction of wheat yield and grain protein content using multitask deep learning from time-series proximal sensing. *Plant Phenomics, 2022*, 9757948. 13 pages, https://doi.org/10.34133/2022/9757948

Tanno, K., & Willcox, G. (2006). The origins of cultivation of Cicer arietinum L. and Vicia faba L.: Early finds from north West Syria (Tell el-Kerkh, late 10th millennium BP). *Vegetation History and Archaeobotany, 15*, 197–204.

Tao, P., Hao, X., Cheng, J., & Chen, L. (2023). Predicting time series by data-driven spatiotemporal information transformation. *Information Sciences, 622*, 859–872.

Thapa, S., Zhao, Z., Li, B., Lu, L., Fu, D., Shi, X., Tang, B., & Qi, H. (2020). Snowmelt-driven streamflow prediction using machine learning techniques (LSTM, NARX, GPR, and SVR). *Water, 12*(6), 1734.

Thimmegowda, M. N., Manjunatha, M. H., Huggi, L., Shivaramu, H. S., Soumya, D. V., Nagesha, L., & Padmashri, H. S. (2023). Weather-based statistical and neural network tools for forecasting rice yields in major growing districts of Karnataka. *Agronomy, 13*(3), 704.

TMO. (2024). Turkish Grain Borad. https://www.tmo.gov.tr/homepage (Last Accessed on 15.08.2024) In Turkish.

TurkStat. (2024). https://biruni.tuik.gov.tr/ (Last Accessed on 15.08.2024) In Turkish.

Wang, D., Jenkins, K., Forstenhäusler, N., Lei, T., Price, J., Warren, R., Jenkins, R., & Guan, D. (2021). Economic impacts of climate-induced crop yield changes: Evidence from agri-food industries in six countries. *Climatic Change, 166*(3), 30.

Wei, M., Ye, M., Li, J. B., Wang, Q., & Xu, X. (2020). State of charge estimation of lithium-ion batteries using LSTM and NARX neural networks. *IEEE Access, 8*, 189236–189245.

Xu, X., & Zhang, Y. (2021). Corn cash price forecasting with neural networks. *Computers and Electronics in Agriculture, 184*, 106120.

Yeşilköy, S., & Demir, I. (2024). Crop yield prediction based on reanalysis and crop phenology data in the agroclimatic zones. *Theoretical and Applied Climatology, 155*, 1–14.

Zencirci, N., & Baran, İ. (1992). Kuru tarım alanları için geliştirilen ekmeklik buğday çeşitleri ile sağlanan genetik ilerleme üzerine bir araştırma. *Tarla Bitkileri Merkez Araştırma Enstitüsü Dergisi, 1*(1), 19–26.

Zohary, D., & Hopf, M. (1988). *Domestication of plants in the Old World*. Clarendon Press.

Chapter 8
Weather Forecasting and State-of-the-Art Soil Analysis in Irrigated Precision Wheat Production

Erhan Akça and Takashi Kume

Abstract Findings from archaeological excavations in Southeastern Anatolia, Türkiye, indicate that humanity began settling the region around 14,000 years ago. Following this, it modified the topography, soil, plants, water resources, and animal life to meet its needs for food, fiber, and shelter. Wheat was reportedly the first plant to be domesticated during the early Holocene, when the environment was favorable for human existence compared to prior geological periods. While wheat grows best in latitudes between 30° and 60°N and 27° and 40°S, it can be harvested in far northern regions, from the Arctic Circle to Tibet, and at elevations of up to 4570 m above sea level. Currently, wheat occupies 22% of all cultivated areas worldwide, making it the most extensively grown plant. Wheat's widespread distribution demonstrates that the majority of the world's population is dependent on this plant, making wheat a crucial plant for local and global food security, as was witnessed wheat shortage in 2022 due to Ukrainian–Russia conflict. Wheat is becoming an increasingly vital plant for strategic purposes. Even in countries like Japan, where rice is the primary food source, the consumption of wheat is on the rise. As a result of technological advancements, all stages of wheat production have been controlled and managed to ensure local and worldwide socioeconomic stability since the mid-twentieth century. First and foremost, significant progress has been made in plant and soil analysis, followed by the establishment of wheat irrigation, fertilization, spraying programs, and production models. Current models and evaluations exhibit a high degree of accuracy and precision, mostly attributed to the abundance of data and the availability of advanced technology. Technological and data capabilities notwithstanding, the presence of climatic instabilities and soil degradation threats requires the gathering of supplementary climate, soil, and plant data, together with software and computers capable of processing it. This underscores the need to

---

E. Akça (✉)
Vocational School of Technical Sciences, Adıyaman University, Adıyaman, Türkiye
e-mail: eakca@adiyaman.edu.tr

T. Kume
Department of Rural Engineering, Faculty of Agriculture, Ehime University, Matsuyama city, Japan

N. Zencirci et al. (eds.), *Empowering Wheat Cultivation with GIS, Digital Approaches and Artificial Intelligence*, https://doi.org/10.1007/978-3-031-99954-3_8

employ artificial intelligence. Utilizing this technology will ensure the secure delivery of wheat from the field to the table, thereby meeting the growing needs of humanity.

Keywords Climate · Wheat cultivation · Rapid soil analyses · Disasters · Soil threats

Abbreviations

APSIM	agricultural production systems sIMulator
CO_2	carbon dioxide
CSM	InfoCrop-Wheat Crop Simulation Model
DRS	diffuse reflectance spectroscopy
EMI	electromagnetic induction
FAO	Food and Agriculture Organization
GCR	Galvanic Contact Resistivity
ITPS	Intergovernmental Technical Panel on Soils
LAI	leaf area index
MCWLA	Crop–Weather Relationship Model Over Larger Areas
NDVI	normalized difference vegetation index
NPP	net primary production
PXRF	portable X-ray fluorescence spectrometry
SOC	soil organic carbon
UNEP	United Nations Environment Programme
US$	United States Dollar
USDA	United States Department of Agriculture
WMO	World Meteorological Organisation

8.1 Introduction

Approximately 10,000 years ago, humans began to establish settlements and modify the environment to suit their needs, resulting in the formation of anthroscapes (Akça et al., 2020). Wheat cultivation was likely the primary crop in the development of anthroscapes (Fig. 8.1). Wheat may thrive in various climate and soil conditions globally; however, peak yields are achieved in Mediterranean climates characterized by 600–800 mm of rainfall occurring in winter and late spring (Lyon, 1907). In numerous dry and semiarid areas, irrigating wheat twice in spring can result in yields that are double those obtained under the same arid conditions (Aykanat & Barut, 2018). Nonetheless, irrigation constitutes a costly endeavor for wheat cultivators in arid and semiarid areas due to the scarcity of water resources and irrigation infrastructure (Rodrigues & Pereira, 2009). Consequently, even if irrigation is possible,

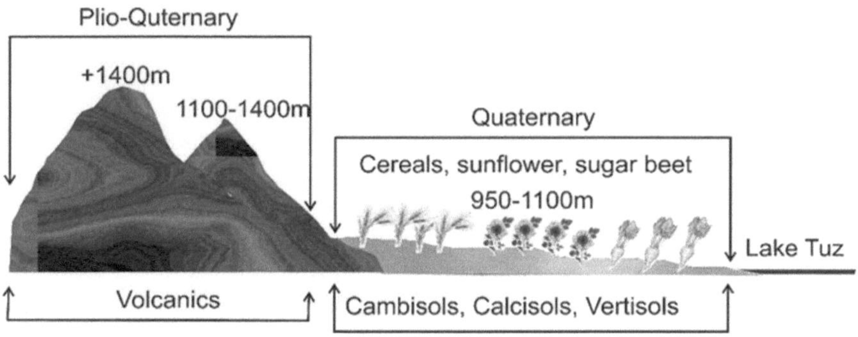

Fig. 8.1 The central Anatolian anthroscape

weather forecasting is essential to avert rising irrigation expenses in wheat cultivation regions.

Agricultural output is contingent upon the interaction of two primary factors: climate and soil. Since the dawn of agriculture, humans have conducted observations and measurements to foresee the impacts of various factors on agricultural practices and to implement appropriate adjustments. For example, a classic Indian and Pakistani quote emphasizes the significance of soil organic matter for crop yield with the saying, "Soil without dung is like a cow without her calf" (Yang et al., 2018). As a result of the accumulation of knowledge gained over the course of thousands of years of climate observations, there are countless proverbs that refer to climate observations. People have been describing the relationship between climate and human health since the time of the ancient Greeks. In his famous work "Air, water, and land," Hippocrates (460 BC–375 BC) discussed the effects of weather on human existence (Yoshino & Miyashita, 2007). For the moment, the impact of climate and soil on agriculture is increasingly diverging from natural processes as agricultural production in fields and orchards is highly susceptible to a multitude of hazards arising from both climate change and human activities. Upon the dawn of agricultural practices, humans commenced forecasting the climatic conditions that predominantly influenced production. Producers and herders employ an integration of meteorological, biological, and astrological signs to predict the weather around them (Radeny et al., 2019). So, many civilizations developed agricultural calendars predicated on river inundations, exemplified by the Nile River, and seasonal precipitation, such as the monsoons of Asia. The remarkable precision of local climate predictions led to their incorporation into societal traditions, ensuring their transmission to subsequent generations due to their dependability (Galacgac & Balisacan, 2009; Ayal et al., 2015). Nonetheless, climate fluctuations that began in the twentieth century have drastically decreased the efficacy of traditional local climate knowledge in properly forecasting weather events. Field and orchard agriculture suffers significant losses as a consequence of climate abnormalities that have become more noticeable and experienced since the 1970s (WMO, 2021). According to the World Meteorological Organization, there has been a fivefold increase in the number

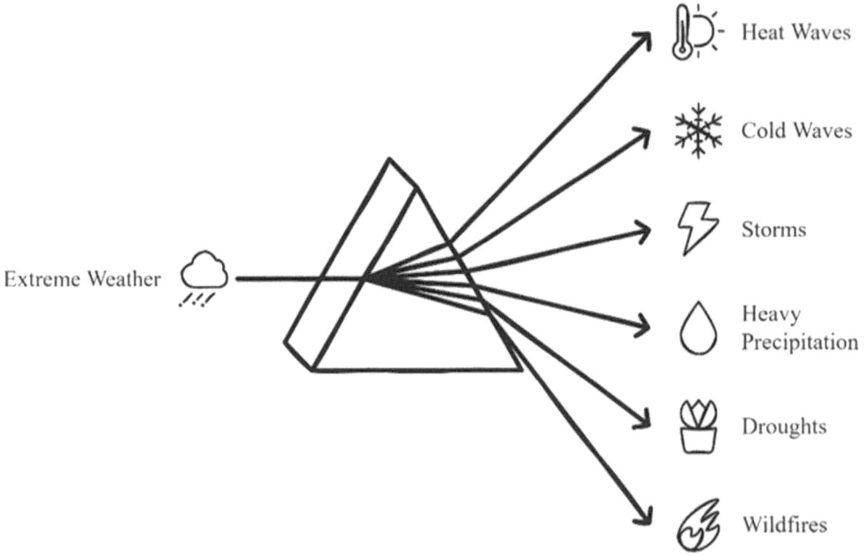

Fig. 8.2 Types of extreme weather events

of reported disaster losses from extreme weather since then. Significant climate anomalies include droughts, storms, hail, severe temperatures (both frosts and heat waves), heavy precipitation, and flooding, resulting in an annual economic loss of US$ 143 billion (Seneviratne et al., 2021) (Fig. 8.2).

The FAO and ITPS (2015) have identified 10 critical threats to agricultural soils: erosion, loss of soil organic carbon, biodiversity, nutrient imbalance, salinity-alkalization, acidity, waterlogging, compaction, soil pollution, and soil sealing (Fig. 8.3). Changes in the soil's physical, chemical, and biological properties are also brought about by these threats (Strawn et al., 2019; Kumar et al., 2019; Nadarajan & Sukumaran, 2021). In dry and semiarid regions, excessive irrigation is causing soil salinization and alkalinization as rainfall is not sufficient to leach salt from the soil profile. A study conducted by Yan et al. (2022) proved that between the years 1980 and 2021, the areas of soil that were salinized and alkalized increased at an annual rate of 1.40% and 0.86%, respectively. Soil productivity faces a grave threat from the anticipated doubling of the present chemical production capacity to 2.3 billion metric tons by 2030, according to the second Global Chemicals Outlook, which was presented at the United Nations Environment Assembly in Nairobi (UNEP, 2019). Herzfeld et al. (2021) have shown that under two distinct climate events, RCP2.6 and RCP8.5, the reserves of agricultural soil organic carbon (SOC) decrease in all simulated scenarios. This is because the breakdown of SOC surpasses the rise in carbon inputs into the soil caused by unsustainable farming methods.

With unforeseen climate events, the magnitude and range of hazards to soil health will escalate to unfavorable levels. The organic matter content of soil is known to decrease as a result of rising temperatures and drought. As a result of the damage that

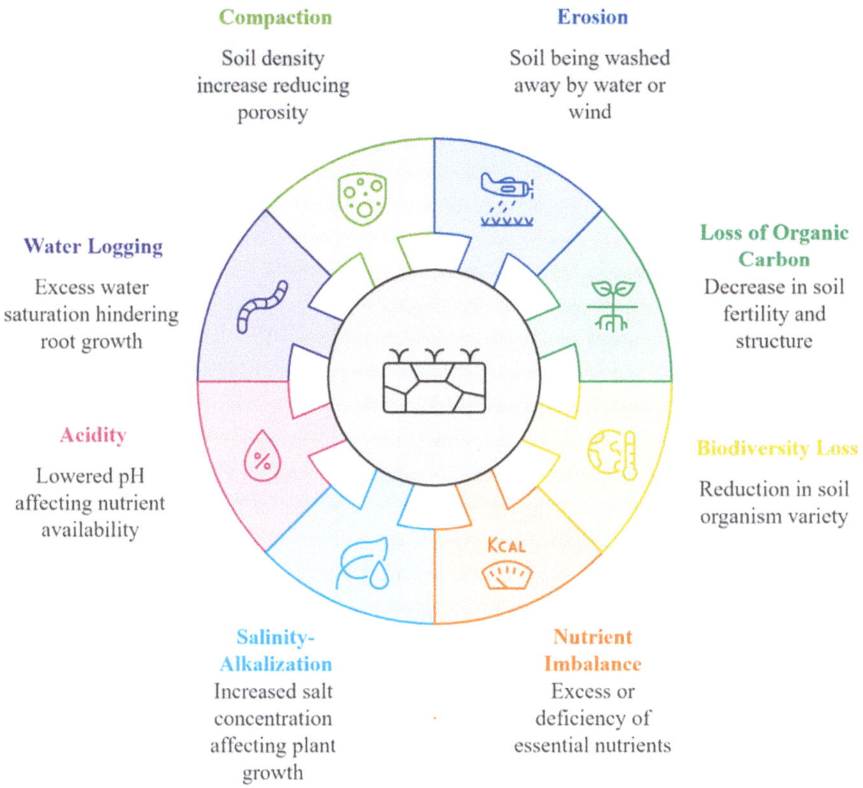

Fig. 8.3 Threats to soils

extreme climate events cause to biomass, the amount of organic matter that is introduced into the soil likewise reduces. In their study, Deng et al. (2021) concluded that drought led to a 59% increase in the accumulation of dissolved organic carbon content. This was attributed to the delayed mineralization and enhanced stability of dissolved organic matter.

Farmers in the majority of wheat-producing nations (Fig. 8.4) are mostly low-income peoples of the society, pushing them to implement increased fertilization and irrigation practices to boost production per unit area. This pressure results in soil degradation, prompting farmers to mitigate productivity losses from declining soil health with intensified fertilization and irrigation, ultimately initiating a process that culminates in land abandonment (Wang et al., 2022).

Wheat lands are currently subject to intensive cultivation, leading to rapid shifts in crop patterns based on income. Department of Agriculture estimated that total wheat cultivation for 2025/26 was 18.615.539 million ha, slightly below the 18.656.008 ha sown in 2024/25 for USA. The report indicated that US farmers are anticipated to expand corn acreage while decreasing soybean and wheat plantings for the forthcoming marketing year (USDA, 2023).

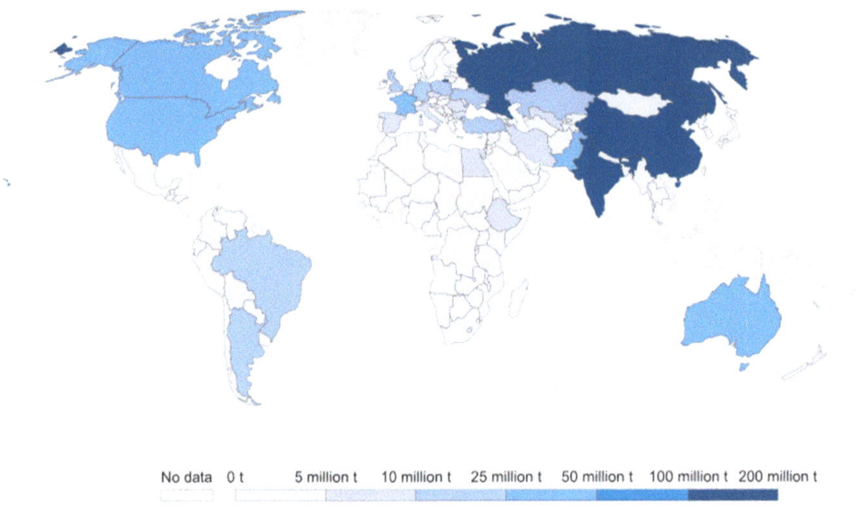

Fig. 8.4 Global wheat production. (FAO, 2022)

So, wheat fields soil properties can change significantly within short time frames, even within a single year, due to the climate effect and land uses. Consequently, climatic change and excessive use of soils necessitate a significantly improved precision, frequency, and quantity of weather forecasts and soil analyses. Classical soil analyses and weather forecasts based on legacy data are inadequate to meet these expectations. There is a demand for immediate weather measures and forecasting that are readily accessible to the public, as well as advanced technology, particularly artificial intelligence, for quicker, cost-effective, and frequent soil analysis.

8.2 Weather Forecasting

Wheat cultivation, practiced across around 240 million hectares globally, yields an average of nearly 800 million tons[1] annually and is distributed over diverse geographical regions due to varying soil and climatic conditions. Consequently, weather forecasts in wheat cultivation regions are a source of crucial information for global food and socioeconomic stability. Moreover, farmers cannot rely solely on rainfall and air temperature figures for agricultural production for the simple reason that knowing the soil's moisture and temperature is essential for effective irrigation, plant nutrition, and pesticide applications in cultivation. General weather forecasts for a province or district are typically insufficient due to climate change and wheat cultivation in marginal regions. The intensity of precipitation varies over short

[1] https://www.statista.com/statistics/267268/production-of-wheat-worldwide-since-1990/

distances, and phenomena such as hail and frost impact limited areas, necessitating weather forecasts at the farm level (Mase & Prokopy, 2014).

In irrigated wheat cultivation, precise weather forecasting will alleviate stress on water resources and inhibit insect proliferation by avoiding excessive soil moisture. For instance, the uncertainty of rainfall following an inaccurate weather forecast results in pre-rain irrigation, which prolongs moisture in the wheat crop and promotes the onset of stripe rust (yellow rust) in wheat fields. However, other than accurate weather forecasting, a more serious issue is the rise in extreme weather events which severely disrupt wheat production (Barlow et al., 2015). The key areas that produce over 50% of Europe's wheat experience a negative event of some kind every 10 years or so, according to Barlow et al. (2015). Using RCP4.5 and RCP8.5 scenarios, the authors found that the likelihood of even a single negative occurrence is expected to rise significantly. In wheat cultivation, weather forecasting, identification of extreme weather events, and implementation of management strategies have become indispensable.

Contemporary weather forecasts now incorporate innovative methodologies, like artificial intelligence, the Internet of Things, and big data, in lieu of traditional forecasting techniques (Ukhurebor et al., 2022). Numerous models are employed for irrigation based on weather forecasts and conditions, including CERES-Wheat (Bannayan et al., 2003), the crop–weather relationship model over larger areas (MCWLA-Wheat model) (Tao & Zhang, 2013), Agricultural Production Systems sIMulator (APSIM)-Wheat (Luo et al., 2018), and the InfoCrop-Wheat crop simulation model (CSM) (Dhakar et al., 2022). To utilize all these models, it is essential to have climatic data, including solar radiation, temperature, precipitation, humidity, and wind speed, in addition to daily long-term temperature and precipitation records (Togliatti et al., 2017). Due to the fact that the majority of this data is archived by states and sold to individual users at exorbitant fees, it is rendered unusable in models as it cannot be integrated with any other definitions. Consequently, models yield unsatisfactory outcomes in certain locations. Despite making use of all available data, research shows that the accuracy of yield predictions decreased as forecast length increased (Togliatti et al., 2017).

8.3 State-of-the-Art Soil Analyses

Wheat is among the most extensively cultivated crops worldwide, and its yield is significantly affected by soil quality. Recent advancements in soil analysis methods have yielded enhanced understanding of soil composition, nutrient availability, and microbial activity, all of which are essential for successful wheat cultivation. Soil testing has shown the soil's health and the necessary maintenance required to preserve crop output. The indicators of soil health are its chemical and physical qualities, as well as plant growth. Consequently, the assessment of soil health can be achieved through the physical, chemical, and biological study of macro- and microplant nutrients, texture, organic matter/carbon, cation exchange capacity, carbonates, pH, electrical conductivity, bulk density, aggregate stability, porosity, CO_2 respiration, and microorganisms. However, due to the cultivation of a single crop in arid regions and the limitations on biomass production from water scarcity, the chemical and physical properties of these soils exhibit minimal changes over time. The application of agricultural chemicals in land cultivation does not significantly change chemical properties, including pH, nutrients, and organic matter, nor does it affect physical features such as soil compaction. In dry and semiarid locations, the impact of any intervention can only be assessed after several years on rainfed soils. For instance, applications designed to enhance organic matter produce results after a decade or longer. Masri and Ryan (2006) investigated the impact of crop rotation involving medic (Medicago sativa) and vetch (Vicia faba) on wheat-fallow management only after 12 years in arid circumstances in Syria. Kapur et al. (2007) measured the impact of crop rotation on soil micromorphology in the same soils studied by Masri and Ryan (2006) in 2008, recognizing the 23rd year of the study launched in 1984. The findings from these studies suggest that changes in nonirrigated wheat soils require a longer period. Consequently, in arid regions of wheat soils, annual chemical and physical studies conducted postharvest through routine soil sampling and subsequent laboratory testing can yield adequate results. When it comes to irrigated areas, however, where two or even three crops are grown annually and intensive agriculture is practiced, the once-a-year soil analyses approach is not appropriate for making plant nutrition programs of the field crops.

In regions characterized by intensive agriculture two or three times a year, submit them to a laboratory and manage plant growth while awaiting analysis results, which can often exceed 15 days. Despite the availability of sophisticated laboratory equipment such as Inductively Coupled Plasma Emission Spectrometry, Inductively Coupled Plasma-Mass Spectroscopy, Neutron Activation Analysis, X-Ray Fluorescence Spectroscopy, Liquid Chromatography, Differential Pulse Voltammetry, Fourier Transform Infrared Spectroscopy, Raman Spectroscopy, Electron Spin Resonance Spectroscopy, X-Ray Photoelectron Spectroscopy, and X-Ray Absorption Fine Structure Spectroscopy, portable devices (sensors and instruments utilizing various wavelengths of light, including X-ray fluorescence) and tracking technologies (drones and satellites) are increasingly valued, as farmers seek prompt and on-site analytical results. In addition to optical and spectroscopic sensors, microelectromechanical system sensors are getting significant popularity for soil quality evaluation due to their mobility, real-time capabilities, quick detection, and in-field mobility. Analyzing soils with this high technology equipment incurs significant time and financial losses due to both analysis costs and the length of time in obtaining results. Currently, as of 2024, a full soil analysis in Türkiye costs approximately 45 US dollars for each sample.[2] Assuming that a minimum of four samples are collected from a 10-ha field and eight analyses are conducted after two crop harvests, the expenditure of 360 dollars will approximate the cost of 1000 kg of

[2]https://arastirma.tarimorman.gov.tr/findik/Menu/26/Analiz-Ucretleri#

Table 8.1 Some soil health indicators (Rayment & Lyons, 2011; Cardoso et al., 2013; Dane & Topp, 2020; Sparks et al., 2020)

Chemical	Physical	Mineralogical	Biological
pH, electrical conductivity, carbon (organic and inorganic), lime, gypsum, macro- (N, P, K, S, Ca) and micronutrients /Fe, Zn; Mn, Cu, Mo, B), cation exchange capacity, base saturation, extractable bases, acid-sulfate soils (extractable S, Ca, Mg, acid neutralizing capacity)	Soil-aggregate stability, size distribution, shear strength, soil erodibility, bulk density, particle density, porosity, specific surface area, penetrability, soil compressibility, water retention, storage, field capacity, wilting point, infiltration, gas diffusity, air permeability, solute content, concentration, transport, etc.	Primary minerals (quartz, feldspar, olivine, amphiboles, etc.), clay mineral abundance and type, weathering, illuviation, accumulation, pore size distribution, pore shape, pore continuity (micromophology), Atterberg limits, coatings (clay, iron oxides, organic matter, etc.)	Enzymes (dehydrogenase, β-glucosidase, cellulase, amylase, urease, glutaminase, asparaginase, phosphatases, Arilsulphatase), micro, meso- and macrofauna analyses, soil respiration

wheat by 2024.[3] Consequently, farmers remain with traditional plant nutrition practices and irrigation rather than adopting fertilization and irrigation strategies derived from soil analysis (Krishna & Sachan, 2016).

When the soil health indicators are enumerated (Table 8.1), it is easy to comprehend why sensor and spectrum analyses are required in soil analysis rather than traditional methods. It is evident that it is extremely challenging and costly to carry out a large number of studies in a single traditional laboratory in a short amount of time. For instance, the majority of soil analyses do not take into account the measurement of biological indicators, such as enzymes, the type and number of microorganisms. These indicators are closely related to the physical and chemical properties of soils, depending on the field application, and they change much more quickly than these two properties. However, the majority of soil analyses do not take into account these indicators because the laboratory opportunities and sample's health (volatilization of soil compounds during transportation) carried to the laboratory are unable to accommodate it. Moreover, farmers often collect soil samples from disparate areas of the field without adequately noting the coordinate information or lacking the necessary equipment for precise location tracking, thus complicating subsequent monitoring research.

The crucial information needed for a farmer to set up a plant nutrition program in irrigated agriculture year-round includes the state of soil nutrients, pH, and salinity values. Consequently, it is essential to perform soil studies swiftly, cost-effectively, and on-site whenever feasible in irrigated wheat fields characterized by intensive agriculture, particularly when 2 or 3 crops are harvested annually. There are now options for doing fast soil analyses, from portable kits for use in the field to more

[3] https://markets.businessinsider.com/commodities/wheat-price (accessed on 10 11 2024).

Table 8.2 The current state-of-the-art soil analyses equipment that can make rapid measurements

Equipment	Analyses
Satellites and Drones	Soil water and carbon contents, texture, aggregates, normalized difference vegetation index (NDVI), net primary production (NPP), soil type.
Soil Color Sensors	Organic matter, Fe, clay content, soil type,
Penetrometer with EC	Soil bulk density, soil moisture, soil texture
Gamma Ray Spectrometry	Soil mappings, available K, texture, peat content, soil erosion mapping
Electromagnetic Induction (EMI) sensor	Salinity, water content, water flow patterns, texture, compaction, organic matter, pH, depth of subsurface horizons, and parent material, soil type mapping
Ion-selective Electrodes	K, nitrate, Na, Ca, Mg, Cl, pH
Galvanic Contact Resistivity (GCR) sensor	Salinity, water content, water flow patterns, texture, compaction, organic matter, pH, depth of subsurface horizons, and parent material, soil type mapping
Gas Sensors	CO_2, SO_2, NH_3, volatile organic compounds
Microcantilever Immunosensor	Insecticides carbofuran) and herbicides (atrazine, diuron, mecoprop, and paraquat)
Diffuse Reflectance Spectroscopy (DRS)	Plant nutrient levels, organic carbon content, clay minerals, soil moisture, and salinity
Portable X-ray Fluorescence Spectrometry (PXRF)	Directly measures total Ca, total Fe, heavy and potentially toxic elements As, Pb, Cu, Zn, Ti, Cr, Cd, Rb, Sr, V, As, and indirectly measures soil cation-exchange capacity, soil reaction, soil salinity

advanced portable equipment. Dimkpa et al. (2017) identified 36 diverse applications in their study of fast analytical kits for soil analysis. For faster soil analysis, sensors and devices utilizing various wavelengths of light are employed instead of the acidic, basic, and saline substances used in traditional laboratory methods. Cutting-edge technologies possess significant potential to address the challenges associated with sampling, storage, transportation, and analysis of soils and are presently employed efficiently for a majority of physical, chemical, mineralogical, and biological assessments. The primary focus in proximal soil sensing has been on soil spectroscopy, conductivity, and ion-specific electrodes, according to Najdenko et al. (2024). Biosensors capable of detecting beneficial and harmful species of diverse sizes in soil, in addition to soil properties, have garnered interest in recent years, exemplified by an exceptionally precise microcantilever immunosensor for the identification of carbofuran pesticide in both soil and plants (Nadporozhskaya et al., 2022). The said techniques have been considered the most fundamental tools for precision agriculture since the 2000s (Table 8.2).

However, it was suggested by Bratoev et al. (2020) that the conditions for a rapid soil test need to have the following properties:

- It should be universally applicable to all types of soils and at any time.
- It should be sensitive and flexible to the management practices for various crops.
- It should be reliable in relation to the information that was put forward.

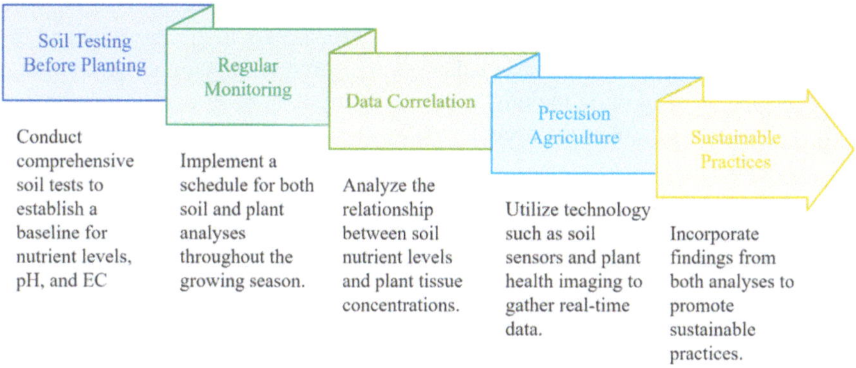

Fig. 8.5 Integrated soil and plant analyses for wheat cultivation

- It should be adaptable to the methods of digital devices in order to select and provide the information that was analyzed for a range of users.

Along with soil analyses to cultivate healthy wheat, soil analyses must be complemented by plant tests. To achieve this, technologically advanced plant analysis instruments, including Chlorophyll Meters for assessing the chlorophyll content, leaf area index (LAI) meters for evaluating leaf development, image sensors for identifying leaf color variations indicative of plant deficiencies, and precise soil analysis tools, must be adjusted in accordance with plant growth advancements. Integrating soil analyses with plant analyses is vital for maximizing wheat production. By understanding the dynamic relationship between soil health and plant nutrition, farmers can implement targeted management strategies that lead to improved yields, better quality crops, and sustainable agricultural practices. This holistic approach, i.e., integrated soil and plant analyses (Fig. 8.5), not only benefits individual farms but also contributes to the overall health of the agricultural ecosystem.

State-of-the-art technologies still suffer from bottlenecks even they are calibrated with sensitive plant tests due to the remarkable variability of soils over short distances; thus, the integration of field measurements with laboratory calibration and the deployment of innovative technologies across extensive wheat cultivated regions are essential for improved production and sustainable soil management.

8.4 Conclusion

Unforeseeable weather events present substantial hazards to the cultivation of wheat, impacting crop production, economic sustainability, and food availability. Through comprehension of these obstacles and use of mitigation strategies, farmers may more effectively manage the unpredictability of climate change and maintain wheat production for future generations. The development of resilient agricultural systems capable of withstanding the impacts of an uncertain climate necessitates ongoing research and collaboration among stakeholders. Free access to climate data and the incorporation of data such as solar radiation, temperature, precipitation, humidity, and wind speed, in addition to daily and even hourly temperature and precipitation, will result in an increase in the accuracy of weather forecasts and models that make use of the data.

Farmers are well aware of the significance of soil analysis, which underpins productivity and sustainable land management in wheat cultivation, alongside weather forecasts. In irrigated wheat cultivation, soil tests must be conducted multiple times year due to the extraction of more than one crop from the field each year. Farmers are reluctant to collect this number of samples. The conditions of sample collection and storage impact the credibility of the analytical results, even if they are conducted. Furthermore, the utilization of classical analytical methods, along with the chemicals employed and the damaged soil samples, results in environmental issues.

Most equipment and sensors that function across different light wavelengths do not induce dangerous processes, such as chemical reactions or heating in ovens during tests, and they are functional without damaging soil samples. These devices first emerged in the 1970s; their limitations are being minimized and they are increasingly becoming more sensitive and reliable with continued usage. However, the intricate and diverse characteristics of soil matrices, even over short distances, can occasionally lead to inaccuracies in the results produced by these devices.

Nearly all soil parameters can be quantified using sophisticated soil testing instruments. Nevertheless, wheat cultivation reliant exclusively on soil assessments may occasionally produce inadequate outcomes. For example, when the lime concentration in the soil exceeds 10%, the plant may struggle to obtain adequate phosphorus (Li et al., 2008), despite the presence of appropriate phosphorus levels in the soil for wheat (app. 90 $kg.ha^{-1}$) (Memon et al., 1991). Consequently, soil analysis outcomes must be corroborated by the observation of plant growth dynamics and foliar assessments. Integrating soil analysis with plant health evaluations establishes a comprehensive strategy for wheat farming. By linking soil nutrient levels with plant tissue analysis, farmers can ascertain specific shortages and customize their fertilization techniques accordingly. This combination results in enhanced resource efficiency and superior crop performance.

Climate change and extreme weather events complicate weather forecasting, while quickly altering soil qualities from heavy irrigation impose significant strain on wheat cultivation. To alleviate these problems, the deployment of rapid and dependable on-site soil measurement instruments should be augmented, and climate data should be gathered from numerous open and accessible locations. These services cannot be rendered by farmers, particularly small-scale farms. Consequently, policymakers must promptly implement choices about the supply of weather forecasting and soil analysis services by cooperatives, producer unions, and the state to ensure the security of wheat production locally and globally.

References

Akça, E., Berberoğlu, S., Nagano, T., & Kapur, S. (2020). Mediterranean anthroscapes: A bioeconomy domain for sustainable land use. In *The bioeconomy approach* (pp. 149–168). Routledge.

Ayal, D. Y., Desta, S., Gebru, G., Kinyangi, J., Recha, J., & Radeny, M. (2015). Opportunities and challenges of indigenous biotic weather forecasting among the Borena herders of Southern Ethiopia. *Springerplus, 4*, 1–11.

Aykanat, S., & Barut, H. (2018). Technical comparison of different planting methods and irrigation in wheat farming. *International Journal of Eastern Mediterranean Agricultural Research, 1*(1), 131–142.

Bannayan, M. N., Crout, N. M. J., & Hoogenboom, G. (2003). Application of the CERES-wheat model for within-season prediction of winter wheat yield in the United Kingdom. *Agronomy Journal, 95*(1), 114–125.

Barlow, K. M., Christy, B. P., O'Leary, G. J., Riffkin, P. A., & Nuttall, J. G. (2015). Simulating the impact of extreme heat and frost events on wheat crop production: A review. *Field Crops Research, 171*, 109–119.

Bratoev, K., Beloev, H., Mitkov, A., & Mitev, G. (2020). On the possibility of conducting fast and reliable soil tests. *Mechanization in agriculture & Conserving of the resources, 66*(2), 71–76.

Cardoso, E. J. B. N., Vasconcellos, R. L. F., Bini, D., Miyauchi, M. Y. H., Santos, C. A. D., Alves, P. R. L., de Paula, A. M., Nakatani, A. S., Pererira, J. M., & Nogueira, M. A. (2013). Soil health: Looking for suitable indicators. What should be considered to assess the effects of use and management on soil health? *Scientia Agricola, 70*, 274–289.

Dane, J. H., & Topp, C. G. (Eds.). (2020). *Methods of soil analysis, part 4: Physical methods* (Vol. 20). Wiley.

Deng, L., Peng, C., Kim, D. G., Li, J., Liu, Y., Hai, X., Liu, Q., Huang, C., Shangguan, Z., & Kuzyakov, Y. (2021). Drought effects on soil carbon and nitrogen dynamics in global natural ecosystems. *Earth-Science Reviews, 214*, 103501.

Dhakar, R., Sehgal, V. K., Chakraborty, D., Sahoo, R. N., Mukherjee, J., Ines, A. V., Kumar, S. N., Shirsath, P. R., & Roy, S. B. (2022). Field scale spatial wheat yield forecasting system under limited field data availability by integrating crop simulation model with weather forecast and satellite remote sensing. *Agricultural Systems, 195*, 103299.

Dimkpa, C., Bindraban, P., McLean, J. E., Gatere, L., Singh, U., & Hellums, D. (2017). Methods for rapid testing of plant and soil nutrients. *Sustainable Agriculture Reviews*, 1–43.

FAO. (2022). *Food and Agriculture Organization of the United Nations (2023)*. https://ourworldindata.org/grapher/wheat-production. Accessed 10 Sept 2024.

FAO and ITPS. (2015). *Status of the world's soil resources (SWSR) – Technical summary*. Food and Agriculture Organization of the United Nations and Intergovernmental Technical Panel on Soils.

Galacgac, E. S., & Balisacan, C. M. (2009). Traditional weather forecasting for sustainable agroforestry practices in Ilocos Norte Province, Philippines. *Forest Ecology and Management, 257*(10), 2044–2053.

Herzfeld, T., Heinke, J., Rolinski, S., & Müller, C. (2021). Soil organic carbon dynamics from agricultural management practices under climate change. *Earth System Dynamics, 12*(4), 1037–1055.

Kapur, S., Ryan, J., Akça, E., Çelik, İ., Pagliai, M., & Tülün, Y. (2007). Influence of mediterranean cereal-based rotations on soil micromorphological characteristics. *Geoderma, 142*(3–4), 318–324.

Krishna, D., & Sachan, H. K. (2016). Farmers attitudes towards adapting climate change through soil testing. *Impact: International Journal of Research in Applied, Natural and Social Sciences, 4*(2026), 185–190.

Kumar, R., Kumar, R., & Prakash, O. (2019). Chapter-5 the impact of chemical fertilizers on our environment and ecosystem. *Chief Ed, 35*(69), 1173–1189.

Li, H., Shen, J., Zhang, F., Clairotte, M., Drevon, J. J., Le Cadre, E., & Hinsinger, P. (2008). Dynamics of phosphorus fractions in the rhizosphere of common bean (Phaseolus vulgaris L.) and durum wheat (Triticum turgidum durum L.) grown in monocropping and intercropping systems. *Plant and Soil, 312*, 139–150.

Luo, Q., Trethowan, R., & Tan, D. K. (2018). Managing the risk of extreme climate events in Australian major wheat production systems. *International Journal of Biometeorology, 62*, 1685–1694.

Lyon, T. L. (1907). The relation of wheat to climate and soil. *Agronomy Journal, 1*(1), 108–125.

Mase, A. S., & Prokopy, L. S. (2014). Unrealized potential: A review of perceptions and use of weather and climate information in agricultural decision making. *Weather, Climate, and Society, 6*(1), 47–61.

Masri, Z., & Ryan, J. (2006). Soil organic matter and related physical properties in a Mediterranean wheat-based rotation trial. *Soil and Tillage Research, 87*(2), 146–154.

Memon, K. S., Puno, H. K., & Fox, R. L. (1991). Phosphate sorption approach for determining phosphorus requirements of wheat in calcareous soils. *Fertilizer Research, 28*, 67–72.

Nadarajan, S., & Sukumaran, S. (2021). Chemistry and toxicology behind chemical fertilizers. In *Controlled release fertilizers for sustainable agriculture* (pp. 195–229). Academic.

Nadporozhskaya, M., Kovsh, N., Paolesse, R., & Lvova, L. (2022). Recent advances in chemical sensors for soil analysis: A review. *Chem, 10*(1), 35.

Najdenko, E., Lorenz, F., Dittert, K., & Olfs, H. W. (2024). Rapid in-field soil analysis of plant-available nutrients and pH for precision agriculture – a review. *Precision Agriculture, 25*, 1–30.

Radeny, M., Desalegn, A., Mubiru, D., Kyazze, F., Mahoo, H., Recha, J., Kimeli, P., & Solomon, D. (2019). Indigenous knowledge for seasonal weather and climate forecasting across East Africa. *Climatic Change, 156*, 509–526.

Rayment, G. E., & Lyons, D. J. (2011). *Soil chemical methods: Australasia* (Vol. 3). CSIRO publishing.

Rodrigues, G. C., & Pereira, L. S. (2009). Assessing economic impacts of deficit irrigation as related to water productivity and water costs. *Biosystems Engineering, 103*(4), 536–551.

Seneviratne, S. I., Zhang, X., Adnan, M., Badi, W., Dereczynski, C., Luca, A. D., Ghosh, S., Iskandar, I., Kossin, J., Lewis, S., Otto, F., Pinto, I., Satoh, M., Vicente-Serrano, S. M., Wehner, M., Zhou, B., & Allan, R. (2021). Weather and climate extreme events in a changing climate. In V. P. Masson-Delmotte, A. Zhai, S. L. Pirani, & C. Connors (Eds.), *Climate change 2021: The physical science basis: Working group I contribution to the sixth assessment report of the intergovernmental panel on climate change* (pp. 1513–1766). Cambridge University Press.

Sparks, D. L., Page, A. L., Helmke, P. A., & Loeppert, R. H. (Eds.). (2020). *Methods of soil analysis, part 3: Chemical methods* (Vol. 14). Wiley.

Strawn, D. G., Bohn, H. L., & O'Connor, G. A. (2019). *Soil chemistry*. Wiley.

Tao, F., & Zhang, Z. (2013). Climate change, wheat productivity and water use in the North China Plain: A new super-ensemble-based probabilistic projection. *Agricultural and forest meteorology, 170*, 146–165.

Togliatti, K., Archontoulis, S. V., Dietzel, R., Puntel, L., & VanLoocke, A. (2017). How does inclusion of weather forecasting impact in-season crop model predictions? *Field Crops Research, 214*, 261–272.

Ukhurebor, K. E., Adetunji, C. O., Olugbemi, O. T., Nwankwo, W., Olayinka, A. S., Umezuruike, C., & Hefft, D. I. (2022). Precision agriculture: Weather forecasting for future farming. In *AI, edge and IOT-based smart agriculture* (pp. 101–121). Academic.

UNEP. (2019). *Global Chemicals Outlook II from Legacies to Innovative Solutions: Implementing the 2030 Agenda for Sustainable Development*. https://wedocs.unep.org/bitstream/handle/20.500.11822/28113/GCOII.pdf?sequence=1&isAllowed=y. Accessed 02 Sept 2024.

USDA. (2023). *Economic Research Report No. (ERR-327)* (eds. B. Williams and G. Pounds-Barnett). https://www.ers.usda.gov/publications/pub-details/?pubid=107995. Accessed 30 Sept 2024.

Wang, Y., Li, J., & Kong, X. (2022). What drives land abandonment in core grain-producing areas? Evidence from China. *International Journal of Environmental Research and Public Health, 19*(9), 5090.

WMO. (2021). *WMO Atlas of mortality and economic losses from weather, climate, and water extremes (1970–2019).* World Meteorological Organization (WMO). https://library.wmo.int/viewer/57564/download?file=1267_Atlas_of_Mortality_en.pdf&type=pdf&navigator=1

Yan, Y., Kayem, K., Hao, Y., Shi, Z., Zhang, C., Peng, J., Liu, W., Zuo, Q., Ji, W., & Li, B. (2022). Mapping the levels of soil salination and alkalization by integrating machining learning methods and soil-forming factors. *Remote Sensing, 14*(13), 3020.

Yang, J. E., Kirkham, M. B., Lal, R., & Huber, S. (2018). *Global soil proverbs: Cultural language of the soil.* CATENA Soil Sciences. ISBN 978-3-510-65431-4.

Yoshino, M., & Miyashita, R. (2007). Studies on bioclimate and weather-health forecasting in Japan. *Global Environmental Research-English Edition, 11*(1), 23.

Chapter 9
How Low-Cost UAVs Could Revolutionize Field Research as a Substitute for Traditional Agronomic Measurements in Crop Phenotyping?

Adrian Gracia-Romero and Marta da Silva Lopes

Abstract Agricultural research is vital for addressing the challenges posed by climate change and ensuring sustainable food production systems. Traditionally, field measurements for crop phenotyping have been labor-intensive, time-consuming, and costly. However, the integration of low-cost unmanned aerial vehicles (UAVs) is revolutionizing field research by offering efficient, accurate, and scalable alternatives to conventional agronomic methods. This chapter explores the evolution of UAV technology in agricultural research, focusing on its applications in crop phenotyping. UAVs equipped with advanced imaging technologies enable high-throughput phenotyping (HTP), capturing key traits such as canopy height, ground cover, and crop phenology with remarkable precision. The use of UAVs not only accelerates the collection of valuable data but also enhances the development and adoption of improved crop varieties and sustainable agronomic practices.

By comparing UAV-derived data with traditional methods, this work demonstrates the effectiveness of drone-based techniques in measuring crucial crop traits. Key examples from cereals (wheat, oats, and barley) illustrate how UAVs enhance the monitoring of plant development, canopy height, and yield prediction. Additionally, this chapter addresses the current limitations and challenges associated with UAV adoption in agricultural research, including the need for expertise in data processing, the high cost of certain equipment, and regulatory concerns.

This chapter concludes by presenting a low-cost, user-friendly protocol for integrating UAVs into crop phenotyping workflows. This protocol simplifies image collection, data analysis, and trait evaluation, making UAV technology more accessible to research teams worldwide. Ultimately, the adoption of UAVs in agricultural research offers a transformative solution to improve the efficiency and

A. Gracia-Romero (✉) · M. S. Lopes (✉)
Field Crops Program, Institute for Food and Agricultural Research and Technology (IRTA), Lleida, Spain
e-mail: adrian.gracia@irta.cat; marta.dasilva@irta.cat

N. Zencirci et al. (eds.), *Empowering Wheat Cultivation with GIS, Digital Approaches and Artificial Intelligence*, https://doi.org/10.1007/978-3-031-99954-3_9

accuracy of field measurements, contributing to the development of more resilient and sustainable cropping systems.

Keywords UAVs (unmanned aerial vehicles) · Crop phenotyping · High-throughput phenotyping (HTP) · Sustainable agriculture · Climate change adaptation

Abbreviations

AI	artificial intelligence
DF	days to flowering
DH	days to heading
DJI	da-jiang innovations (drone manufacturer)
DEM	digital elevation model
DM	days to maturity
DSM	digital surface model
DTM	digital terrain model
GIS	geographic information system
HSI	hyperspectral imaging
HTP	high throughput phenotyping
HTPPs	high-throughput phenotyping platforms
LAI	leaf area index
LiDAR	light detection and ranging
NDVI	normalized difference vegetation index
NGRDI	normalized green-red difference index
NIR	near-infrared reflectance
PH	canopy height
R^2	coefficient of determination
RGB	red green blue
RMSE/RSME	root mean squared error
SfM	structure-from-motion
UAV	unmanned aerial vehicle
UNCTAD	united nations conference on trade and development
VI	vegetation index
WOFOST	world food studies crop model

9.1 Introduction

Agricultural research delivers quantitative and objective data to define optimal agronomic practices and develop new crop varieties tailored to specific regions or those with similar climates. As the global population continues to grow, the demand for food production increases, necessitating innovative solutions that ensure food security while preserving the environment (UNCTAD, 2017). Research institutions worldwide, encompassing both private and public organizations, are at the forefront of advancing crop breeding, sustainable agronomy, and agroecological transitions. Their goal is to develop a more efficient food production system that minimizes waste and reduces environmental impact.

Through rigorous scientific investigation, including extensive field comparisons of microplots with various treatments and experimental units across different environments, research programs develop improved agronomic practices and plant varieties. These new varieties are better adapted to changing climatic conditions and innovative cropping systems, exhibiting crucial traits such as drought and heat tolerance, pest and disease resilience, and enhanced nutritional profiles, all while minimizing environmental impact. The data generated from these research efforts are crucial for translating scientific discoveries into practical applications. Effective communication and targeted training workshops enable farmers to quickly adopt these technologies, thereby making sustainable food production systems a reality. This entire process of testing, analyzing, and communicating is costly and requires years of dedication, well-prepared research teams, appropriate infrastructure, and ongoing maintenance. Traditionally, agricultural research has relied on labor-intensive and time-consuming methods to collect data on crop performance. However, the use of remote sensing applications at different scales and, recently, the integration of low-cost drones and advanced imaging technologies marks a significant advancement, enabling researchers to gather accurate data more efficiently. This chapter examines the benefits, protocols, and applications of using drone imagery in agricultural research, highlighting its role in accelerating the development and adoption of improved crop varieties and agronomic practices. These innovations are designed to create cropping systems that require fewer inputs and are more resilient to the unpredictable nature of global climate conditions.

9.2 The Evolution of Drone Imagery Technology in Crop Phenotyping for Agricultural Research

9.2.1 Historical Background: Overview of the Technological Advancements Leading to the Current State of UAV Technology

The use of aerial imagery in agriculture dates back to the early twentieth century when photographs taken from balloons and aircraft were used for surveying and mapping purposes. These early efforts laid the groundwork for more sophisticated applications in agriculture. However, the earliest recorded instance of aerial application of agricultural materials occurred in 1906, when John Clervaux Chaytor dispersed seeds over a flooded valley floor at his family farm, "Marshlands," in Wairau, New Zealand, utilizing a hot air balloon with movable tethers. Not long after, in 1921, the US Department of Agriculture, in conjunction with the US Army Signal Corps research station in Ohio, conducted crop dusting using an airplane (Johnson, 2002).

In the 1970s, the advent of multispectral imaging technology enabled the capture of data across different wavelengths of light, beyond the visible spectrum. This development was pivotal for agricultural applications, as it allowed for the detection of various plant stressors, including water deficiency, pest infestations, and nutrient deficiencies. The Landsat program, launched in 1972, was particularly influential, providing continuous satellite imagery that could be used for agricultural monitoring (Tucker et al., 1980). Tucker et al. (1980) highlighted the relationship between spectral data and grain yield variation, emphasizing the potential of these technologies in crop phenotyping.

The 1980s and 1990s saw significant advancements in remote sensing technologies, including the development of high-resolution satellite imagery and the increased use of geographic information systems (GIS). These tools enabled more precise mapping and analysis of agricultural lands. Researchers began using these technologies for precision agriculture, which aimed to optimize field-level management based on crop variability (Mulla, 2013).

In the early 2000s, the introduction of unmanned aerial vehicles (UAVs), commonly known as drones, marked a major shift in agricultural imaging. Drones equipped with multispectral, hyperspectral, and thermal cameras provided high-resolution, real-time data at a lower cost than satellite imagery. This technology allowed for more frequent monitoring and more detailed assessments of crop health and phenotyping (Zhang & Kovacs, 2012). The integration of machine learning and data analytics with aerial imagery has further enhanced crop field phenotyping. These technologies enable the automated analysis of large datasets, improving the accuracy and efficiency of phenotyping. Lelong et al. (2008) demonstrated the use of UAV imagery for quantitative monitoring of wheat crops, illustrating the potential of combining advanced imaging with data analytics.

9.2.2 *Crop Phenotyping Advancements. Recent Integrations of UAV Imagery in Crop Phenotyping*

Crop phenotyping, defined as the assessment of observable plant traits, has significantly evolved with advancements in technology. Crop phenotyping in research institutions involves the systematic study and measurement of plant characteristics and traits across various environments and conditions. This process is crucial for identifying and developing optimal sustainable agronomic practices and crop varieties. Specifically, crop phenotyping at institutions that test several thousands of microplots in the field includes: (i) extensive field testing: research institutions utilize microplots, which are small, controlled field plots, to conduct experiments. These microplots allow for the testing of numerous treatments (e.g., crop varieties, agronomic practices) and experimental units under a variety of environmental conditions. (ii) Data collection and analysis: researchers gather comprehensive data on plant traits such as growth rate, yield, disease resistance, drought and heat tolerance, pest resilience, and nutritional content. Advanced technologies, such as drones, sensors, and imaging techniques, are often employed to enhance the precision and efficiency of data collection. (iii) Trait evaluation: the collected data are analyzed to evaluate the performance of different crop varieties and agronomic practices. This analysis helps in understanding how various factors, including environmental conditions and management practices, affect plant growth and productivity. (iv) Development of improved practices: based on the findings from these field trials, researchers develop improved agronomic practices that are more sustainable and efficient. These practices aim to maximize yield, enhance crop resilience, and reduce the environmental impact of farming. The final step involves translating quantitative research findings into practical applications and demonstrations to farmers. Effective communication and targeted training workshops are conducted to help farmers adopt these new technologies and practices, ensuring the transition to sustainable food production systems.

In this context, the use of UAVs has become particularly impactful in crop phenotyping, providing detailed, high-resolution data for calculating crop attributes. This technological integration is especially beneficial in breeding and agronomy experimental designs. Since the 2010s, advances in sensors, aeronautics, and high-performance computing have transformed the measurement of challenging field traits, which are often time-consuming and resource-intensive, into high-throughput phenotyping platforms (HTPPs), making them more efficient and affordable (Araus & Cairns, 2014). Araus and Cairns (2014) very well described the transition from proximal to remote sensing in agricultural research. At that stage of field phenotyping, various remote sensing and imaging techniques have been proposed, such as visible and near-infrared spectroradiometry, thermal imaging, and RGB photography, which facilitated noninvasive and detailed trait evaluation in field conditions. Aerial platforms, including UAVs, have become increasingly popular for their ability to cover large areas quickly and with high resolution. Laboratory analyses, including NIRS, complement field phenotyping by providing detailed information on grain characteristics and stress adaptation traits. Addressing spatial variability and incorporating real-time environmental monitoring were highlighted as crucial for accurate phenotypic data. Advanced data management and modeling tools were detected as necessary to handle the vast amounts of data generated by HTPPs and to integrate genotypic, phenotypic, environmental, and management data effectively. The Araus and Cairns (2014) paper highlighted that while technological advances in HTPPs were promising, their integration into the breeding pipeline and the development of user-friendly data management tools were critical for their widespread adoption. Efficient use of HTPPs will significantly enhance crop genetic improvement efforts, ensuring food security for future generations.

9.2.3 Recent Integrations of UAV Imagery in Crop Phenotyping

The integration of UAVs has revolutionized crop phenotyping, offering unprecedented accuracy, efficiency, and scalability in agricultural research. Recent studies highlight a range of innovative applications and methodologies that showcase the transformative potential of these technologies. Gracia-Romero et al. (2023) have shown a 50% improvement in grain yield prediction modeling aerial normalized difference vegetation index (NDVI) with canopy height, phenology, and ear density from experimental trials. Guo et al. (2024) demonstrated a significant improvement in maize yield estimation by integrating UAV-based leaf area index (LAI) measurements into the WOFOST crop model. This approach bridged the gap between remote sensing data and crop modeling, enhancing the precision of yield predictions and providing timely data for decision-making (Guo et al., 2024). In the realm of wheat phenotyping, Sharma et al. (2024) utilized UAV remote sensing to monitor wheat collections under water stress conditions. By employing machine learning, Sharma et al. (2024) established a strong correlation between UAV-derived canopy height and manual measurements. They utilized reductions in canopy height to predict yield under water stress conditions, paving the way for advanced predictive modeling of crop yields under various environmental stresses. Dua et al. (2024) developed an automated pipeline for generating multipolygon shapefiles for phenotyping and precision agriculture applications. Crop breeders traditionally depended on manual and detailed extraction, processing, and analysis of high-resolution images to derive results. One of the major difficulties was pinpointing plot locations within multispectral imagery, defining plot boundaries, and assigning spatial references to each plot. Consequently, this study aimed to develop an automated method for generating multiple polygon shapefiles with unique identifiers. This method facilitates overlaying drone imagery data with centimeter-level accuracy to obtain zonal statistics for each phenotype plot. This innovation streamlines data processing from UAV imagery, facilitating detailed crop monitoring and management practices (Dua et al., 2024). Shepard et al. (2024) employed deep learning techniques for high-throughput phenotyping of maize tasseling using UAV imagery across different environments. This study highlighted the potential of combining UAVs with AI to capture critical phenotypic traits with high accuracy and consistency, significantly advancing high-throughput phenotyping capabilities (Shepard et al., 2024). Exploring the feasibility of substituting UAVs with satellite imagery, Shrestha et al. (2024) found that plot-level satellite imagery could effectively complement UAV data in assessing maize phenotypes across multistate field trials. This scalable solution is particularly valuable for large-scale agricultural research (Shrestha et al., 2024). In winter wheat phenotyping, Kaur et al. (2024) combined hyperspectral imaging with machine learning to predict various crop traits. The use of UAV-based hyperspectral measurements in breeding programs represented a significant advancement in high-throughput phenotyping highlighting the importance of utilizing appropriate machine learning algorithms for improved prediction of plant traits (Kaur et al.,

2024). (Mohammadi et al. 2024) integrated UAV imaging and machine learning to enhance phenotyping efficiency in faba bean breeding. UAV data were utilized to estimate traditional phenotypic measures like canopy height, improving the efficiency and accuracy of breeding processes (Mohammadi et al. 2024). Balota et al. (2024) used aerial remote sensing and crop index data to phenotype peanut plants under drought stress. UAVs proved to be valuable tools for assessing drought tolerance, supporting genetic improvements for drought resistance in peanut crops (Balota et al., 2024). Finally, Gillham (2024) assessed maize height patterns in intercropping systems using UAV imagery. This study highlighted the role of UAVs in monitoring crop dynamics in complex agricultural systems, aiding in the optimization of intercropping practices (Gillham, 2024).

In conclusion, advancements in crop phenotyping using UAVs have markedly enhanced the accuracy, efficiency, and scalability of agricultural research across various crops. These developments have been pivotal in promoting the widespread application of UAVs in plant phenotyping. Our team is currently focused on creating user-friendly pipelines that utilize low-cost drones, cameras, and algorithms. This initiative aims to enable research teams globally to adopt these technologies, significantly reducing their workload.

9.3 Current Utilization and Challenges of UAVs Technology in Agricultural Research

A comprehensive survey was conducted among researchers and users involved in crop research programs to understand their perceptions, experiences, and the current usage of UAV systems in their work. The survey included a mix of closed- and open-ended questions and received responses from 160 participants from 47 different countries. Most of the responses were obtained from researchers at public institutions and universities, with a significant portion having experience with UAVs and varying levels of perceived effectiveness in their application (Fig. 9.1).

In this section, and derived from the survey responses, the actual main uses of UAVs in agricultural research will be determined, while the reasons why scientists and technical staff are still reluctant to widely adopt this technology will be discussed.

The participants were asked to rate the feasibility of using UAV systems to assess various parameters related to crop establishment, phenology, stress, and production, based on their personal experience and opinions. For crop establishment, respondents had a positive perception of using UAVs to measure emergence and early vigor, particularly when evaluating indicators like crop cover or leaf area index (LAI). Crop phenology was also a commonly monitored trait using UAVs, although there was a lower perceived feasibility in evaluating developmental stages prior to grain filling, such as heading and flowering. In contrast, assessing traits related to crop maturity and senescence is considered much more feasible. Respondents

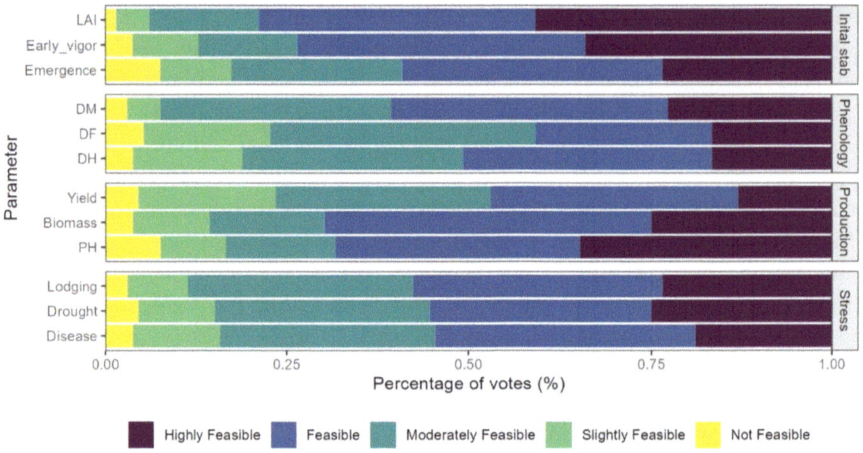

Fig. 9.1 Perception on the feasibility of UAV-derived measures estimating crop initial establishment, phenology, production, and stress response for agricultural research. LAI leaf area index, DM days to maturity, DF days to flowering, DH days to heading, PH canopy height

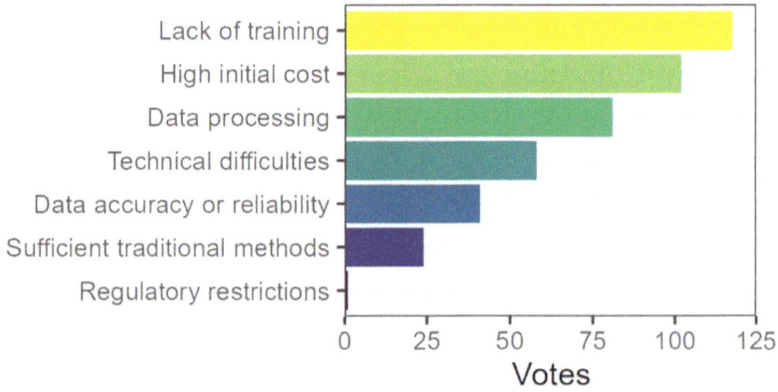

Fig. 9.2 Perception of the most important limitations found by the agricultural research community for the use of UAV systems

frequently mentioned the use of UAVs to assess crop production traits, with biomass and canopy height being well-established parameters. However, there was more uncertainty about yield forecasting, as opinions on its feasibility were divided. Finally, in terms of stress monitoring, respondents expressed strong confidence in the feasibility of UAVs, regardless of the type of stress. Additionally, lodging evaluation was also considered feasible by most participants.

While there was generally a positive opinion regarding the use of UAVs in crop monitoring, participants widely agreed on the main limitations for applying these technologies (Fig. 9.2). The most commonly cited restriction was the lack of training or expertise, particularly in processing and analyzing the data. Operating drones efficiently and safely requires specific skills to understand and control flight mechanisms. However, programming and executing autonomous flights with a preset flight route plan and waypoints is making data collection more consistent and repeatable. Flight mission planning software facilitates fully automated flights from take-off to landing, by defining routes and calculating the UAV's speed to ensure precise image mapping. Despite these technological advancements, the primary concern among participants was the lack of expertise in data processing and analysis. UAVs generate large volumes of data, which users often find difficult to manage and interpret effectively. Processing this data typically requires expertise in GIS and image processing software to derive accurate insights. The surveys also highlighted that, even with the right tools, insufficient expertise can lead to suboptimal data and poor research results. To address this gap, universities are increasingly incorporating UAV protocols for crop monitoring into their curricula, such as precision agriculture and phenotyping. In addition, training opportunities are increasingly available outside universities, such as those offered by research organizations or through funding projects and networks.

Another frequently cited limitation was the high initial cost of UAVs. However, drone and camera costs have significantly decreased over the past few decades due to technological advancements, growing demand, and broader applications across various industries. Improvements in battery life, GPS technology, and camera quality have made UAVs more affordable and capable, especially for agricultural management. Currently, entry-level UAVs with the necessary features to capture high-quality phenomic data can be purchased for less than 1500€. This downward trend in pricing is expected to continue, making UAVs an increasingly cheap and essential tool for different users. Some companies now offer UAVs-as-a-service, reducing the need for large upfront investments and the hiring of specialized expertise within agricultural research teams. This makes it much easier for agricultural researchers to focus on addressing critical questions and developing solutions, rather than being concerned about how to measure basic agronomic traits.

In addition to the platform costs, participants also expressed concerns about the high expenses associated with computer and software requirements. A more affordable solution is offered by cloud-based applications, which minimize upfront investments, provide efficient use of computational resources, and operate on a usage-based cost model (Ampatzidis et al., 2020).

Finally, concerns were raised regarding legal restrictions on drone use in certain countries of specific areas where test fields are located. Additionally, compliance with regulatory authorities often requires enrolling in certification courses. To overcome these challenges, standardized protocols for both drone operation and data analysis are essential. Developing user-friendly tools that simplify data processing and analysis will help bridge the expertise gap. These tools should feature intuitive interfaces and automate complex analytical tasks, making them accessible to a wider range of users.

9.4 Transitioning from Traditional Field Measurements to UAV-Based Techniques

UAV-based phenotyping has the potential to streamline routine field work, reducing the associated effort, costs, and time. However, the absence of affordable, standardized, and user-friendly protocols for UAV operation and data processing is hindering widespread adoption within the crop research community. In this section, detailed methods for using UAVs to measure key crop traits such as canopy cover, canopy height, and to monitor crop growth stages will be presented.

9.4.1 Ground Cover Evaluation in Plant Phenotyping by UAV Systems

One of the most popular applications of UAV technology in agriculture is the evaluation of ground cover. Field surveys of this trait using sensors or cameras have become highly reliable, effectively replacing visual scoring and destructive measurements. Various efficient techniques exist for estimating the quantity and condition of aboveground biomass, most of which rely on spectral indices. These indices are based on the characteristic low reflectance in the visible spectrum (400–700 nm) and high reflectance in the near-infrared (NIR) spectrum (700–1100 nm) (Hassan et al., 2019). The normalized difference vegetation index (NDVI) is the most widely used, but other variations have been developed to minimize the effects of soil interference or saturation in dense canopies (Tucker, 1979). In addition to vegetation indices, another commonly used approach for evaluating crop cover involves classifying image pixels as vegetation or nonvegetation through thresholding techniques (Myint et al., 2011). A notable example can be found in Duan et al. (2016), which applied segmentation methods to accurately estimate ground cover for different crop species in field experiments (cotton: $R^2 = 0.97$, sorghum: $R^2 = 0.98$, sugarcane: $R^2 = 0.84$). High-resolution RGB images have also improved the accuracy of ground cover assessments (Fernandez-Gallego et al., 2018). Currently, advanced machine learning algorithms further enhance classification accuracy and automate the process, making it more efficient and reliable (Hu et al., 2021). Depending on the timing of measurements, UAV-estimated crop cover can be linked to various key traits, such as emergence establishment (Feng et al., 2020; Li et al., 2019), crop vigor (Aharon et al., 2020; Tefera et al., 2022), or, at later stages, to senescence (Makanza et al., 2018).

Spatial resolution (i.e., pixel size) is a significant factor influencing the accuracy of ground cover estimation (Waldner & Defourny, 2017), with higher spatial resolution providing more detailed vegetation information. In aerial systems, flight altitude is a key determinant of spatial resolution; higher altitudes result in lower image resolution (Hu et al., 2019). This is particularly important when assessing early vigor traits (Sankaran et al., 2015), where differences in canopy cover are minimal and require greater precision.

9.4.2 Canopy Height Measurement: Techniques for Using UAVs to Accurately Measure Canopy Height Across Different Crop Species

Monitoring crops by measuring aboveground biomass and canopy height using UAV systems has been a primary goal in plant phenotyping. Advances of light detection and ranging (LiDAR) technology for three-dimensional analysis of canopy shape and structure have brought this goal closer to reality, with several reports of its application in field crops (Jimenez-Berni et al., 2018; Jin et al., 2021; Zolkos et al., 2013). However, despite improvements in smaller, lighter sensors with increased resolution and sensitivity (Remondino & El-hakim, 2006), the routine use of LiDAR remains limited by the high cost of sensors and challenges in standardizing data processing.

Alternatively, readily available consumer-level UAV systems equipped with high-resolution cameras are being used to reconstruct the 3D structure of crop canopies through the structure-from-motion (SfM) method. This method generates a point cloud model by identifying common points in overlapping images. The steps for generating the crop surface model include: (i) mosaicking the collected images into a geo-referenced orthophoto, (ii) generating the point cloud model, and (iii) exporting both digital surface and terrain models (DSM and DTM, respectively). The canopy height model is then derived by subtracting the DTM, representing ground elevation, from the DSM, representing the height of the highest surfaces at each location. In phenotyping field trials where the ground is visible between plots, the SfM technique generally provides a good description of both DSM and DTM layers (Holman et al., 2016; Khanna et al., 2015) except if the plants are still in the early stages of development (low height). The accuracy of the 3D model retrieval improves as the number of images acquired from different angles increases (Walter et al., 2018), and an adequate canopy description can be achieved in flight plans with the camera in a zenithal position. Madec et al. (2017) compared UAV-derived canopy height to the LiDAR measurements, demonstrating a similar degree of accuracy, although canopy height derived from SfM tends to be systematically lower than that from LiDAR.

To implement the SfM methodology correctly, the available literature offers different approaches. Canopy height for each plot can be retrieved from the mean or median values, or from specific percentiles (Watanabe et al., 2017). For example, Khan et al. (2018) designated canopy height as the 98th percentile of the estimated plot's height distribution, reporting correlations of $R^2 = 0.85$ and RMSE $= 6.64$ cm throughout the crop season. Other authors (Hassan et al., 2019) proposed improving the methodology by assessing the canopy height as the average of the highest and lowest points on the canopy's upper boundary, reporting high correlations during midgrain filling ($R^2 = 0.96$).

9.4.3 *Heading and Maturity Dates: Methods for Determining the Phenological Stages of Crops Using UAV Imagery*

Selecting cultivars based on their phenological timing is critical for optimizing adaptation to specific environmental conditions and production strategies (Gulino & Lopes, 2024; Hunt et al., 2019; Lopes et al., 2018; Rezaei et al., 2018). Traditionally, tracking the initiation of phenological stages relies on human observation, often requiring routine measurements (every 1–2 days) to ensure data precision, as well as well-trained personnel to maintain consistent criteria. Recently, HTP-UAV has emerged as a way to monitor the phenological dynamics in crop development (Herr et al., 2023). However, interpreting a single VI value during the growing season can be challenging, as it may reflect various crop stress parameters (e.g., water stress, a disease, lodging, etc.), which introduce uncertainty in phenology monitoring (Feng et al., 2024; Gao et al., 2023). As a result, predicting crop phenology using UAV platforms remains challenging, often relying on postseasonal forecasting through analysis of temporal changes in VIs and canopy height time series.

One of the most common approaches found in the available literature focuses on monitoring senescence. Stay green is considered a key trait for extending grain filling duration and improving crop performance (Lopes & Reynolds, 2012; Pinto et al., 2016). This is achievable because canopy greenness is easily monitored using VIs, as it transitions from green to yellow. Repeated measurements of canopy reflectance indices have been widely used to capture the onset of senescence (Anderegg et al., 2020; Christopher et al., 2014; Kipp et al., 2014). Increasingly, spectral sensors are being replaced by more affordable RGB cameras (Cao et al., 2021; Hassan et al., 2018; Li et al., 2023). Canopy color properties measured using different color models, such as the hue–saturation–intensity (HSI) space (Zaman-Allah et al., 2015) or the a* parameter from the CIELab space (Gracia-Romero et al., 2020), are being used to monitor yellowing tonalities as indicators of senescence.

On the other hand, the determination of other phenological stages, such as heading, remains more challenging. Some studies have used computer vision technology to detect heading by recognizing ears (Bai et al., 2018; Zhu et al., 2016), but this approach faces limitations when applied to UAVs due to image resolution constraints (Fernandez-Gallego et al., 2018). A more practical solution at an aerial scale involves tracking the temporal changes in VIs and canopy height time series, often related to growth stages, as demonstrated by Zhao et al. (2021) with wheat and Taniguchi et al. (2022) with rice. The point when canopy height reaches its maximum has also been used as a proxy of the flowering date (Madec et al., 2017).

Recent studies have shown the feasibility of using postheading measurements to assess earlier phenological stages in wheat, based on the principle that an early-heading variety is likely to mature earlier in its growth cycle (Gracia-Romero & Lopes, 2024; Lopes et al., 2012; Zhao et al., 2022).

9.5 Exploring the Application of Low-Cost UAVs as a Replacement for Traditional Agronomic Measurements in Crop Phenotyping: A Case Study

The various applications proposed in this chapter, using RGB images captured by a compact UAV system to evaluate two key parameters for selection—crop phenology and canopy height—were tested and validated in a case study. The experiment involved a panel of 200 modern varieties of bread wheat (*Triticum aestivum L.*) grown under irrigated conditions at the experimental station of "la Melusa" in Tamarite de Litera (Huesca, Spain; 41°47′N, 00°22′E, 215 m a.s.l) during the 2022/2023 crop season. The environmental conditions of the study area are characterized by a temperate semiarid climate with cool, wet winters, and dry and hot spring to summer seasons. The study consisted of a nonreplicated augmented design with two replicated checks, the modern cultivars "Artur Nick" and "Chambo," in plots of 9.60 m^2 with eight rows spaced 0.15 m apart and border rows were strategically employed to mitigate the impact of border effects.

The unmanned aerial vehicle used in this study was a DJI Mavic 2 Enterprise Dual (DJI, Shenzhen, China), featuring a standard RGB camera with a 1/2″ CMOS 48-megapixel image sensor. The drone was operated at an altitude of 30 m above ground level. All flight campaigns were conducted under clear skies and in conditions of low wind speed. The flight missions were meticulously planned using DJI Pilot software, ensuring a front overlap ratio of 80% for image collection. A total area of 5.36 Ha was surveyed in 34 min. Aerial observations were scheduled approximately every 2–3 weeks, as detailed in Table 9.1.

Table 9.1 Dates, corresponding phenological stages, and days after sowing (DAS) for each of the UAV surveys performed after stem elongation

UAV Survey	Date	Phenological stage	DAS
1	2023-04-13	Stem elongation	91
2	2023-04-24	Booting	102
3	2023-05-04	Ear emergence	112
4	2023-05-10	Ear emergence	118
5	2023-05-17	Anthesis	125
6	2023-05-24	Milk development	132
7	2023-05-30	Milk development	138
8	2023-06-07	Dough development	146
9	2023-06-14	Dough development	153
10	2023-06-21	Ripening	160

Table 9.2 Indices derived from the RGB images

Parameter	Model	Phenological stage	Trait	Reference
Green area (GA)	RGB orthomosaic; HIS model	$60^\circ < \text{Hue} < 180^\circ$	Vegetation cover	Casadesús et al. (2007)
*a**	RGB orthomosaic; CIElab model	–	Senescence	Gracia-Romero et al. (2020)
Normalized green red index (NGRDI)	RGB orthomosaic; RGB model	$\frac{(\text{DN Green} - \text{DN Red})}{(\text{DN Green} + \text{DN Red})}$	Plot greenness	Haboudane et al. (2002)
PHmax	SfM model	–	Height	

HSI Hue–Saturation–Intensity color space, *DN* digital number

Preprocessed aerial images were combined to create two orthomosaics, one generated from the RGB images and the other from the digital elevation model exported from the RGB images. Both mosaics were produced using the cloud-based platform *WebOpenDroneMap* (*WebODM*), an open-source toolchain written in Python for performing photogrammetric workflows (OpenDroneMap Authors ODM, 2020). For each orthomosaic, regions of interest corresponding to each plot were segmented and exported using the MosaicTool (Shawn C. Kefauver, https://www.integrativecropecophysiology.com/software-development/mosaictool/, https://gitlab.com/sckefauver/MosaicTool, University of Barcelona, Spain) integrated as a plugin for the open-source image analysis platform FIJI (Fiji is Just ImageJ; http://fiji.sc/Fiji). Four different vegetation indices based on different image color properties were calculated and used in this study, together with maximum canopy height points within each plot derived from the SfM model (Table 9.2).

9.5.1 Canopy Height

First, UAV-derived parameters were compared with ground truth measurements taken at physiological maturity. The UAV-SfM modeled crop heights showed better correlation with ground measurements than any VIs through the crop cycle (Fig. 9.3a). The manual measurements were performed near maturity in 10 main stems per plot, measuring from the tillering node to the top of the spike, excluding the awns, using a ruler. Plant height derived from the UAV surveys was estimated using the SfM method and subtracting the DTM from the DSM with the maximum height values for each plot being recorded. The correlation between UAV-derived and manual measurements was conducted closer to maturity, with the best model obtained during the tenth UAV survey yielding an R^2 coefficient of 0.801 and a root mean square error of 3.41 cm (Fig. 9.3b).

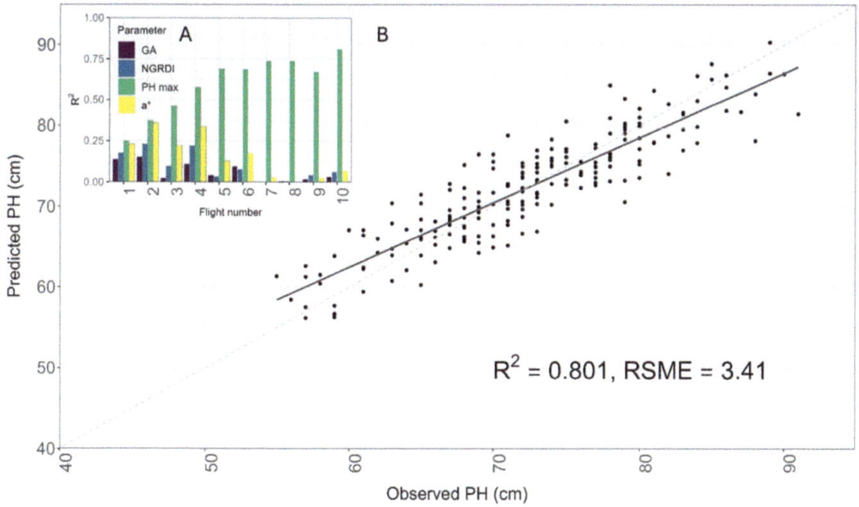

Fig. 9.3 (**a**) Pearson coefficient of determination (R^2) of the relationships between the UAV-derived parameters and canopy height (PH) measured in the field at physiological maturity across each UAV survey conducted after stem elongation. (**b**) Scatter plot of the observed versus predicted PH values using the best linear regression model, based on the PH_{max} measurements from the tenth UAV survey

9.5.2 Phenology

UAV-SfM modeled canopy height and RGB-derived vegetation indices were used to predict days to heading (DH) and days to maturity (DM) as indicators of crop phenology. To monitor these developmental stages accurately in the field, a continuous observation protocol was implemented throughout the growth cycle, involving frequent assessments of individual plot developmental stages every 2–3 days (Pask et al., 2012). Days to heading (DH) was defined as the number of days from sowing to the point when 50% of spikes emerged in a given plot [Zadoks Stage 59 (Zadoks et al., 1974)], while days to maturity (DM) was defined as the time span between sowing and when 50% of the peduncles exhibited ripeness, characterized by a distinct yellow hue [Zadoks Stage 87 (Zadoks et al., 1974)]. The correlations of both phenology parameters followed similar trends throughout the crop cycle: the correlation between SfM-derived canopy height and phenology peaked at survey eight and then declined, while the RGB indices showed increasing correlation until the UAV survey 13, after which the correlations sharply decreased due to the crop ripening (Figs. 9.4a and 9.5a). Overall, the correlations with DM were stronger than those with DH. These findings align with previous results (Gracia-Romero & Lopes, 2024), where VIs measured during the grain-filling stage of irrigated bread wheat accurately predicted DH and DM. This period is crucial because canopies transition from green to yellow, and cultivars with early heading dates tend to reach maturity sooner in their growth cycle. However, it becomes impractical to use VIs at stages

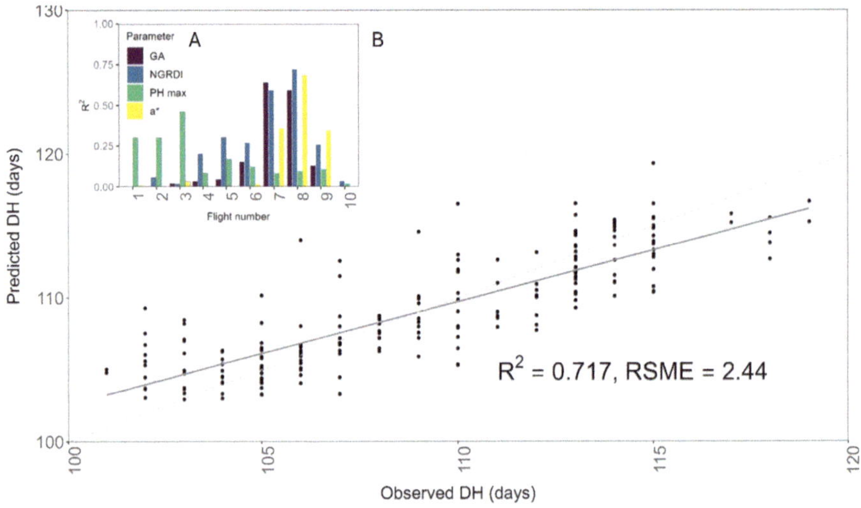

Fig. 9.4 (**a**) Pearson coefficient of determination (R^2) for the relationships between the UAV-derived parameters and days to heading (DH) across each UAV survey performed after stem elongation. (**b**) Scatter plot of the observed versus predicted DH values using the best linear regression model, based on NGRDI measures from the eighth UAV survey

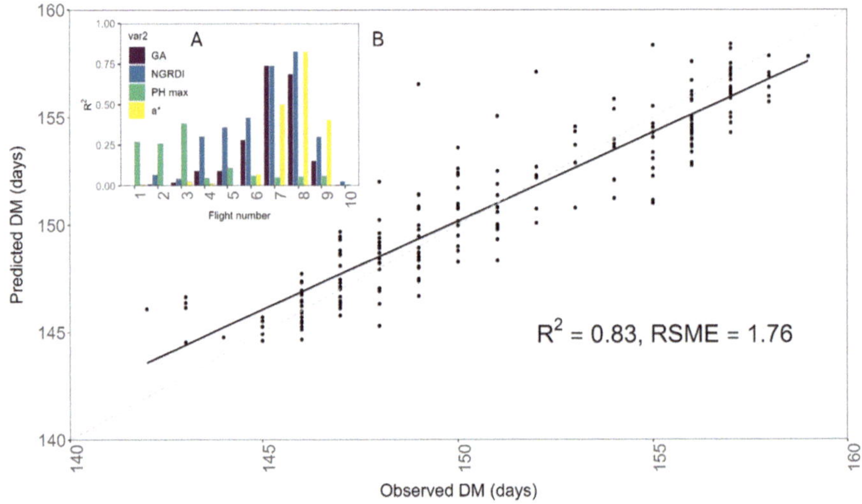

Fig. 9.5 (**a**) Pearson coefficient of determination (R^2) of the relationships between UAV-derived parameters and days to maturity (DM) across each UAV survey performed after stem elongation. (**b**) Scatter plot of the observed versus predicted DM values using the best linear regression model corresponding to the a* measurements from the eighth UAV survey

when crops exhibit very vigorous canopies and saturation (Gracia-Romero et al., 2023) becomes a major impediment for adequate predictions.

The best models for predicting DH and DM (Figs. 9.4b and 9.5b), using values from the eighth UAV survey, were validated. Measurements of plot greenness from the NGRDI index reported the best predictions for DH (R^2 = 0.717, RSME = 2.44 days), while the senescence index a* was most effective to predict DM (R^2 = 0.819, RSME = 1.81 days).

9.5.3 *Protocol Design for UAV-Based Phenological Stage and Canopy Height Assessment*

Based on the results from the case study, a protocol for assessing key phenological stages and canopy height development was established (Fig. 9.6). Before conducting any UAV survey, it is essential to ensure flight permissions and safety procedures are in place. Autonomous flight planning is crucial for capturing the entire experimental area with sufficient image resolution and overlap to produce high-quality data for accurate assessment of canopy height and coverage. While lower flight altitudes provide higher pixel resolution, they cover less area per unit of time. However, the results demonstrated that with current camera quality in commonly used commercial drones (e.g., 1/2″ CMOS 48-megapixel image sensor on the DJI Mavic 2 Enterprise Advanced), flying at an altitude of 30 m produced images with adequate resolution to detect differences among cultivars.

A common concern among researchers was the technical difficulty of processing large volumes of images and performing subsequent analysis. Many widely used high-throughput image analysis pipelines are based on mapping software, which requires operational expertise to accurately transform aerial images into precise, georeferenced 3D models. In addition to the cost of software licenses, running these programs requires a powerful computer with a strong CPU, RAM, and graphics card. An affordable alternative is the use of cloud-based photogrammetry services, which simplify the process. These services handle orthophoto and elevation model generation through basic georeferencing, dense cloud generation, meshing, and texturing without the need for costly processing equipment. This is especially important when generating three-dimensional point clouds, where technical limitations are a significant challenge.

For the segmentation steps, there are numerous software packages available to analyze microplot schemes on orthomosaics to extract information for each plot. For example, in the workflow presented, the segmentation process was carried out using the MosaicTool plugin for FIJI ImageJ. This tool exported 200 plots in about 15 min and also calculated VIs. Additionally, the same software was used to compute canopy height and perform calculations for the maximum height values of each plot. It is also compatible with multispectral and thermal images. While many of

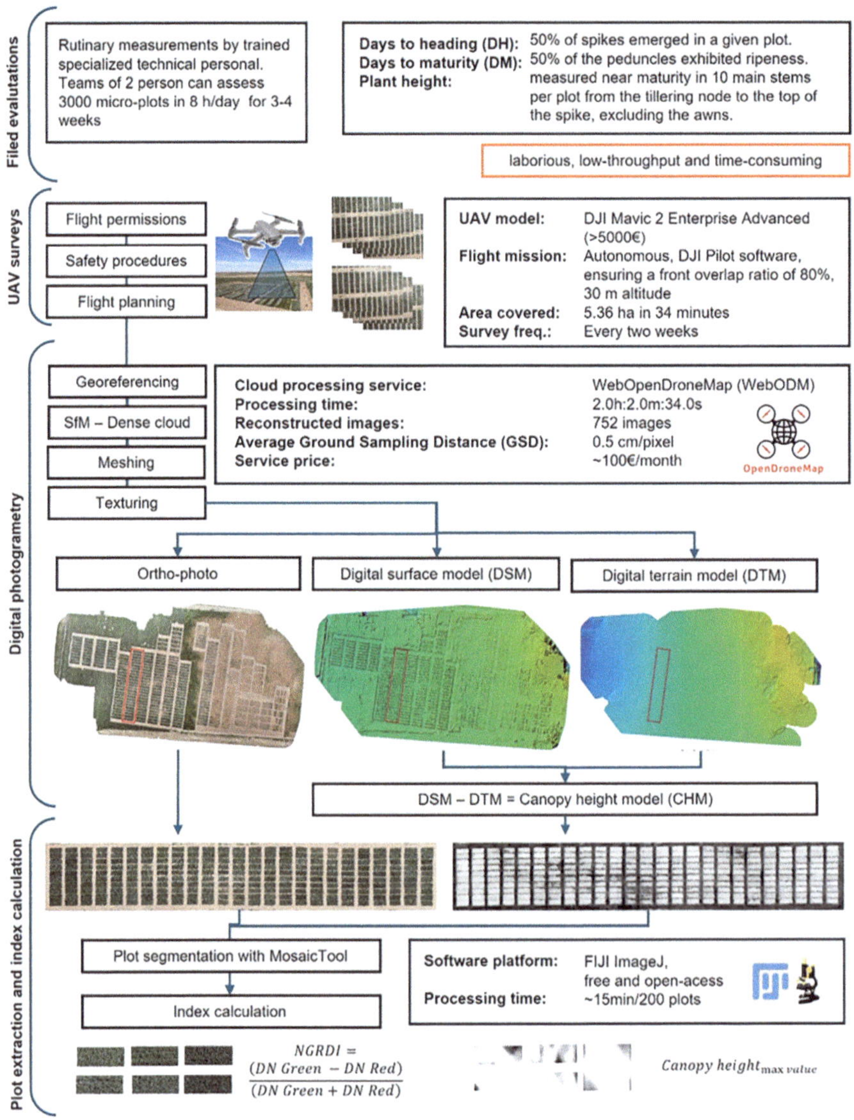

Fig. 9.6 Workflow summarizing the steps followed as a low-cost, simple, and straightforward protocol to implement commercial UAVs as phenotyping platforms to evaluate canopy height and crop phenology

these plugins are designed to handle basic and commonly used indices, some custom coding work may be required for more advanced calculations.

Once the workflow is defined, the recommended protocols could be defined as follows:

Canopy Height Protocol

(i) Conduct a UAV flight during grain filling. A single UAV evaluation has been reported to be sufficient for an accurate assessment of the final canopy height.
(ii) Use the structure-from-motion (SfM) method to construct the 3D structure of the crop canopy. The canopy height model is generated by subtracting the digital terrain model (DTM) from the digital elevation model (DEM).
(iii) Calculate the maximum height points from each canopy. Mean values may not be as effective for accurate assessments.

Crop Phenology Protocol

(i) Perform weekly UAV flights during the grain-filling stage. Multiple surveys during this period will help capture the time when the evaluated varieties exhibit the most significant differences in canopy color.
(ii) Generate orthomosaics from the RGB images and calculate greenness and senescence indices for predicting days to heading and to maturity, respectively. While greenness indices perform well for both phenological stages, the a* value provides a more precise evaluation of green tones turning yellow as plots approach maturity.
(iii) Develop training models incorporating at least one replicate for each experimental condition, ensuring comprehensive ground phenotyping. Visual in-field evaluation of a representative subset is still necessary to capture the heterogeneity of experimental conditions. Employ the collected data to create and refine the most accurate model, emphasizing the attainment of the highest R^2 (coefficient of determination) and the lowest RMSE (root mean square error). With this model, DH and DM predictions can be calculated for each plot.

9.6 Challenges and Future Directions

While UAVs were once associated with complex piloting skills and high costs, they have now evolved into user-friendly platforms, with many affordable models available in the market. However, there are still technical limitations and challenges in adopting drone technology for agricultural research. These challenges include the need for high-resolution sensors, data processing capabilities, and understanding how to integrate UAV-collected data into phenotyping, breeding, and agronomy research pipelines. Additionally, regulatory hurdles and weather conditions can also impact the widespread use of UAVs in agriculture. However, UAVs are increasingly being used for a wide range of applications, from providing a comprehensive view of entire fields to pinpointing specific issues for targeted interventions. Future developments may include better integration of AI for real-time analysis, enhanced multispectral sensors for more accurate crop assessments, and improved automation for flight and data processing. The UAV procedures presented herein, including a literature review and a case study, highlight the potential of UAV systems to assess key traits that assist in the crop selection process and agronomic evaluations. However, it is important to note that these methods cannot fully replace field evaluations. Visual inspections and manual sampling remain critical for collecting data to train models tailored to specific seasonal conditions. Nevertheless, UAV technology can significantly reduce the amount of fieldwork required for precision phenotyping. By streamlining the measurement of key routine traits, researchers in the future will have more time to focus on developing innovative solutions for a more sustainable and climate-resilient agriculture, which is essential for ensuring food security and safety.

References

Araus, J. L., & Cairns, J. E. (2014). Field high-throughput phenotyping: The new crop breeding frontier. *Trends in Plant Science, 19*, 52–61.

Ampatzidis, Y., Partel, V., & Costa, L. (2020). Agroview: Cloud-based application to process, analyze and visualize UAV-collected data for precision agriculture applications utilizing artificial intelligence. *Computers and Electronics in Agriculture, 174*, 105457. https://doi.org/10.1016/j.compag.2020.105457

Anderegg, J., Yu, K., Aasen, H., Walter, A., Liebisch, F., & Hund, A. (2020). Spectral vegetation indices to track senescence dynamics in diverse wheat germplasm. *Frontiers in plant science, 10*, 1749. https://www.frontiersin.org/article/10.3389/fpls.2019.01749

Aharon, S., Peleg, Z., Argaman, E., Ben-David, R., & Lati, R. N. (2020). Image-based high-throughout phenotyping of cereals early vigor and weed-competitiveness traits. *Remote Sensing, 12*(23), 1–17. https://doi.org/10.3390/rs12233877

Balota, M., Sarkar, S., Bennett, R. S., & Burow, M. D. (2024). Phenotyping peanut drought stress with aerial remote-sensing and crop index data. *Agriculture*. Retrieved from https://www.mdpi.com/2077-0472/14/4/565

Bai, X., Cao, Z., Zhao, L., Zhang, J., Lv, C., Li, C., & Xie, J. (2018). Rice heading stage automatic observation by multi-classifier cascade based rice spike detection method. *Agricultural and Forest Meteorology, 259*, 260–270. https://doi.org/10.1016/j.agrformet.2018.05.001

Cao, X., Liu, Y., Yu, R., Han, D., & Su, B. (2021). A comparison of UAV RGB and multispectral imaging in phenotyping for stay green of wheat population. *Remote Sensing, 13*(24). https://doi.org/10.3390/rs13245173

Casadesús, J., Kaya, Y., Bort, J., Nachit, M. M., Araus, J. L., Amor, S., Ferrazzano, G., Maalouf, F., Maccaferri, M., Martos, V., Ouabbou, H., & Villegas, D. (2007). Using vegetation indices derived from conventional digital cameras as selection criteria for wheat breeding in water-limited environments. *Annals of applied biology, 150*(2), 227–236. https://doi.org/10.1111/j.1744-7348.2007.00116.x

Christopher, J. T., Veyradier, M., Borrell, A. K., Harvey, G., Fletcher, S., & Chenu, K. (2014). Phenotyping novel stay-green traits to capture genetic variation in senescence dynamics. *Functional Plant Biology, 41*(11), 1035–1048. https://doi.org/10.1071/FP14052

Dua, A., Sharda, A., & Schapaugh, W. T. (2024). *Automated pipeline for multi-polygon shapefile generation for phenotype and precision agriculture applications*. Proceedings of SPIE - The International Society for Optical Engineering. Retrieved from https://www.spiedigitallibrary.org/conference-proceedings-of-spie/13053/1305302/Automated-pipeline-for-multi-polygon-shapefile-generation-for-phenotype-and/10.1117/12.3013542.short

Duan, T., Zheng, B., Guo, W., Ninomiya, S., Guo, Y., & Chapman, S. C. (2016). Comparison of ground cover estimates from experiment plots in cotton, sorghum and sugarcane based on images and ortho-mosaics captured by UAV. *Functional Plant Biology, 44*(1), 169-183. https://doi.org/10.1071/FP16123

Fernandez-Gallego, J., Kefauver, S., Kerfal, S., & Araus, J. (2018). Comparative canopy cover estimation using RGB images from UAV and ground. https://doi.org/10.1117/12.2501531

Feng, A., Zhou, J., Vories, E., & Sudduth, K. A. (2020). Evaluation of cotton emergence using UAV-based imagery and deep learning. *Computers and Electronics in Agriculture, 177*, 105711. https://doi.org/10.1016/j.compag.2020.105711

Feng, Z., Cheng, Z., Ren, L., Liu, B., Zhang, C., Zhao, D., Sun, H., Feng, H., Long, H., Xu, B., Yang, H., Song, X., Ma, X., Yang, G., & Zhao, C. (2024). Real-time monitoring of maize phenology with the VI-RGS composite index using time-series UAV remote sensing images and meteorological data. *Computers and Electronics in Agriculture, 224*. https://doi.org/10.1016/j.compag.2024.109212

Gillham, L. (2024). *Crop monitoring in intercropping systems: Assessing Maize Height patterns using UAV-imagery*. Utrecht University. Retrieved from https://studenttheses.uu.nl/handle/20.500.12932/46414

Gao, S., Zhong, R., Yan, K., Ma, X., Chen, X., Pu, J., Gao, S., Qi, J., Yin, G., & Myneni, R. B. (2023). Evaluating the saturation effect of vegetation indices in forests using 3D radiative transfer simulations and satellite observations. *Remote Sensing of Environment, 295*, 113665. https://doi.org/10.1016/j.rse.2023.113665

Gracia-Romero, A., & Lopes, M. S. (2024). Heading and maturity date prediction using vegetation indices: A case study using bread wheat, barley, and oat crops. *European Journal of Agronomy, 160*(127), 330.

Gracia-Romero, A., Kefauver, S. C., Vergara-Díaz, O., et al. (2020). Leaf versus whole-canopy remote sensing methodologies for crop monitoring under conservation agriculture: A case study with maize in Zimbabwe. *Scientific Reports, 10*, 16008. https://doi.org/10.1038/s41598-020-73110-3

Gracia-Romero, A., Rufo, R., Gómez-Candón, D., Soriano, J. M., Bellvert, J., Yannam, V. R. R., Gulino, D., & Lopes, M. S. (2023). Improving in-season wheat yield prediction using remote sensing and additional agronomic traits as predictors. *Frontiers in Plant Science, 14*, 1063983.

Gulino, D., & Lopes, M. S. (2024). Phenological adaptation of wheat varieties to rising temperatures: Implications for yield components and grain quality. *Plants, 13*(20), 2929. https://doi.org/10.3390/plants13202929

Guo, Y., Hao, F., Zhang, X., He, Y., & Fu, Y. H. (2024). Improving maize yield estimation by assimilating UAV-based LAI into WOFOST model. *Field Crops Research*. Retrieved from https://www.sciencedirect.com/science/article/pii/S0378429024002302

Haboudane, D., Miller, J. R., Tremblay, N., Zarco-Tejada, P. J., & Dextraze, L. (2002). Integrated narrow-band vegetation indices for prediction of crop chlorophyll content for application to precision agriculture. *Remote sensing of environment, 81*(2), 416–426. https://doi.org/10.1016/S0034-4257(02)00018-4

Hassan, M. A., Yang, M., Rasheed, A., Jin, X., Xia, X., Xiao, Y., & He, Z. (2018). Time-series multispectral indices from unmanned aerial vehicle imagery reveal senescence rate in bread wheat. *Remote Sensing, 10*(6). https://doi.org/10.3390/rs10060809

Hassan, M. A., Yang, M., Fu, L., Rasheed, A., Zheng, B., Xia, X., Xiao, Y., & He, Z. (2019). Accuracy assessment of plant height using an unmanned aerial vehicle for quantitative genomic analysis in bread wheat. *Plant Methods, 15*(1), 37. https://doi.org/10.1186/s13007-019-0419-7

Herr, A. W., Adak, A., Carroll, M. E., Elango, D., Kar, S., Li, C., Jones, S. E., Carter, A. H., Murray, S. C., Paterson, A., Sankaran, S., Singh, A., & Singh, A. K. (2023). Unoccupied aerial systems imagery for phenotyping in cotton, maize, soybean, and wheat breeding. *Crop Science*. https://doi.org/10.1002/csc2.21028

Holman, F. H., Riche, A. B., Michalski, A., Castle, M., Wooster, M. J., & Hawkesford, M. J. (2016). High throughput field phenotyping of wheat plant height and growth rate in field plot trials using UAV based remote sensing. *Remote Sensing, 8*(12). https://doi.org/10.3390/rs8121031

Hu, P., Chapman, S. C., & Zheng, B. (2021). Coupling of machine learning methods to improve estimation of ground coverage from unmanned aerial vehicle (UAV) imagery for high-throughout phenotyping of crops. *Functional Plant Biology, 48*(8), 766–779. https://doi.org/10.1071/FP20309

Hunt, J. R., Lilley, J. M., Trevaskis, B., Flohr, B. M., Peake, A., Fletcher, A., Zwart, A. B., Gobbett, D.,& Kirkegaard, J. A. (2019). Early sowing systems can boost Australian wheat yields despite recent climate change. *Nature Climate Change, 9*(3), 244–247. https://doi.org/10.1038/s41558-0190417-9

Jimenez-Berni, J. A., Deery, D. M., Rozas-Larraondo, P., Condon, A. T. G., Rebetzke, G. J., James, R. A., et al. & Sirault, X. R. (2018). High throughput determination of plant height, ground cover, and above-ground biomass in wheat with LiDAR. *Frontiers in plant science, 9*, 237

Jin, S., Sun, X., Wu, F., Su, Y., Li, Y., Song, S., Xu, K., Ma, Q., Baret, F., Jiang, D., Ding, Y., & Guo, Q. (2021). Lidar sheds new light on plant phenomics for plant breeding and management: Recent advances and future prospects. *ISPRS Journal of Photogrammetry and Remote Sensing, 171*, 202–223. https://doi.org/10.1016/j.isprsjprs.2020.11.006

Johnson, A. A. (2002). *The history of McCook Field 1917–1927* (pp. 190–191). Landfall Press. ISBN 0-913428-84-1.

Kaur, S., Kakani, V. G., & Carver, B. (2024). Hyperspectral imaging combined with machine learning for high-throughput phenotyping in winter wheat. *Plant Phenome J.* Retrieved from https://acsess.onlinelibrary.wiley.com/doi/abs/10.1002/ppj2.20111

Khan, Z., Chopin, J., Cai, J., Eichi, V. R., Haefele, S., & Miklavcic, S. J. (2018). Quantitative estimation of wheat phenotyping traits using ground and aerial imagery. *Remote Sensing, 10*(6). https://doi.org/10.3390/rs10060950

Kipp, S., Mistele, B., & Schmidhalter, U. (2014). Identification of stay-green and early senescence phenotypes in high-yielding winter wheat, and their relationship to grain yield and grain protein concentration using high-throughput phenotyping techniques. *Functional Plant Biology, 41*(3), 227–235. https://doi.org/10.1071/FP13221

Khanna, R., Möller, M., Pfeifer, J., Liebisch, F., Walter, A., & Siegwart, R. (2015). Beyond point clouds-3D mapping and field parameter measurements using UAVs. *In 2015 IEEE 20th conference on Emerging Technologies & Factory Automation (ETFA)* (pp. 1–4)

Lelong, C. C., Burger, P., Jubelin, G., Roux, B., Labbé, S., & Baret, F. (2008). Assessment of unmanned aerial vehicles imagery for quantitative monitoring of wheat crop in small plots. *Sensors, 8*(5), 3557–3585.

Lopes, M. S., & Reynolds, M. P. (2012). Stay-green in spring wheat can be determined by spectral reflectance measurements (normalized difference vegetation index) independently from phenology. *Journal of Experimental Botany, 63*(10), 3789–3798. https://doi.org/10.1093/jxb/ers071

Lopes, M. S., Reynolds, M. P., Jalal-Kamali, M. R., Moussa, M., Feltaous, Y., Tahir, I. S. A., Barma, N., Vargas, M., Mannes, Y., & Baum, M. (2012). The yield correlations of selectable physiological traits in a population of advanced spring wheat lines grown in warm and drought environments. *Field Crops Research, 128*, 129–136. https://doi.org/10.1016/j.fcr.2011.12.017

Lopes, M. S., Royo, C., Alvaro, F., Sanchez-Garcia, M., Ozer, E., Ozdemir, F., Karaman, M., Roustaii, M., Jalal-Kamali, M. R., & Pequeno, D. (2018). Optimizing winter wheat resilience to climate change in rain-fed crop systems of Turkey and Iran. *Frontiers in Plant Science, 9*, 563. https://doi.org/10.3389/fpls.2018.00563

Li, B., Xu, X., Han, J., Zhang, L., Bian, C., Jin, L., & Liu, J. (2019). The estimation of crop emergence in potatoes by UAV RGB imagery. *Plant Methods, 15*(1). https://doi.org/10.1186/s13007-019-0399-7

Li, L., Hassan, M. A., Song, J., Xie, Y., Rasheed, A., Yang, S., Li, H., Liu, P., Xia, X., He, Z., & Xiao, Y. (2023). UAV-based RGB imagery and ground measurements for high-throughput phenotyping of senescence and QTL mapping in bread wheat. *Crop Science, 63*(6), 3292–3309. https://doi.org/10.1002/csc2.21086

Mulla, D. J. (2013). Twenty five years of remote sensing in precision agriculture: Key advances and remaining knowledge gaps. *Biosystems engineering, 114*(4), 358–371. https://doi.org/10.1016/j.biosystemseng.2012.08.009

Mohammadi, S., Uhlen, A. K., Lillemo, M., Ergon, Å., & Shafiee, S. (2024). Enhancing phenotyping efficiency in faba bean breeding: integrating UAV imaging and machine learning. *Precision Agriculture, 25*(3), 1502–1528. https://doi.org/10.1007/s11119-024-10121-4

Myint, S. W., Gober, P., Brazel, A., Grossman-Clarke, S., & Weng, Q. (2011). Per-pixel vs. object-based classification of urban land cover extraction using high spatial resolution imagery. *Remote sensing of environment, 115*(5), 1145–1161. https://doi.org/10.1016/j.rse.2010.12.017

Madec, S., Baret, F., De Solan, B., Thomas, S., Dutartre, D., Jezequel, S., Hemmerlé, M., Colombeau, G., & Comar, A. (2017). High-throughput phenotyping of plant height: comparing unmanned aerial vehicles and ground LiDAR estimates. *Frontiers in plant science, 8*. https://doi.org/10.3389/fpls.2017.02002

Makanza, R., Zaman-Allah, M., Cairns, J. E., Magorokosho, C., Tarekegne, A., Olsen, M., & Prasanna, B. M. (2018). High-throughput phenotyping of canopy cover and senescence in maize field trials using aerial digital canopy imaging. *Remote Sensing, 10*(2). https://doi.org/10.3390/rs10020330

Pinto, R. S., Lopes, M. S., Collins, N. C., & Reynolds, M. P. (2016). Modelling and genetic dissection of staygreen under heat stress. *Theoretical and Applied Genetics, 129*(11), 2055–2074. https://doi.org/10.1007/s00122-016-2757-4

Pask, A., Pietragalla, J., Mullan, D., & Reynolds, M. (2012). *Physiological Breeding: II, A field guide to wheat phenotyping*. CIMMYT

Remondino, F., & El-Hakim, S. (2006). Image-based 3D modelling: a review. *In photogrammetric record, 21*(115), 269–291. https://doi.org/10.1111/j.1477-9730.2006.00383.x

Rezaei, E. E., Siebert, S., Hüging, H., & Ewert, F. (2018). Climate change effect on wheat phenology depends on cultivar change. *Scientific reports, 8*(1). https://doi.org/10.1038/s41598-018-23101-2

Sharma, V., Honkavaara, E., Hayden, M., & Kant, S. (2024). UAV remote sensing phenotyping of wheat collection for response to water stress and yield prediction using machine learning. *Plant Stress*. Retrieved from https://www.sciencedirect.com/science/article/pii/S2667064X24001180

Sankaran, S., Khot, L. R., Espinoza, C. Z., Jarolmasjed, S., Sathuvalli, V. R., Vandemark, G. J., Miklas, P. N., Carter, A. H., Pumphrey, M. O., Knowles, R. R. N., & Pavek, M. J. (2015). Low-altitude, high-resolution aerial imaging systems for row and field crop phenotyping: A review. *European journal of agronomy, 70*, 112–123. https://doi.org/10.1016/j.eja.2015.07.004

Shepard, N. R., DeSalvio, A. J., Arik, M., Adak, A., & Murray, S. C. (2024). Deep learning-based high-throughput phenotyping of Maize (Zea mays L.) tasseling from UAS imagery across environments. *bioRxiv*. Retrieved from https://www.biorxiv.org/content/10.1101/2024.06.24.600506.abstract

Shrestha, N., Powadi, A., Davis, J., Ayanlade, T. T., & Liu, H. (2024). Plot-level satellite imagery can substitute for UAVs in assessing maize phenotypes across multistate field trials. *agriRxiv*. Retrieved from https://www.cabidigitallibrary.org/doi/abs/10.31220/agriRxiv.2024.00251

Tucker, C. J. (1979). Red and photographic infrared linear combinations for monitoring vegetation. *Remote sensing of Environment, 8*(2), 127–150. https://doi.org/10.1016/0034-4257(79)90013-0

Tucker, C. J., Holben, B. N., Elgin, J. H., & McMurtrey, J. E. (1980). Relationship of spec-tral data to grain yield variation. *Photogrammetric Engineering and Remote Sensing, 46*(5), 657–666

Taniguchi, S., Sakamoto, T., Imase, R., Nonoue, Y., Tsunematsu, H., Goto, A., Matsushita, K., Ohmori, S., Maeda, H., Takeuchi, Y., Ishii, T., Yonemaru, J. I., & Ogawa, D. (2022). Prediction of heading date, culm length, and biomass from canopy-height-related parameters derived from time-series UAV observations of rice. *Frontiers in Plant Science, 13*. https://doi.org/10.3389/fpls.2022.998803

Tefera, A. T., Banerjee, B. P., Pandey, B. R., James, L., Puri, R. R., Cooray, O. O., Marsh, J., Richards, M., Kant, S., Fitzgerald, G.,& Rosewarne, G. M. (2022). Estimating early season growth and biomass of field pea for selection of divergent ideotypes using proximal sensing. *Field Crops Research, 277*. https://doi.org/10.1016/j.fcr.2021.108407

United Nations Conference on Trade and Development (UNCTAD). (2017). *The role of science, technology and innovation in ensuring food security*. United Nations Conference on Trade and Development. Retrieved from https://unctad.org/system/files/official-document/dtlstict2017d5_en.pdf

Waldner, F., & Defourny, P. (2017). Where can pixel counting area estimates meet user-defined accuracy requirements?. *International Journal of Applied Earth Observation and Geoinformation, 60*, 1–10. https://doi.org/10.1016/j.jag.2017.03.014

Walter, J., Edwards, J., McDonald, G., & Kuchel, H. (2018). Photogrammetry for the estimation of wheat biomass and harvest index. *Field Crops Research, 216*, 165–174 https://doi.org/10.1016/j.fcr.2017.11.024

Watanabe, K., Guo, W., Arai, K., Takanashi, H., Kajiya-Kanegae, H., Kobayashi, M., Yano, K., Tokunaga, T., Fujiwara, T., Tsutsumi, N., & Iwata, H. (2017). High-throughput phenotyping of sorghum plant height using an unmanned aerial vehicle and its application to genomic prediction modeling. *Frontiers in plant science, 8*. https://doi.org/10.3389/fpls.2017.00421

Zhang, C., & Kovacs, J. M. (2012). The application of small unmanned aerial systems for precision agriculture: A review. *Precision Agriculture, 13*, 693–712.

Zaman-Allah, M., Vergara, O., Araus, J. L., Tarekegne, A., Magorokosho, C., Zarco-Tejada, P. J., Hornero, A., Albà, A. H., Das, B., Craufurd, P., Olsen, M., Prasanna, B. M., & Cairns, J. (2015). Unmanned aerial platform-based multi-spectral imaging for field phenotyping of maize. *Plant methods, 11*(1), 35. https://doi.org/10.1186/s13007-015-0078-2

Zhao, L., Guo, W., Wang, J., Wang, H., Duan, Y., Wang, C., Wu, W., & Shi, Y. (2021). An efficient method for estimating wheat heading dates using UAV images. *Remote Sensing, 13*(16). https://doi.org/10.3390/rs13163067

Zhao, F., Yang, G., Yang, H., Long, H., Xu, W., Zhu, Y., Meng, Y., Han, S., & Liu, M. (2022). A method for prediction of winter wheat maturity date based on MODIS time series and accumulated temperature. *Agriculture, 12*(7). https://doi.org/10.3390/agriculture12070945

Zhu, Y., Cao, Z., Lu, H., Li, Y., & Xiao, Y. (2016). In-field automatic observation of wheat heading stage using computer vision. *Biosystems Engineering, 143*, 28–41.https://doi.org/10.1016/j.biosystemseng.2015.12.015

Zadoks, J. C., Chang, T. T., & Konzak, C. F. (1974). A decimal code for the growth stages of cereals. *Weed research, 14*(6), 415–421

Zolkos, S. G., Goetz, S. J., & Dubayah, R. (2013). A meta-analysis of terrestrial aboveground biomass estimation using lidar remote sensing. *Remote sensing of environment, 128*, 289–298. https://doi.org/10.1016/j.rse.2012.10.017

Chapter 10
Perfecting Wheat Quality, Quality Analysis, and Production Processes via Digital and Artificial Intelligence Approaches

Asuman Kaplan Evlice and Buket Cetiner

Abstract Wheat as a staple crop supports the dietary needs of billions worldwide, making its quality vital for both nutritional value and industrial processing. High-quality wheat is essential for producing diverse end products such as bread, pasta, and other baked goods, which require specific grain and flour properties. Quality traits such as protein content, gluten strength, and kernel hardness directly influence milling performance and product quality.

The integration of digital tools and artificial intelligence (AI) in grain storage, wheat quality assessment, and wheat-based production processes is transforming the wheat industry, enabling faster, more accurate, and scalable quality analysis. Technologies such as near-infrared spectroscopy (NIR), machine vision, and machine learning (ML) models are revolutionizing how wheat quality is evaluated, offering numerous advantages over traditional methods. The Internet of things (IoT) and ML are used on grain storage for better moisture control and spoilage prevention in silos.

Predictive maintenance of milling and processing equipment is a proactive approach in the production sector that leverages technologies like IoT to optimize maintenance costs, reduce downtime, and enhance operational efficiency. With the advent of new technologies, particularly digital and AI approaches, the ways of analysing quality and managing production processes have been transformed. The 3D/4D printing can be used to produce wheat-based food items tailored to specific contexts and individuals with specific needs.

This chapter highlights the role of digital and AI technologies in wheat quality evaluation, storage, and production processes, paving the way for a smarter and more sustainable future in the agriculture and food sectors.

A. K. Evlice (✉)
Faculty of Agricultural Sciences and Technology, Department of Field Crops, Sivas Science and Technology University, Sivas, Türkiye
e-mail: asuman.kaplanevlice@sivas.edu.tr

B. Cetiner
Department of Quality and Technology, Field Crops Central Research Institute, Ankara, Türkiye

N. Zencirci et al. (eds.), *Empowering Wheat Cultivation with GIS, Digital Approaches and Artificial Intelligence*, https://doi.org/10.1007/978-3-031-99954-3_10

Keywords Wheat · Quality · Artificial intelligence (AI) · 3D/4D printing

Abbreviations

ABC	artificial bee colony
AFB1	aflatoxin B1
AI	artificial intelligence
ANN	artificial neural network
BPO	benzoyl peroxide
CFS	correlation-based feature selection
CNN	convolutional neural network
CSVM	cubic support vector machine
CVS	computer vision systems
DF	deep forest
DLT	distributed ledger technology
DON	deoxynivalenol
DRIFTS	diffuse reflectance infrared fourier transform spectroscopy
DSIFT	dense scale-invariant feature transform
DT	decision tree
DT	digital twins
ELM	extreme learning machine
EMCVS	Monte Carlo variable selection
EW	effective wavelengths
FL	fuzzy logic
FT-MIR	Fourier transform mid-infrared
FTIR	Fourier transform infrared spectroscopy
Gcforest	multi-grained cascaded forest
HI	hardness index
HSI	Hyperspectral Imaging
Iot	Internet of things
KNN	K-nearest neighbours
LDA	linear discriminant analysis
LF-NMR	low-field nuclear magnetic resonance
LR	logistic regression
LSVM	linear support vector machine
MCW	marker-controlled watershed
ML	machine learning
MLP	multilayer perceptron
NB	naïve Bayes
NFC	near field communication
NIR	near-infrared
NIRS	near-infrared spectroscopy
PCA	principal component analysis

PLS	partial least squares
PLS-DA	partial least-squares discriminant analysis
QDA	quadratic discriminant analysis
QSVM	quadratic support vector machine
R^2	coefficient of determination
RACNN	residual attention convolutional neural network
RF	random forest
RFID	radio frequency identification
RMSEP	root mean square error of prediction
R_p^2	coefficient of determination for prediction
RPD	predictive residual deviation
SACNN	symmetric all-convolutional neural network
SAE	sparse autoencoder
SIAP	solid-phase latex microsphere immunochromatography platform
SLS	selective laser sintering
SR-IMS	synchrotron radiation infrared microspectroscopy
SVM	support vector machine
SWIR	short-wave infrared
ZEN	Zearalenone

10.1 Introduction

Wheat, a cornerstone of global agriculture, is one of the most extensively cultivated crops worldwide. Its significance extends beyond nutrition, impacting economies, cultures, and industries worldwide. Wheat accounts for approximately 20% of the total caloric intake globally, serving as a dietary staple for 40% of the world's population (Acevedo et al., 2018).

As a key source of carbohydrates, protein, and micronutrients, wheat significantly contributes to food security, particularly in regions heavily reliant on cereal-based diets (An et al., 2023). It is used to produce a variety of food products, including bread, pasta, noodles, and pastries, catering to different cultural preferences (Ciudad-Mulero et al., 2020). The quality of these products is heavily influenced by the quality of the raw material. Consequently, assessing quality is important to the farmers, breeders, food industry, and the scientific community that supports these sectors (Caporaso et al., 2018).

Traditional methods for wheat quality analysis, including grain grading and laboratory-based testing such as protein content, gluten strength, and starch composition, have served the industry well over the years. These methods provide valuable insights into the wheat's suitability for different applications, from milling to baking (Patrício & Rieder, 2018). However, these techniques are often regarded as time-intensive, requiring sample preparation, laborious procedures, reliance on chemicals, and specialized technical expertise to operate the equipment. Besides, as the global demand for wheat grows, there is an increasing need for more efficient, accurate, and scalable methods to assess wheat quality (Vithu & Moses, 2016; Patrício & Rieder, 2018). Considering this, there has been a shift towards employing green analytical chemistry methods for quality detection (An et al., 2023).

The integration of digital technologies such as artificial intelligence (AI), the Internet of things (IoT), and cloud computing in the agri-food sector holds the promise of transforming the industry while enhancing sustainability (Ben Ayed & Hanana, 2021). These technologies can empower farmers, breeders, manufacturers, and distributors by enabling data-driven decision-making, increasing efficiency, minimizing waste, and contributing to improved food security and sustainable practices (Taneja et al., 2023).

AI, a branch of computer science, is recognized as one of the most advanced technologies of the twenty-first century, alongside genetic engineering and nanoscience. This recognition stems from the continuous advancements in computing power, the development of robust algorithms, and the accessibility of these algorithms through open-source platforms, which have collectively accelerated the growth of AI and its application across various scientific domains (Ayres et al., 2021). AI has evolved into an independent discipline encompassing both theoretical and practical frameworks. It is an expansive field that includes machine learning (ML), computer vision, and natural language processing, among others (Abioye et al., 2021; An et al., 2023).

AI utilizes a variety of algorithms such as reinforcement learning (Pantazi et al., 2016), swarm intelligence, cognitive science, expert systems, fuzzy logic (FL), artificial neural networks (ANNs), and logic programming (Kumar & Thakur, 2012; Taneja et al., 2023). Each algorithm has distinct strengths and limitations, with their selection tailored to the specific task or challenge being addressed (Taneja et al., 2023). Overall, the primary objective of AI research is to enable machines to undertake complicated tasks that traditionally need human intelligence (Chen et al., 2020).

AI has become an integral part of cereal science, particularly in wheat research, offering transformative solutions for both qualitative and quantitative analyses. Qualitative analysis focuses on tasks such as classification of variety, adulteration detection, unsound kernel recognition, and other classification challenges like determining geographical origin and maturity levels. On the other hand, quantitative analysis aims to predict the internal nutrient content of kernels, including moisture, protein, starch, oil, and minerals, alongside regression tasks such as hardness index (HI) and toxin content (An et al., 2023).

With increasing technological advancements, particularly in digital and AI domains, the agricultural and food sectors are undergoing a transformation, redefining traditional practices, including wheat quality analysis and production processes. This makes the process much more accurate, faster, cheaper, and objective (Yasar, 2024). This chapter explores the key AI techniques currently used in wheat quality assessment, their applications, and their potential impact on the wheat industry.

10.2 Application of AI in the Assessment of Wheat Quality

Wheat is one of the most important staple crops globally, with a diverse range of end products including bread, pasta, noodles, bulgur, and baked goods. Wheat quality is multifaceted, encompassing a range of physical, chemical, and functional attributes that influence its suitability for different end products. Understanding these quality traits is essential for both wheat breeders and processors to meet the demands of the food industry and to optimize production processes.

In recent years, there has been a significant shift towards the integration of digital tools and AI in wheat quality assessment. These advancements aim to address the challenges posed by conventional methods such as subjectivity, time consumption, and scalability issues. AI-driven technologies such as near-infrared spectroscopy (NIR), image analysis, machine vision, and ML models used in the assessment of wheat quality are given below.

10.2.1 Kernel Classification

The physicochemical properties of wheat exhibit considerable variation across different varieties, which directly influences the quality and taste of the end products. AI can be used in the classification of wheat varieties, facilitating advancements in this field (An et al., 2023).

Sabanci et al. (2017) developed a computer vision system employing a simplified classification approach with high accuracy. This system classifies wheat grains of *Triticum aestivum* and *Triticum durum* species based on their visual attributes, utilizing an artificial neural network (ANN) of the multilayer perceptron (MLP) type.

ML techniques, including artificial neural networks (ANNs), the dense scale-invariant feature transform (DSIFT) algorithm, and support vector machines (SVMs), have been widely applied in the agricultural sector for the classification and identification of grains and other products. ANNs are employed to differentiate wheat species (e.g., durum and common wheat) by analysing visual features such as shape, size, and colour (Sabanci et al., 2017). The DSIFT algorithm, a computer vision method, detects features like grain size, shape, and texture, enabling the classification of wheat grains into distinct categories (Patrício & Rieder, 2018). Additionally, SVM is an effective ML technique for categorizing wheat grains and identifying germinated grains, enhancing precision in quality assessment (Taneja et al., 2023).

Olgun et al. (2016) introduced an ML classification system designed to distinguish 40 different wheat varieties. The system employed a support vector machine (SVM) classifier, which utilized features extracted through the dense scale-invariant feature transform (DSIFT) algorithm, achieving an accuracy of 88.3%. Similarly, Khatri et al. (2022) developed a classification model incorporating the K-nearest neighbour (KNN), naïve bayes (NB), decision tree (DT), and an Ensemble method to classify three wheat varieties. Among these, the ensemble method attained the highest classification accuracy of 95%, based on shape features. Fazel-Niari et al. (2022) proposed an ML-based system for classifying wheat grains and seven different wheat components. This system utilized quadratic discriminant analysis (QDA), linear support vector machine (LSVM), quadratic support vector machine (QSVM), and cubic support vector machine (CSVM) algorithms. The QSVM algorithm achieved the highest accuracy of 97.6%, relying on shape, colour, and texture features. Furthermore, Gulzar et al. (2020) introduced a convolutional neural network (CNN) model for the classification of 14 commonly used seed types. The CNN model outperformed traditional classifiers such as KNN, DT, NB, random forest (RF), and logistic regression (LR), achieving an impressive classification accuracy of 99.9% (Agarwal & Bachan, 2023). These studies highlight the growing efficacy of ML and deep learning methods in enhancing precision in agricultural classification tasks.

Spoladore et al. (2021) developed a partial least-squares discriminant analysis (PLS-DA) model utilizing NIR data, successfully achieving 100% accuracy in classifying wheat grains from five distinct genotypes. In more recent studies, deep learning techniques have been applied to classify wheat grains using image data (Parrenin et al., 2023). Unlersen et al. (2022) introduced a convolutional neural network (CNN)-based model designed to differentiate between four wheat varieties by analysing images of wheat lot samples.

10.2.2 Unsound Kernel Identification

Unsound wheat kernels, such as those that are broken, sprouted, mouldy, diseased, infected by fungi, damaged by insect, heat, or frost, mainly occur during production, storage, and transportation. These unsound kernels both diminish the nutritional quality of wheat and pose considerable health risks to consumers (Sendin et al., 2018; Caporaso et al., 2018). Implementing effective detection and sorting methods to identify and remove unsound kernels is crucial for ensuring the safety and quality of wheat products. Advancements in technologies such as deep learning and image processing have shown promise in automating the identification of these compromised grains.

High moisture levels before or during harvesting can cause premature germination of wheat grains. Sprouting increases the grain's susceptibility to insect damage and microbial contamination (Sendin et al., 2018), while also significantly altering the starch, protein, sugar, and dry matter content of the grain, ultimately impacting its quality (Barbedo et al., 2018). Slightly sprouted kernels closely resemble unsound kernels, making them harder to differentiate. A previous study introduced a deep learning model called deep forest (DF), designed for small sample sets, which achieved over 89% accuracy in identifying slightly sprouted wheat kernels (Zhang et al., 2020). Shrestha et al. (2016) proposed a computer vision system using two cameras to separate and identify germinated wheat grains in a controlled environment. This system classifies the grains into three categories: healthy, damaged by germination, and severely damaged by germination. The system includes a top-mounted camera and a base-mounted camera, using the marker-controlled watershed (MCW) technique for segmentation. The watershed algorithm helps create accurate object contours and is computationally efficient. An artificial neural network uses 16 input parameters for classification. Zhu et al. (2023) created CNN models to distinguish the sound wheat kernels and slightly sprouted kernels using hyperspectral imaging. The presence of sprouted grains in wheat milling negatively affects the flour quality and yield. This imaging technique offers potential for monitoring grain quality and sorting varieties for milling purposes. Although applied to a limited number of wheat varieties, it holds promise for selecting wheat grains for milling (Parrenin et al., 2023).

Many methods in the literature are proposed to detect damaged wheat grains. These studies typically involve extracting different grain features to apply AI techniques (Sabanci et al., 2022). Singh et al. (2009) employed near-infrared hyperspectral images of wheat to detect insect-damaged kernels using statistical discriminant classifiers, achieving classification accuracy ranging from 85% to 100%. In a subsequent study, Singh et al. (2010) combined short-wave near-infrared hyperspectral and digital colour images of wheat, increasing classification accuracy to between 91% and 100%. Shahin and Symons (2011) achieved 92% accuracy in detecting damaged wheat kernels using visible/near-infrared hyperspectral images. Boniecki et al. (2014) detected damaged kernels with 100% accuracy using X-ray images. Turhal and Turhal (2014) proposed an automatic system utilizing an artificial neural network (ANN) to identify wheat grains damaged by sunn pest (*Eurygaster* spp.), with the histogram of grain images serving as the input to the ANN. Their method achieved a success rate of 75%. Basati et al. (2018) classified wheat grains using 25 features, including nine colour features, ten morphological features, and six textural features, achieving a classification accuracy of 90.2% with a decision tree algorithm and 87.5% accuracy with an ANN classifier. Sabanci et al. (2022) implemented an artificial bee colony (ABC) optimization-based ANN and extreme learning machine (ELM) algorithms to classify wheat grains damaged by sunn pest, using a dataset of 300 wheat grains (150 healthy and 150 damaged). They also applied a correlation-based feature selection (CFS) technique to identify the most effective among 17 grain features. This study significantly improved detection accuracy for damaged wheat grains due to the enhanced ANN algorithm and the CFS approach.

Grains affected by *Fusarium* have low market demand, resulting in a reduced final sales price. These diseased kernels also pose a potential health risk due to concerns about food safety, as *Fusarium* contamination can lead to the presence of mycotoxins in the grain. A recent study introduced a residual attention convolutional neural network (RACNN), improved by increasing its width and depth, and integrating channel attention and residual modules for detecting different levels of *Fusarium* head blight infection in wheat, reaching a classification accuracy of 98.1% (Weng et al., 2021). Cereals are particularly vulnerable to toxic fungi, which produce secondary metabolites like aflatoxin B1 and Deoxynivalenol

(DON), both of which are highly carcinogenic (Kimuli et al., 2018; Femenias et al., 2020). Liang et al. (2020) found that sparse autoencoder (SAE) surpasses SVM in detecting DON-contaminated wheat flour and kernel. Additionally, they observed that the Vis-NIR region is more effective for identifying wheat samples, whereas the short-wave infrared (SWIR) region is more suitable for distinguishing wheat flour samples. SWIR-hyperspectral imaging (HIS) was used to identify aflatoxin B1 (AFB1) in wheat, combined with principal component analysis (PCA) and SVM. Images were collected in the wavelength range of 1000–2500 nm and characteristic wavelengths were noted as 1317, 1459, 1865, 1934, and 2274 nm for AFB1 detection, achieving classification rates of 83.75% for the training set and 82.50% for the test set (Chu et al., 2017). Shen et al. (2022) proposed a local partial least-squares (PLS) model based on global PLS scores (LPLS-S), which predicted DON content in wheat kernels with a coefficient of determination (R_p^2) of 0.81 and a root mean square error of prediction (RMSEP) of 40.25 mg/kg, yielding better results than the SVR and PLSR models. De Girolamo et al. (2019) evaluated the effectiveness of Fourier transform near-infrared (FT-NIR) and Fourier transform mid-infrared (FT-MIR) spectroscopy techniques, integrated with multivariate classification models, for screening unprocessed durum wheat samples that were naturally contaminated with ochratoxin A. FT-NIR and FT-MIR, both, demonstrated comparable and satisfactory results, with FT-NIR achieving 94% accuracy and FT-MIR achieving 96%. Notably, the classifier using FT-MIR data showed slight improvement over the one based on FT-NIR data.

Destructive climates, like frost, can alter the chemical composition of wheat, resulting in a reduction in nutritional value, nutrient availability, and the quality of grinding and baking. These changes in the chemical structure are considered a primary factor in this decline (An et al., 2023). Synchrotron radiation infrared microspectroscopy (SR-IMS) was utilized to examine the microchemical structure of both normal and frost-damaged wheat kernels. The findings revealed that frost caused significant alterations in the spectral distribution in the kernel. Frost-damaged kernels exhibited significant reductions ($p < 0.05$) in spectral intensities, particularly in the area and height ratios of amide I to amide II bands, as well as in spectral parameters related to carbohydrate-associated functional groups such as SCHO, CHO, and NSCHO, when compared to normal kernels (Xin et al., 2013).

Imaging-based technologies have proven to be effective techniques for monitoring the quality, discrimination, authentication, and prediction of cereals (Zareef et al., 2021). A study reported the detection of broken wheat kernels using NIR-HSI at various percentages (0%, 3%, 6%, 9%, 12%, and 15%). Images in the wavelength range of 1000–1600 nm were collected, and both PCR and PLSR models were used for prediction, while linear discriminant analysis (LDA) and quadratic discriminant analysis (QDA) models were applied for classification. The results showed that both classification and quantification models were effectively used for detecting broken kernels (Ravikanth et al., 2016).

10.2.3 Adulteration Recognition

Food adulteration is a widespread issue globally, posing significant threats to consumer health and safety and representing a major challenge in ensuring food security. Economic motives often drive cereal adulteration, including species or variety substitution and fraudulent organic labelling. Commonly, low-quality grains are blended with high-quality ones to maximize profits. In more sophisticated cases, high-quality flour is blended with other substances, making the adulteration difficult to detect visually. The spectral data of adulterated cereals typically exhibit complex nonlinear components. Additionally, variations in light scattering caused by differences in particle size, shape, and distribution within a sample can negatively affect classification accuracy (Yu et al., 2018; An et al., 2023).

Wheat is used for producing various wheaten foods depending on their different functions. For instance, the use of common wheat is banned in the productions of Italian (Vermeulen et al., 2018) and Turkish pastas. A prior study developed a PLS-DA model that combined morphological features with NIR spectral profiles, achieving 99% accuracy in detecting fraud in durum wheat classification (Vermeulen et al., 2018). Another study utilized FT-NIR and FT-MIR spectroscopy alongside LDA and PLS-DA models to identify common wheat adulteration in durum wheat samples sourced from Argentina and Italy (De Girolamo et al., 2020). In this research, samples were divided into two scenarios: one with three classes (1%, 1–5%, and >5% common wheat) and another with two classes, based on a 2% common wheat threshold. The PLS-DA model demonstrated superior performance, achieving classification accuracies of 85–97%. The three-class scenario yielded better results in the NIR range, while the two-class scenario offered similar accuracy across both spectral ranges.

Flour adulteration, whether intentional or accidental, is a common challenge that complicates the assurance of flour quality in industrial production. For instance, peanut flour, walnut flour, and wheat flour are frequently used ingredients, and cross-contamination can occur when shared production equipment is involved, leading to wheat flour becoming unintentionally contaminated with peanut or walnut flour. This can be life-threatening for individuals with allergies. Additionally, some illegal traders add harmful substances like benzoyl peroxide (BPO) to achieve a whitening effect (An et al., 2023). A multi-grained cascaded forest (gcForest) model with a non-neural DL architecture to fuse image and spectral features from HSI was utilized for detecting adulterants in wheat flour by Zheng et al. (2022). The study identified adulteration types (walnut, peanut, or BPO) and their different concentrations. Effective wavelengths (EW) were extracted using ensemble Monte Carlo variable selection (EMCVS), and image features were obtained through a symmetric all-convolutional neural network (SACNN) based on EW. The gcForest-related models, which combined EW and SACNN image features, achieved over 94% accuracy, surpassing the performance of models using only EW spectral features or SACNN image features. Visualization techniques were highlighted as a valuable tool for providing intuitive insights into cereal adulteration.

Deep learning algorithms have been used to detect and quantify adulterants, even at trace levels, in grain or flour (Pradana-López et al., 2022a; Sharma et al., 2024). An ML-supported Fourier transform infrared spectroscopy (FTIR) used to detect and quantify cross-contact between gluten-free and gluten-rich flours revealed contamination of cornflour (gluten-free) with gluten-containing flours, namely wheat, barley, and rye at different concentrations (Okeke, 2020). An allergen can be detected quickly using smartphones equipped with convolutional deep learning algorithms like Resnet, which utilizes transfer learning to detect trace amounts of wheat with 99.1% accuracy (Pradana-López et al., 2022a). Additionally, a real-time AI-based approach was applied to detect adulterated lentil flour containing trace levels of wheat (gluten) or pistachios (nuts), achieving an accuracy of 96.4% (Pradana-López et al., 2022b). These advancements highlight the potential of AI and deep learning in improving food safety and ensuring product integrity.

10.2.4 Functional Quality

The quality of wheat-based products relies heavily on the quality of the wheat itself. Therefore, evaluating the composition, purity, and physicochemical properties of wheat is essential for the food industry, breeders, farmers, and the scientific community that supports these sectors (Caporaso et al., 2018).

Moisture content significantly impacts the storage duration of wheat and serves as a key parameter for assessing its quality (Zhang & Guo, 2020). NIR is a non-destructive and rapid analytical technique that has gained popularity due to its ability to quickly assess key quality attributes of wheat, such as moisture, protein, starch contents (Caporaso et al., 2018). Recent research revealed that the NIR region provides more information about grain moisture content compared to the Vis-NIR region (Zhang & Guo, 2020).

Protein content is one of the most important quality parameters that influences the technological characteristics of baked foods such as bread, pasta (Sendin et al., 2018). Developing predictive models with strong generalization capabilities requires considering factors like geographical origin, variety, and agronomic conditions. One notable study collected 190 wheat varieties from Canada and various European countries, establishing a PLSR model that demonstrated strong predictive performance (coefficient of determination for prediction $R_p^2 = 0.82$; RMSEP = 0.86%) for estimating protein content in individual wheat kernels (Caporaso et al., 2018). Cui and Fearn (2018) compared PLSR and CNN models for predicting protein content using datasets of wheat kernels and wheat flour, respectively. The CNN model demonstrated superior performance with a lower noise level compared to the PLSR model. FTIR was employed to measure total protein and wet gluten in wheat flour, achieving an external validation $R^2 = 0.82$. The study concluded that NIR spectroscopy could effectively predict the gluten protein content of wheat flour (Liu et al., 2022). Furthermore, recent research used feature selection techniques to identify the most relevant wavelength ranges for detecting gluten in flour samples.

The best accuracy, 84.4%, was achieved by focusing on the 1089–1325 nm wavelength range and applying a random forest classifier (Jossa-Bastidas et al., 2022).

NIR is particularly effective in measuring protein and moisture content in wheat, which are critical factors for milling and baking quality. Dowell et al. (2006) demonstrated the ability of NIR data to predict wheat grain quality, flour quality, and milling yield. The study concluded that intrinsic properties derived from NIR data are valuable for assessing wheat quality. Today, NIR is widely used for wheat quality analysis due to its reliability, speed, and non-destructive nature (Caporaso et al., 2018). Shi and Yu (2017) compared the effectiveness of NIR and MIR spectroscopy in predicting thr moisture and protein contents of wheat using 10 different spectral preprocessing methods in combination with PLSR modelling. For moisture content, the NIR-based model outperformed the MIR-based model, achieving $R^2 = 0.86$; predictive residual deviation (RPD) = 2.64, compared to $R^2 = 0.72$, RPD = 1.87 for MIR. Similarly, for protein content, the NIR-based model showed superior prediction performance ($R^2 = 0.97$, RPD = 6.09) compared to the MIR-based model ($R^2 = 0.90$, RPD = 3.14). In a subsequent study, Shi et al. (2019) applied the same methodology to predict protein content in a different batch of wheat kernels. The findings aligned with those of Shi and Yu (2017), confirming that the NIR-based model ($R_p^2 = 0.98$, RPD = 8.00) outperformed the MIR-based model ($R_p^2 = 0.96$, RPD = 4.90) in terms of predictive ability. These results highlight the performance of NIR spectroscopy in wheat quality assessment.

Kernel hardness significantly influences the quality and yield of flour and the energy consumption of milling equipment. Erkinbaev et al. (2019) investigated the prediction of the hardness index in wheat kernel using PLSR and ANN models. The study revealed that the ANN model outperformed the PLSR model, achieving an R_p^2 of 0.90 and an RMSEP of 6.59. In contrast, the PLSR model achieved an R_p^2 of 0.80 and an RMSEP of 12.90. These results highlight the predictive performance of ANN models for assessing wheat kernel hardness.

In recent years, interest in resistant starch has increased significantly due to its resistance to breakdown in the small intestine, its role in promoting satiety, and its benefits in regulating intestinal metabolism, managing body weight, and lowering blood glucose levels. A recent study by Wang et al. (2021) demonstrated that a model based on MIR ($Rp^2 = 0.83$; RMSEP = 0.28%) outperformed the one based on NIR ($Rp^2 = 0.55$; RMSEP = 0.46%) for predicting resistant starch content in wheat. For confirming the MIR as an effective alternative method to predict the resistant starch content of wheat, the researchers identified two mutant wheats with high resistant starch content from a mutant wheat library using the MIR-calibrated model, finding no significant differences between chemically measured and predicted values. In another valuable study, Nieto-Ortega et al. (2022) analysed around 400 samples of various cereals, including wheat, barley, maize, rye, oats, and sorghum, collected over several years. They developed a global NIR calibration model using PLSR, and achieved high accuracy in predicting total starch content ($R^2 = 0.99$), monosaccharides ($R^2 > 0.97$), lignin ($R^2 = 0.94$), and cellulose ($R^2 = 0.70$) in these cereals.

Micronutrients in cereals like iron, calcium, manganese, magnesium, and zinc are vital for human health and play a crucial role in regulating metabolic activities. Accurately predicting these micronutrients rapidly and non-destructively can significantly reduce both time and costs. Recent studies indicate that HSI is a viable method for predicting micronutrient content in cereals. Hu et al. (2021) developed a partial least-squares regression (PLSR) model using the spectral reflectance of wheat kernels, achieving strong predictive performance for Zn, Mg, and Ca ($Rp^2 > 0.70$). The study also highlighted that grinding wheat kernels tends to dilute micronutrient concentrations, resulting in lower prediction accuracy for wheat flour compared to whole wheat kernels.

Neural networks, a branch of AI, utilize computer algorithms for tasks such as classification, pattern recognition, and parameter estimation. Habib and Farzad (2016) employed artificial neural networks with appropriate models to assess the impact of hydrocolloids on baguette staling. Another study applied NIR spectroscopy to bread, processing the spectral data using PCA and PLS for model development. The results were highly satisfactory, with R^2 values ranging from 0.90 to 0.96, demonstrating the effectiveness of NIR spectroscopy in bread analysis (Chakravartula et al., 2019). In a recent study, NIRS technology combined with ML models was used to evaluate the quality traits of sourdough bread flour made from six different flour sources. The ML models successfully classified the wheat type with an accuracy of 96.3% and precision of 99.4%, utilizing NIRS and a low-cost electronic nose (Gonzalez Viejo et al., 2022). Additionally, the moisture content in pasta was analysed using DRIFTS-NIR. Spectral data collected in the 4000–9000 cm^{-1} range yielded optimal results when processed with PLS (Czaja et al., 2018).

10.3 New Technologies for Moisture Control and Spoilage Prevention on Grain Storage

Cereal grains, oilseeds, and pulses, collectively known as grains, constitute a significant part of the daily diet for both humans and domesticated animals, while also carrying substantial economic importance for producers and the grain industry (Anukiruthika & Jayas, 2024). Developing cost-effective strategies for microbial decontamination of cereals is of significant industrial importance. Such approaches play a vital role in minimizing post-harvest crop losses and mitigating consumer health risks associated with microbial contamination. By reducing spoilage and ensuring food safety, these strategies enhance storage stability, product quality, and sustainability in cereal production and distribution systems. Monitoring stored grain is a critical aspect of the post-harvest phase in the food production chain. Preserving the characteristics of grains after harvest is crucial to ensure high-quality products for consumers (Ziegler et al., 2021). Grains are typically stored in large metal structures known as bins or silos. However, grains are susceptible to spoilage during storage, which can render them unfit for use. Consequently, effective monitoring of grain bins and silos is vital to identify and address conditions that may lead to spoilage. Unfortunately, many existing grains bin monitoring methods lack sufficient sensitivity, as they only detect spoilage once the conditions exceed a certain threshold (Asefi et al., 2015). The absence of proper grain storage structures and inadequate control and monitoring of the stored grain mass significantly contribute to quantitative and qualitative losses (Lutz & Coradi, 2022). The research has shown that, globally, up to 30% of grain produced annually is lost due to poor grain storage and management (Singh & Jayas, 2024).

Grain moisture content, temperature, and the relative humidity of intergranular air and the surrounding environment are crucial variables that must be monitored during storage to maintain grain quality (Vilas-Boas et al., 2023; Lutz & Coradi, 2022). Harvested grain can be stored anywhere from a few weeks to several years before being utilized. A key concern during this storage period is the potential for significant deterioration that renders the grain unacceptable. Insects and mould in grain bins can create localized high-temperature areas, known as hotspots, which accelerate the deterioration of the stored grain (Muir & White, 2000). For instance, insect activity can generate pockets of heat reaching up to 42 °C, while mould can produce mycotoxins that also lead to localized heating. Ensuring a consistent moisture content and temperature during storage, along with early hot-spot detection to allow for timely intervention before significant losses occur, is crucial. Various methods have been developed and implemented in silo management to monitor grain quality during the storage period and detect the potential contaminations (Asefi et al., 2015).

Technological advancements including smart sensors, the IoT, ML, AI, digital twins (DT), distributed ledger technology (DLT), and hyperspectral imaging have been proposed to monitor grain conditions and analyse collected data, thereby enhancing decision-making in grain silos (Vilas-Boas et al., 2023; Aviara et al., 2022; Hashim et al., 2022). The integration of sensors with IoT principles and ML algorithms offers advanced methods for monitoring and predicting grain quality during storage. These technologies enable real-time data collection and analysis, facilitating proactive decision-making to maintain optimal storage conditions and preserve grain quality.

10.3.1 Applications of IoT in Grain Storage

The IoT offers innovative solutions to enhance the efficiency and effectiveness of grain storage management. By integrating smart sensors and connected devices, the use of IoT technologies provides the monitoring and control of sensitive environmental variables, from remote sensing, and enhances grain yield. IoT enables real-time monitoring and control of critical parameters such as grain moisture content, temperature, and intergranular air humidity (Lutz & Coradi, 2022).

Radio frequency identification (RFID), near field communication (NFC), and DLT are enabling technologies for IoT. The integration of IoT technologies, such as RFID devices, facilitates the automatic identification of grain batches during storage, thereby minimizing input errors. The advancement of automatic identification technologies has enabled the differentiation of various wheat sources and the establishment of data relationships between processing and logistics units. Technologies like barcodes and RFID can be incorporated into traceability systems to enhance tracking and monitoring capabilities (Qian et al., 2012). The integration of IoT-based electronic identification technologies, including RFID and barcodes, into traceability systems has become a research hotspot in recent years, aiming to ensure real-time, safe, and reliable data tracking in the grain supply chain. Another alternative proposed for identification is NFC. When a batch of grains enters a storage silo, the first step is to identify it. This identification process is performed using optical, RFID, or NFC readers to recognize the tags applied to the wheat batches or their carriers. DLT refers to a computer-based system that utilizes ledger-based data structures to enhance trust, service availability, resilience, and security in digital systems, while also enabling distributed storage, computation, and control (Nurgazina et al., 2021). The environmental conditions within the warehouse, such as temperature and humidity, are monitored and recorded on the DLT, associating this data with the wheat batches until they leave the storage facility. The identification process must also be conducted at the point of exit to ensure traceability and quality control (Vilas-Boas et al., 2023).

The IoT technology provides several key applications:

- Real-time data monitoring: IoT sensors continuously track storage conditions, providing instant alerts for any deviations that could lead to spoilage or pest infestations.
- Predictive maintenance: By analysing data trends, IoT systems can predict potential risks, such as equipment malfunctions or environmental changes, allowing for timely intervention to prevent losses.
- Automation of storage processes: IoT can automate ventilation, aeration, and temperature regulation systems to optimize storage conditions and reduce manual labour.
- Improved traceability: Data collected through IoT devices can be logged and analysed, ensuring traceability of grain storage conditions throughout the supply chain.
- Energy efficiency: IoT-driven insights help minimize energy use in storage operations by precisely controlling equipment like fans and dryers based on actual needs.

IoT applications in grain storage improve quality preservation, reduce losses, and enhance overall operational efficiency.

10.3.2 Applications of ML in Grain Storage

ML techniques are recognized as predicting or data classification methods, offering a wide range of applications in the post-harvest sector. The growing volume of data created through communication between the objects has highlighted the significance of ML. As a branch of AI, ML enables systems to learn from data. It employs various algorithms that learn iteratively from data to make predictions, with training data used to guide this process. This approach allows for the development of increasingly accurate models based on the available data (Karimi et al., 2017; Lutz & Coradi, 2022).

ML offers transformative potential for optimizing grain storage by leveraging advanced data analysis and predictive capabilities. By processing large datasets generated from storage environments, ML algorithms can identify patterns, predict risks, and optimize management processes. Key applications of ML in grain storage include:

- Spoilage prediction: ML models analyse data from sensors monitoring moisture, temperature, and humidity to predict spoilage risks before they occur, enabling proactive interventions.
- Pest detection and control: Using image recognition and environmental data, ML algorithms can detect signs of pest infestations early and recommend appropriate control measures.
- Condition optimization: ML-driven systems can optimize storage conditions by learning from historical data to maintain ideal moisture, temperature, and ventilation levels for different grain types.
- Energy efficiency: ML can predict the most energy-efficient times for aeration and drying processes based on environmental conditions and grain state, reducing operational costs.
- Anomaly detection: ML algorithms can detect unusual patterns or outliers in sensor data, indicating potential equipment malfunctions or storage issues that require attention.
- Quality prediction: By analysing historical data on storage conditions, ML models can predict the quality and shelf life of stored grains, assisting in inventory management and marketing strategies.
- Integrated decision support: ML-powered systems provide actionable insights and recommendations, helping operators make informed decisions to maximize grain preservation and minimize losses.

ML enhances the precision, efficiency, and sustainability of grain storage management through these applications, reducing waste and improving the overall quality of stored grains.

10.4 New Technologies for Quality and Production Processes in the Wheat-Based Industry

10.4.1 Predictive Maintenance of Milling and Processing Equipment

In the era of Industry 4.0, AI is reshaping the manufacturing sector (Traini et al., 2019; Zonta et al., 2020). Industry 4.0 represents a new industrial model that defines the Fourth Industrial Revolution. This advanced manufacturing paradigm is characterized by intelligent, digital, and virtual operations within large-scale industries, marking a significant shift from the previous three industrial revolutions. It encompasses an integrated factory-wide structure and cutting-edge technologies applied across various industrial activities. These technologies form the foundation of Industry 4.0's design principles, which are essential for driving the innovative capabilities and performance of this new industrial era (Carvalho & Cazarini, 2020). With the integration of the IoT and ML techniques, manufacturing systems can now monitor physical processes and make intelligent decisions in real time by collaborating with humans, machines, sensors, and other devices. AI empowers manufacturers to minimize equipment downtime, detect production defects, optimize supply chains, and accelerate design processes through ML algorithms that adapt based on experience. A key recent advancement in this field is the development of predictive maintenance systems (Traini et al., 2019).

Maintenance has an important role in industrial production. It represents a meaningful percentage of the production costs (Bevilacqua & Braglia, 2000; Nunes et al., 2023). Effective maintenance strategies may help to reduce unexpected production stops, decrease the costs, and increase the lifetime of machines. The common strategy is corrective maintenance, also known as "run-to-failure". It consists of repairing or replacing a part only when it is impaired and the equipment has failed to operate without assistance (Nunes et al., 2023). Corrective maintenance turned to proactive methods because of the high expenses associated with unplanned production halts (Nunes et al., 2023). Predictive maintenance has become a crucial approach in the manufacturing industry, offering solutions to minimize downtime, optimize maintenance costs, and enhance operational efficiency (Lukito et al., 2025). Preventive maintenance involves periodical checks conducted by specialists and the replacement of components before the failure occurs (Nunes et al., 2023). This replacement occurs at equal intervals or after a predetermined number of operating cycles and is often provided by the equipment's manufacturer (Nunes et al., 2023). This could result in component replacement sooner or later. In the first scenario, replacing parts that are still in good condition and capable of many more operational cycles leads to increased maintenance costs. Conversely, the second scenario may have more serious consequences, as it necessitates corrective maintenance actions (Hao et al., 2010; Nunes et al., 2023).

In the last decade, research has been focused on fault detection and the rise of ML concepts with the technological advances in the field of Industry 4.0 and the development of the IoT (Nunes et al., 2023; Schwendemann et al., 2021). Minimizing the unscheduled downtime of machines due to the outages of machine constituents in automated production lines is very crucial (Schwendemann et al., 2021). Inspections previously conducted by technicians and experts have been replaced by automated systems utilizing sensors and devices that measure, monitor, and process signals representing various physical parameters of industrial equipment, including acoustic signals, current, voltage, temperature, force, and vibration (Mobley, 2002). This method allows interventions to be driven by sensor readings, with actions initiated whenever values exceed predefined limits (Nunes et al., 2023). These systems combine Industrial IoT with ML to accurately predict when manufacturing equipment will require maintenance, enabling proactive problem resolution and timely decision-making (Traini et al., 2019).

10.4.2 Digital Technologies in Quality Control

Quality control is a fundamental pillar of manufacturing, guaranteeing that products adhere to quality and reliability standards. Maintaining product quality and integrity is crucial in the fast-changing world of industrial manufacturing. However, conventional quality control methods, which largely depend on manual inspections and basic automated systems, are often inefficient, susceptible to human error, and inadequate for identifying intricate defects. While traditional quality control methods provide some effectiveness, they often fall short in meeting the growing demands for efficiency, precision, and adaptability required in modern production environments (Islam et al., 2024). The emergence of deep learning and computer vision, RFID, NFC devices, and sensor-based systems technologies has introduced transformative opportunities for automated defect detection, paving the way for a significant shift in how industries handle quality control and inspection processes (Cocco et al., 2021; Islam et al., 2024).

Computer vision systems (CVS) have become a powerful tool for analysing various appearance parameters such as shape, size, colour, and texture from digitized images. Compared to traditional destructive methods, CVS offers significant advantages, including higher speed, cost-efficiency, and flexibility. These benefits make CVS a reliable and non-invasive technique for various food analyses, such as classification and shelf life evaluation. The application of CVS extends to the quality assessment of diverse food products, including cereal-based items and bakery goods (Amani et al., 2022).

The widespread availability of smartphones equipped with advanced sensors, substantial processing capabilities, network connectivity, and user-friendly interfaces positions them as effective tools for non-invasive, portable food quality assessment. When combined with recent advancements in the IoT, deep learning algorithms, and cloud computing, smartphones offer the potential to develop accessible and sustainable analytical methods applicable at every stage of cereal-based food production and distribution (Kalinowska et al., 2021). For instance, smartphones have been utilized in food safety data collection for various purposes, including quality assessment, due to their portability and connectivity (Jin et al., 2020). Additionally, integrating IoT-based food quality systems with smartphones enables real-time monitoring of critical factors like temperature and humidity throughout the food supply chain, ensuring optimal conditions and facilitating informed decision-making. Furthermore, smartphone-based food diagnostic technologies have been developed, combining biosensors with IoT to issue immediate warnings about food contamination before purchase, enhancing consumer safety (Rateni et al., 2017).

Zearalenone (ZEN) is a mycotoxin produced by Fusarium fungi, known for its teratogenic effects and strong reproductive toxicity. It primarily contaminates cereals such as corn, rice, barley, wheat, and animal feed (Llorent-Martinez et al., 2019; Li et al., 2019). Li et al. (2019) developed a portable smartphone-based lateral flow analyser integrated with a solid-phase latex microsphere immunochromatography platform (SIAP) for the on-site detection of ZEN in cereals and feed. This innovative system enables sensitive and quantitative detection of ZEN, achieving results highly consistent with those obtained using both commercial kits and LC-MS/MS. The device utilizes a 3D-printed component equipped with lenses and a camera, coupled with a user-friendly Android application for data analysis and reporting. This advancement represents a significant step towards accessible and efficient monitoring of mycotoxins in food products, facilitating rapid on-site testing and enhancing food safety protocols. The study concluded that such smartphone-based detection systems have great potential for widespread on-site monitoring of various hazard factors, including veterinary drugs, antibiotics, and illegal additives (Li et al., 2019).

RFID, NFC devices, and sensor-based systems are anticipated to play a significant role in tracking goods throughout the supply chain and monitoring product quality. Integrating Blockchain technology with IoT devices significantly enhances supply chain integrity by improving scalability, security, immutability, auditing, information flows, traceability, and system interoperability. This integration enables real-time monitoring and secure data sharing across the supply chain, ensuring transparency and trust among stakeholders. The deployment of blockchain technology in combination with the IoT infrastructure can benefit modern supply chains by enhancing transparency and increasing business-to-business trust. By leveraging these technologies, organizations can achieve more efficient and secure supply chain operations, leading to improved product quality and customer satisfaction (Rejeb et al., 2019; Cocco et al., 2021).

10.4.3 Possible 3D/4D Printing Applications in the Wheat-Based Food Sector

The purpose of applying additive manufacturing technology to food is not simply industrial processing; it is needed to adapt to personalized consumption. 3D printing, also referred to as additive manufacturing, has garnered significant attention for its potential applications in the food industry, including the development of customized food products, enhancement of supply chain efficiency, personalization of nutrient profiles, and expansion of food accessibility (Kumar et al., 2024). In other words, the food industry must customize the appearance, colour, and nutrition according to the needs of different consumer groups (Teng et al., 2021). Recent advancements in 3D and 4D printing technologies have significantly impacted the food industry, offering innovative solutions for food design, production, and personalization. These technologies enable the creation of complex, customized food structures with precise control over texture, flavour, and nutritional content (Demei et al., 2022). For instance, 3D food printing allows for the fabrication of novel shapes and internal structures, facilitating personalized nutrition and dietary preferences (Chen et al., 2022; Derossi et al., 2024). Building upon this technology, 4D printing represents an advanced iteration wherein intelligent materials are printed to enable dynamic changes in shape, colour, nutritional content, and flavour in response to external stimuli such as pH, temperature, moisture absorption, and microwave heating (Singh et al., 2024). Several types of printing technologies have been investigated, including extrusion-based printers, binder jetting-based printers, selective laser sintering (SLS) printers, and inkjet printers. These technologies influence the accuracy and resolution of printed structures, as determined by machine parameters and material properties. Analytical methods such as computer numerical simulations, low-field nuclear magnetic resonance (LF-NMR), and morphological identification techniques have been employed to evaluate the structural integrity and characteristics of printed food products. The application of 3D printing has been extensively studied for producing food items tailored to specific contexts, such as space exploration, military operations, desserts, and nutrient-rich, texture-modified foods for elderly populations (Kumar et al., 2024).

3D printing technology has evolved from rapid moulding technology and may control complex processes using digital operations. It possesses several advantages such as high precision and speed, low cost, simplicity, and convenience in use. The rapid development of 3D printing technology in recent years is indicative of the arrival of the "Fourth Industrial Revolution". Applications of 3D printing technology have been effective in a variety of fields, including biology, medicine, electronics, mechanics, aerospace, and food. In the view of the food industry, applications of 3D printing technology range from spaceships to homes and restaurants. For instance, 3D printing technology enables astronauts to create personalized meals for extended space travel (Chen et al., 2022). Furthermore, 4D printing has introduced innovative possibilities, such as the creation of blooming edible flowers designed to captivate children by releasing colour and nutrients during consumption (Kumar et al., 2024). 4D food printing is an extension of 3D printing and it permits the customization of new and innovative food products. 4D printing technology has attracted attention in industrial sectors, although commercialization still needs additional research and development studies (Teng et al., 2021). Despite these advancements, the digestibility of printed foods may be reduced due to the formation of smooth, compact, and ordered molecular structures achieved during the printing process. Continued research is required to optimize these technologies, ensuring their scalability and functionality while addressing challenges related to the nutritional and sensory quality of printed food products (Kumar et al., 2024).

Wheat is one of the most extensively studied cereals in the field of 3D food printing. This prominence can be attributed to the wide range of foods and snacks produced from wheat and its derivatives, which form a significant portion of global diets. As a primary ingredient in numerous culinary applications, wheat offers versatile functional properties, making it an ideal candidate for exploring innovative manufacturing techniques like 3D food printing (Sundarsingh et al., 2024). Maniglia et al. (2020) stated that dry heating treatment is a relevant process to improve the properties of hydrogels based on wheat starch, making this ink suitable for 3D printing application. Habuš (2022) investigated the possibility of 3D printing of oat snack and breakfast cereals enriched with wheat bran, focusing on bran pre-processing while defining 3D printing conditions and a post-processing method. It is suggested that 3D printing technology could provide a diverse range of healthier snack products with bran addition (Habuš, 2022).

Zheng et al. (2021) investigated the effects of material composition on the quality of 3D-printed food using wheat flour, wheat starch, and whole meal as base materials. Wheat flour showed the poorest printability due to its tendency to adhere to the walls of the 3D printer, causing disruptions during the printing process. The results demonstrated that samples produced using wheat starch and whole meal exhibited superior print accuracy and overall quality compared to those made with wheat flour (Zheng et al., 2021). The possible reason is that the presence of additional components such as protein, fat, and fibre facilitated the development of various complexes in the whole meal. Whole meal, owing to its enriched compositional profile, outperforms wheat starch and wheat flour in achieving high-quality 3D-printed food structures.

3D printing has attracted growing interest due to its unique capability to fabricate geometrically complex designs. This technology holds promise not only for mass manufacturing applications but also for delivering significant environmental and economic advantages. By minimizing material waste and optimizing resource utilization, 3D printing contributes to sustainable production practices while enabling cost-effective customization in food manufacturing (Leontiou et al., 2023).

Another study explored the impact of material composition on 3D-printed food quality using wheat flour, freeze-dried mango powder, olive oil, and water. The findings suggest that the combination of these ingredients influences the rheological properties of the dough, which in turn affects the printability and structural characteristics of the printed food (Liu et al., 2019). Furthermore, the selection of flour and fat types has been shown to affect the flow properties and dynamic moduli of the dough, thereby influencing the 3D printing process and the quality of the final product (Lisovska & Harasym, 2023).

The development of biodegradable packaging from renewable resources is a crucial step towards addressing the environmental challenges posed by synthetic plastics. Materials derived from polysaccharides, anthocyanins, and essential oils have garnered significant attention for their potential in food packaging due to their safety, sustainability, and wide availability. These natural components not only provide a renewable alternative to conventional plastics but also offer functional properties such as improved barrier performance, antioxidant activity, and antimicrobial effects, enhancing the shelf life and quality of packaged foods (Leontiou et al., 2023). Advancements in three-dimensional (3D) printing have revolutionized intelligent food packaging by enabling the creation of multifunctional, self-indicating components. These components are fabricated using biocompatible, non-toxic substances, offering a cost-effective alternative to traditional manufacturing methods. This innovation enhances the functionality and sustainability of food packaging, contributing to the reduction of food waste and ensuring that consumers receive high-quality products (Tracey et al., 2022; Leontiou et al., 2023).

The integration of 3D and 4D printing technologies in the food sector holds promise for creating innovative products, improving food accessibility, and addressing specific dietary needs. Ongoing research and development in this area continues to explore the potential applications and benefits of these technologies in food science and technology (Srivastava et al., 2023).

10.5 Conclusions

The integration of digital tools and AI is revolutionizing the wheat industry by enhancing quality analysis and production processes. Technologies such as NIR spectroscopy, image analysis, machine vision, and ML models are enabling faster, more accurate, and scalable assessments of wheat quality, surpassing traditional methods. Innovations in 3D and 4D printing technologies offer the potential to produce customized wheat-based food products tailored to specific dietary needs and preferences. These advancements enable the creation of personalized nutrition solutions, aligning with individual health requirements.

The integration of predictive maintenance within the wheat industry's digital transformation underscores the critical role of AI and IoT in modernizing production facilities. As these technologies advance, they will continue to drive improvements in efficiency, product quality, and sustainability across the sector.

As digital and AI technologies continue to evolve, their integration into the wheat industry holds immense potential for further improvements. These tools are poised to streamline quality control in storage, milling, and baking, enable real-time monitoring of wheat quality, optimize production processes, and reduce waste. The ongoing development and adoption of these technologies promise a more efficient, sustainable, and high-quality future for the wheat industry.

References

Abioye, S. O., Oyedele, L. O., Akanbi, L., et al. (2021). Artificial intelligence in the construction industry: A review of present status, opportunities and future challenges. *Journal of Building Engineering, 44*, 103299. https://doi.org/10.1016/j.jobe.2021.103299

Acevedo, M., Zurn, J. D., Molero, G., et al. (2018). The role of wheat in global food security. In U. S. Nagothu (Ed.), *Agricultural development and sustainable intensification* (1st ed., pp. 81–110). Taylor & Francis Group, Routledge Imprint.

Agarwal, D., & Bachan, P. (2023). Machine learning approach for the classification of wheat grains. *Smart Agricultural Technology, 3*, 100136.

Amani, H., Badak-Kerti, K., & Mousavi Khaneghah, A. (2022). Current progress in the utilization of smartphone-based imaging for quality assessment of food products: A review. *Critical Reviews in Food Science and Nutrition, 62*(13), 3631–3643.

An, D., Zhang, L., Liu, Z., et al. (2023). Advances in infrared spectroscopy and hyperspectral imaging combined with artificial intelligence for the detection of cereals quality. *Critical Reviews in Food Science and Nutrition, 63*(29), 9766–9796.

Anukiruthika, T., & Jayas, D. S. (2024). Mathematical modeling for management of stored-grain ecosystems: Approaches, opportunities, and research needs. *Journal of Stored Products Research, 106*, 102304.

Asefi, M., Jeffrey, I., LoVetri, J., et al. (2015). Grain bin monitoring via electromagnetic imaging. *Computers and Electronics in Agriculture, 119*, 133–141.

Aviara, N. A., Liberty, J. T., Olatunbosun, O. S., et al. (2022). Potential application of hyperspectral imaging in food grain quality inspection, evaluation, and control during bulk storage. *Journal of Agriculture and Food Research, 8*, 100288.

Ayres, L. B., Gomez, F. J. V., Linton, J. R., et al. (2021). Taking the leap between analytical chemistry and artificial intelligence: A tutorial review. *Analytica Chimica Acta, 1161*, 338403. https://doi.org/10.1016/j.aca.2021.338403

Barbedo, J. G. A., Guarienti, E. M., & Tibola, C. S. (2018). Detection of sprout damage in wheat kernels using NIR hyperspectral imaging. *Biosystems Engineering, 175*, 124–132. https://doi.org/10.1016/j.biosystemseng.2018.09.012

Basati, Z., Rasekh, M., & Abbaspour-Gilandeh, Y. (2018). Using different classification models in wheat grading utilizing visual features. *IntAgrophys, 32*, 225–235.

Ben Ayed, R., & Hanana, M. (2021). Artificial intelligence to improve the food and agriculture sector. *Journal of Food Quality, 2021*(1), 5584754.

Bevilacqua, M., & Braglia, M. (2000). Analytic hierarchy process applied to maintenance strategy selection. *Reliability Engineering and System Safety, 70*(1), 71–83. https://doi.org/10.1016/S0951-8320(00)00047-8

Boniecki, P., Piekarska-Boniecka, H., Swierczy, K., et al. (2014). Detection of the granary weevil based on X-ray images of damaged wheat kernels. *Journal of Stored Products Research, 56*, 38–42.

Caporaso, N., Whitworth, M. B., & Fisk, I. D. (2018). Near-infrared spectroscopy and hyperspectral imaging for non-destructive quality assessment of cereal grains. *Applied Spectroscopy Reviews, 53*, 667–687. https://doi.org/10.1080/05704928.2018.1425214

Carvalho, N. G. P., & Cazarini, E. W. (2020). Industry 4.0—What is it? In *Industry 4.0—Current status and future trends*. IntechOpen..

Chakravartula, S. S. N., Cevoli, C., Balestra, F., et al. (2019). Evaluation of drying of edible coating on bread using NIR spectroscopy. *Journal of Food Engineering, 240*, 29–37.

Chen, J., Zhang, M., Xu, B., et al. (2020). Artificial intelligence assisted technologies for controlling the drying of fruits and vegetables using physical fields: A review. *Trends in Food Science and Technology, 105*, 251–260. https://doi.org/10.1016/j.tifs.2020.08.015

Chen, X., Zhang, M., Teng, X., et al. (2022). Recent progress in modeling 3D/4D printing of foods. *Food Engineering Reviews, 14*, 1–14.

Chu, X., Wang, W., Yoon, S.-C., et al. (2017). Detection of aflatoxin B1 (AFB1) in individual maize kernels using short wave infrared (SWIR) hyperspectral imaging. *Biosystems Engineering, 157*, 13–23.

Ciudad-Mulero, M., Barros, L., Fernandes, A., et al. (2020). Potential health claims of durum and bread wheat flours as functional ingredients. *Nutrients, 12*(2), 504.

Cocco, L., Mannaro, K., Tonelli, R., et al. (2021). A blockchain-based traceability system in agri-food SME: Case study of a traditional bakery. *IEEE Access, 9*, 62899–62915.

Cui, C., & Fearn, T. (2018). Modern practical convolutional neural networks for multivariate regression: Applications to NIR calibration. *Chemometrics and Intelligent Laboratory Systems, 182*, 9–20. https://doi.org/10.1016/j.chemolab.2018.07.008

Czaja, T., Kuzawi'nska, E., Sobota, A., et al. (2018). Determining moisture content in pasta by vibrational spectroscopy. *Talanta, 178*, 294–298.

De Girolamo, A., von Holst, C., Cortese, M., et al. (2019). Rapid screening of ochratoxin a in wheat by infrared spectroscopy. *Food Chemistry, 282*, 95–100. https://doi.org/10.1016/j.foodchem.2019.01.008

De Girolamo, A., Arroyo, M. C., Cervellieri, S., et al. (2020). Detection of durum wheat pasta adulteration with common wheat by infrared spectroscopy and chemometrics: A case study. *LWT, 127*, 109368. https://doi.org/10.1016/j.lwt.2020.109368

Demei, K., Zhang, M., Phuhongsung, P., et al. (2022). 3D food printing: Controlling characteristics and improving technological effect during food processing. *Food Research International, 156*, 111120.

Derossi, A., Spence, C., Corradini, M. G., et al. (2024). Personalized, digitally designed 3D printed food towards the reshaping of food manufacturing and consumption. *NPJ Science of Food, 8*(1), 54.

Dowell, F. E., Maghirang, E. B., Xie, F., et al. (2006). Predicting wheat quality characteristics and functionality using near-infrared spectroscopy. *Cereal Chemistry, 83*, 529–536. https://doi.org/10.1094/CC-83-0529

Erkinbaev, C., Derksen, K., & Paliwal, J. (2019). Single kernel wheat hardness estimation using near infrared hyperspectral imaging. *Infrared Physics & Technology, 98*, 250–255. https://doi.org/10.1016/j.infrared.2019.03.033

Fazel-Niari, Z., Afkari-Sayyah, A. H., Abbaspour-Gilandeh, Y., et al. (2022). Quality assessment of components of wheat seed using different classifications models. *Applied Sciences, 12*(9), 4133. https://doi.org/10.3390/app12094133

Femenias, A., Gatius, F., Ramos, A. J., et al. (2020). Use of hyperspectral imaging as a tool for *Fusarium* and deoxynivalenol risk management in cereals: A review. *Food Control, 108*, 106819. https://doi.org/10.1016/j.foodcont.2019.106819

Gonzalez Viejo, C., Harris, N. M., & Fuentes, S. (2022). Quality traits of sourdough bread obtained by novel digital technologies and machine learning modelling. *Fermentation, 8*, 516.

Gulzar, Y., Hamid, Y., & Soomro, A. B. (2020). A convolution neural network-based seed classification system. *Symmetry, 12*(12), 2018. https://doi.org/10.3390/sym12122018

Habib, V., & Farzad, A. (2016). Modelling the effect of adding hydrocolloids on baguette bread staling using neural networks. *Acta Medica, 32*, 1325.

Habuš, M. (2022). *Development of 3D-printed cereal-based products enriched with pre-processed wheat bran*. Doctoral dissertation, University of Zagreb.

Hao, Q., Xue, Y., & Shen, W., et al. (2010). A decision support system for integrating corrective maintenance, preventive maintenance, and condition-based maintenance. In: *Proceedings of the Constr Res Congr ASCE, Banff, Alberta, 08 May 2010.*

Hashim, N., Onwude, D. I., & Maringgal, B. (2022). Technological advances in postharvest management of food grains. *Research and Technological Advances in Food Science*, 371–406.

Hu, N., Li, W., Du, C., et al. (2021). Predicting micronutrients of wheat using hyperspectral imaging. *Food Chemistry, 343*, 128473. https://doi.org/10.1016/j.foodchem.2020.128473

Islam, M. R., Zamil, M. Z. H., Rayed, M. E., et al. (2024). *Deep learning and computer vision techniques for enhanced quality control in manufacturing processes*. IEEE Access.

Jin, C., Bouzembrak, Y., Zhou, J., et al. (2020). Big data in food safety—A review. *Current Opinion in Food Science, 36*, 24–32.

Jossa-Bastidas, O., Sainz Lugarezaresti, U., Osa Sanchez, A., et al. (2022). Gluten analysis composition using NIR spectroscopy and artificial intelligence techniques. *Telematique, 21*, 7487–7496.

Kalinowska, K., Wojnowski, W., & Tobiszewski, M. (2021). Smartphones as tools for equitable food quality assessment. *Trends in Food Science and Technology, 111*, 271–279.

Karimi, N., Ramin, R., & Tohid, A. (2017). An intelligent system for quality measurement of Golden bleached raisins using two comparative machine learning algorithms. *Measurement, 107*, 68–76. https://doi.org/10.1016/j.measurement.2017.05.009

Khatri, A., Agrawal, S., & Chatterjee, J. M. (2022). Wheat seed classification: Utilizing ensemble machine learning approach. *Scientific Programming, 626868*, 1. https://doi.org/10.1155/2022/2626868

Kimuli, D., Wang, W., Lawrence, K. C., et al. (2018). Utilisation of visible/near-infrared hyperspectral images to classify aflatoxin B1 contaminated maize kernels. *Biosystems Engineering, 166*, 150–160. https://doi.org/10.1016/j.biosystemseng.2017.11.018

Kumar, K., & Thakur, G. S. M. (2012). Advanced applications of neural networks and artificial intelligence: A review. *International Journal of Information Technology and Computer Science, 4*, 57–68.

Kumar, S., Karunanithi, S., Rout, S., et al. (2024). 3D and 4D structure food printing. In *Structured foods* (pp. 64–91). CRC Press.

Leontiou, A., Georgopoulos, S., Karabagias, V. K., et al. (2023). Three-dimensional printing applications in food industry. *Nano, 3*(1), 91–112.

Li, X., Wang, J., Yi, C., et al. (2019). A smartphone-based quantitative detection device integrated with latex microsphere immunochromatography for on-site detection of zearalenone in cereals and feed. *Sensors and Actuators B: Chemical, 290*, 170–179.

Liang, K., Huang, J., He, R., et al. (2020). Comparison of Vis-NIR and SWIR hyperspectral imaging for the non-destructive detection of DON levels in *Fusarium* head blight wheat kernels and wheat flour. *Infrared Physics & Technology, 106*, 103281. https://doi.org/10.1016/j.infrared.2020.103281

Lisovska, T., & Harasym, J. (2023). 3D printing progress in gluten-free food – Clustering analysis of advantages and obstacles. *Applied Science, 13*(22), 12362.

Liu, Y., Liang, X., Saeed, A., et al. (2019). Properties of 3D printed dough and optimization of printing parameters. *Innovative Food Science and Emerging Technologies, 54*, 9–18.

Liu, Q., Zhang, W., Zhang, B., et al. (2022). Determination of total protein and wet gluten in wheat flour by fourier transform infrared photoacoustic spectroscopy with multivariate analysis. *Journal of Food Composition and Analysis, 106*, 104349.

Llorent-Martinez, E. J., Fernandez-Poyatos, M. P., & Ruiz-Medina, A. (2019). Automated fluorimetric sensor for the determination of zearalenone mycotoxin in maize and cereals feedstuff. *Talanta, 191*, 89–93.

Lukito, T., Herlianti, R., Mayanti, M., et al. (2025). Implementation of predictive maintenance in various industry: A review. *TEKNOSAINS: Jurnal Sains, Teknologi dan Informatika, 12*(1), 133–144.

Lutz, É., & Coradi, P. C. (2022). Applications of new technologies for monitoring and predicting grains quality stored: Sensors, internet of things, and artificial intelligence. *Measurement, 188*, 110609.

Maniglia, B. C., Lima, D. C., da Matta, J. M., et al. (2020). Dry heating treatment: A potential tool to improve the wheat starch properties for 3D food printing application. *Food Research International, 137*, 109731.

Mobley, R. K. (2002). *An introduction to predictive maintenance*. Elsevier. https://doi.org/10.1016/B978-0-7506-7531-4.X5000-3

Muir, W. E., & White, N. (2000). Microorganisms in stored grain. In W. E. Muir (Ed.), *Manitoba: Grain preservation biosystems* (pp. 1–17). University of Manitoba.

Nieto-Ortega, B., Arroyo, J. J., Walk, C., et al. (2022). Near infrared reflectance spectroscopy as a tool to predict non-starch polysaccharide composition and starch digestibility profiles in common monogastric cereal feed ingredients. *Animal Feed Science and Technology, 285*, 115214. https://doi.org/10.1016/j.anifeedsci.2022.115214

Nunes, P., Santos, J., & Rocha, E. (2023). Challenges in predictive maintenance: A review. *CIRP Journal of Manufacturing Science and Technology, 40*, 53–67.

Nurgazina, J., Pakdeetrakulwong, U., Moser, T., et al. (2021). Distributed ledger technology applications in food supply chains: A review of challenges and future research directions. *Sustainability, 13*(8), 4206.

Okeke, A. G. (2020). *Fourier transform infrared spectroscopy (as a rapid method) coupled with machine learning approaches for detection and quantification of gluten contaminations in grain-based foods*. Master thesis, University of Kentucky.

Olgun, M., Onarcan, A. O., Özkan, K., et al. (2016). Wheat grain classification by using dense SIFT features with SVM classifier. *Computers and Electronics in Agriculture, 122*, 185–190. https://doi.org/10.1016/j.compag.2016.01.033

Pantazi, X.-E., Moshou, D., & Bravo, C. (2016). Active learning system for weed species recognition based on hyperspectral sensing. *Biosystems Engineering, 146*, 193–202.

Parrenin, L., Danjou, C., Agard, B., et al. (2023). Future trends in organic flour milling: The role of AI. *AIMS Agriculture and Food, 8*(1), 48–77.

Patrício, D. I., & Rieder, R. (2018). Computer vision and artificial intelligence in precision agriculture for grain crops: A systematic review. *Computers and Electronics in Agriculture, 153*, 69–81.

Pradana-López, S., Pérez-Calabuig, A. M., Cancilla, J. C., et al. (2022a). Single-digit ppm quantification of melamine in powdered milk driven by computer vision. *Food Control, 131*, 108424. https://doi.org/10.1016/j.foodcont.2021.108424

Pradana-López, S., Pérez-Calabuig, A. M., Otero, L., et al. (2022b). Is my food safe?-AI-based classification of lentil flour samples with trace levels of gluten or nuts. *Food Chemistry, 386*, 132832.

Qian, J. P., Yang, X. T., Wu, X. M., et al. (2012). A traceability system incorporating 2D barcode and RFID technology for wheat flour mills. *Computers and Electronics in Agriculture, 89*, 76–85.

Rateni, G., Dario, P., & Cavallo, F. (2017). Smartphone-based food diagnostic technologies: A review. *Sensors, 17*(6), 1453. https://doi.org/10.3390/s17061453

Ravikanth, L., Chelladurai, V., Jayas, D. S., et al. (2016). Detection of broken kernels content in bulk wheat samples using near-infrared hyperspectral imaging. *Agricultural Research, 5*(3), 285–292. https://doi.org/10.1007/s40003-016-0227-5

Rejeb, A., Keogh, J. G., & Treiblmaier, H. (2019). Leveraging the internet of things and blockchain technology in supply chain management. *Future Internet, 11*(7), 161.

Sabanci, K., Kayabasi, A., & Toktas, A. (2017). Computer vision-based method for classification of wheat grains using artificial neural network. *Journal of the Science of Food and Agriculture, 97*, 2588–2593.

Sabanci, K., Aslan, M. F., Ropelewska, E., et al. (2022). A novel convolutional-recurrent hybrid network for Sunn pest-damaged wheat grain detection. *Food Analytical Methods, 15*(6), 1748–1760.

Schwendemann, S., Amjad, Z., & Sikora, A. (2021). A survey of machine-learning techniques for condition monitoring and predictive maintenance of bearings in grinding machines. *Computers in Industry, 125*, 103380.

Sendin, K., Williams, P. J., & Manley, M. (2018). Near infrared hyperspectral imaging in quality and safety evaluation of cereals. *Critical Reviews in Food Science and Nutrition, 58*(4), 575–590. https://doi.org/10.1080/10408398.2016.1205548

Shahin, M. A., & Symons, S. J. (2011). Detection of *Fusarium* damaged kernels in Canada Western Red Spring wheat using visible/near-infrared hyperspectral imaging and principal component analysis. *Computers and Electronics in Agriculture, 75*, 107–112.

Sharma, L., Rahman, F., & Sharma, R. A. (2024). The emerging role of biotechnological advances and artificial intelligence in tackling gluten sensitivity. *Critical Reviews in Food Science and Nutrition*, 1–17.

Shen, G., Cao, Y., Yin, X., et al. (2022). Rapid and nondestructive quantification of deoxynivalenol in individual wheat kernels using near-infrared hyperspectral imaging and chemometrics. *Food Control, 131*, 108420. https://doi.org/10.1016/j.foodcont.2021.108420

Shi, H., & Yu, P. (2017). Comparison of grating-based near-infrared (NIR) and Fourier transform midinfrared (ATR-FT/MIR) spectroscopy based on spectral preprocessing and wavelength selection for the determination of crude protein and moisture content in wheat. *Food Control, 82*, 57–65. https://doi.org/10.1016/j.foodcont.2017.06.015

Shi, H., Lei, Y., Prates, L. L., et al. (2019). Evaluation of near-infrared (NIR) and Fourier transform mid-infrared (ATR-FT/MIR) spectroscopy techniques combined with chemometrics for the determination of crude protein and intestinal protein digestibility of wheat. *Food Chemistry, 272*, 507–513. https://doi.org/10.1016/j.foodchem.2018.08

Shrestha, B. L., Kang, Y.-M., Yu, D., et al. (2016). A two-camera machine vision approach to separating and identifying laboratory sprouted wheat kernels. *Biosystems Engineering, 147*, 265–273. https://doi.org/10.1016/j.biosystemseng.2016.04.008

Singh, C. B., & Jayas, D. S. (2024). Grain storage and management. In *Unit operations in food grain processing* (pp. 161–174). Academic.

Singh, C. B., Jayas, D. S., Paliwal, J., et al. (2009). Detection of insect-damaged wheat kernels using near-infrared hyperspectral imaging. *Journal of Stored Products Research, 45*, 151–158.

Singh, C. B., Jayas, D. S., Paliwal, J., et al. (2010). Identification of insect-damaged wheat kernels using short-wave near-infrared hyperspectral and digital colour imaging. *Computers and Electronics in Agriculture, 73*, 118–125.

Singh, P., Pandey, V. K., Tripathi, A., et al. (2024). 4D food printing technology: Structural changes to culinary art and beyond. *Journal of Food Process Engineering, 47*(1), e14535.

Spoladore, S. F., Scholz, M. D. S., & Bona, E. (2021). Genotypic classification of wheat using near-infrared spectroscopy and PLS-DA. *Applied Food Research, 1*(2), 100019. https://doi.org/10.1016/j.afres.2021.100019

Srivastava, S., Pandey, V. K., Singh, R., et al. (2023). Recent insights on advancements and substantial transformations in food printing technology from 3 to 7D. *Food Science and Biotechnology, 32*(13), 1783–1804.

Sundarsingh, A., Zhang, M., Mujumdar, A. S., et al. (2024). Research progress in printing formulation for 3D printing of healthy future foods. *Food and Bioprocess Technology, 17*(11), 3408–3439.

Taneja, A., Nair, G., Joshi, M., et al. (2023). Artificial intelligence: Implications for the agri-food sector. *Agronomy, 13*(5), 1397.

Teng, X., Zhang, M., & Mujumdar, A. S. (2021). 4D printing: Recent advances and proposals in the food sector. *Trends in Food Science and Technology, 110*, 349–363.

Tracey, C. T., Predeina, A. L., Krivoshapkina, E. F., et al. (2022). A 3D printing approach to intelligent food packaging. *Trends in Food Science and Technology, 127*, 87–98.

Traini, E., Bruno, G., D'Antonio, G., et al. (2019). Machine learning framework for predictive maintenance in milling. *IFAC-PapersOnLine, 52*(13), 177–182. https://doi.org/10.1016/j.ifacol.2019.11.172

Turhal, Ü. Ç., & Turhal, K. (2014). Determininination of effects of sunn pest on wheat grain by artificial neural networks. *Trakya University Journal of Natural Sciences, 15*, 25–30.

Unlersen, M. F., Sonmez, M. E., Aslan, M. F., et al. (2022). CNN–SVM hybrid model for varietal classification of wheat based on bulk samples. *European Food Research and Technology, 248*, 2043–2052. https://doi.org/10.1007/s00217-022-04029-4

Vermeulen, P., Suman, M., Fernández Pierna, J. A., et al. (2018). Discrimination between durum and common wheat kernels using near infrared hyperspectral imaging. *Journal of Cereal Science, 84*, 74–82. https://doi.org/10.1016/j.jcs.2018.10.001

Vilas-Boas, J. L., Rodrigues, J. J., & Alberti, A. M. (2023). Convergence of distributed ledger technologies with digital twins, IoT, and AI for fresh food logistics: Challenges and opportunities. *Journal of Industrial Information Integration, 31*, 100393.

Vithu, P., & Moses, J. A. (2016). Machine vision system for food grain quality evaluation: A review. *Trends in Food Science and Technology, 56*, 13–20.

Wang, R., Wei, X., Wang, H., et al. (2021). Development of attenuated total reflectance mid-infrared (ATR-MIR) and near-infrared (NIR) spectroscopy for the determination of resistant starch content in wheat grains. *Journal of Analytical Methods in Chemistry*, 5599388. https://doi.org/10.1155/2021/5599388

Weng, S., Han, K., Chu, Z., et al. (2021). Reflectance images of effective wavelengths from hyperspectral imaging for identification of Fusarium head blight-infected wheat kernels combined with a residual attention convolution neural network. *Computers and Electronics in Agriculture, 190*, 106483. https://doi.org/10.1016/j.compag.2021.106483

Xin, H., Zhang, X., & Yu, P. (2013). Using synchrotron radiation-based infrared microspectroscopy to reveal microchemical structure characterization: Frost damaged wheat vs. normal wheat. *International Journal of Molecular Sciences, 14*(8), 16706–16718. https://doi.org/10.3390/ijms140816706

Yasar, A. (2024). Analysis of selected deep features with CNN-SVM-based for bread wheat seed classification. *European Food Research and Technology, 250*(6), 1551–1561.

Yu, Y., Yu, H., Guo, L., et al. (2018). Accuracy and stability improvement in detecting Wuchang rice adulteration by piece-wise multiplicative scatter correction in the hyperspectral imaging system. *Analytical Methods, 10*(26), 3224–3231. https://doi.org/10.1039/C8AY00701B

Zareef, M., Arslan, M., Hassan, M. M., et al. (2021). Recent advances in assessing qualitative and quantitative aspects of cereals using nondestructive techniques: A review. *Trends in Food Science and Technology, 116*, 815–828.

Zhang, Y., & Guo, W. (2020). Moisture content detection of maize seed based on visible/near-infrared and near-infrared hyperspectral imaging technology. *International Journal of Food Science and Technology, 55*(2), 631–640. https://doi.org/10.1111/ijfs.14317

Zhang, L., Sun, H., Rao, Z., et al. (2020). Non-destructive identification of slightly sprouted wheat kernels using hyperspectral data on both sides of wheat kernels. *Biosystems Engineering, 200*, 188–199. https://doi.org/10.1016/j.biosystemseng.2020.10.004

Zheng, L., Liu, J., Liu, R., et al. (2021). 3D printing performance of gels from wheat starch, flour, and whole meal. *Food Chemistry, 356*, 129546.

Zheng, L., Bao, Q., Weng, S., et al. (2022). Determination of adulteration in wheat flour using multi-grained cascade forest-related models coupled with the fusion information of hyperspectral imaging. *Spectrochimica Acta. Part A, Molecular and Biomolecular Spectroscopy, 270*, 120813. https://doi.org/10.1016/j.saa.2021.120813

Zhu, J., Li, H., Rao, Z., et al. (2023). Identification of slightly sprouted wheat kernels using hyperspectral imaging technology and different deep convolutional neural networks. *Food Control, 143*, 109291. https://doi.org/10.1016/j.foodcont.2022.109291

Ziegler, V., Paraginski, R. T., & Ferreira, C. D. (2021). Grain storage systems and effects of moisture, temperature and time on grain quality—A review. *Journal of Stored Products Research, 91*, 101770.

Zonta, T., Da Costa, C. A., da Rosa, R. R., et al. (2020). Predictive maintenance in the industry 4.0: A systematic literature review. *Computers and Industrial Engineering, 150*, 106889.

Chapter 11
Infrared Sensing for Wheat Biochemical Analysis: Methods and Protocols

Ceren Orak, Işıl İlter Baysan, and Ulaş Baysan

Abstract In this chapter, the infrared (IR) sensing methods used for the biochemical analysis of wheat were introduced. In this context, commonly used IR sensing methods are near-infrared spectroscopy (NIRS), mid-infrared spectroscopy, Fourier transform infrared (FTIR) spectroscopy, attenuated total reflectance (ATR) spectroscopy, as well as chemometric techniques. Their working principles and intended uses were explained in detail. In addition, their application protocols were discussed, and some examples from the literature were provided, since proper calibration and validation are crucial to ensure the accuracy and reliability of analysis results. The main parts of this chapter are introduction, including wheat, wheat analyses, biochemical analyses, spectroscopic methods and types; basis of infrared spectroscopy [near-infrared spectroscopy (NIRS); Fourier transform infrared (FTIR) spectroscopy; attenuated total reflectance (ATR) spectroscopy]; and use of these methods in the biochemical analysis of wheat.

Keywords Wheat · NIRS · FTIR · ATR

Abbreviations

ANN	artificial neural network
ATR	attenuated total reflectance
CNN	convolutional neural network

C. Orak (✉)
Sivas University of Science and Technology, Department of Chemical Engineering, Sivas, Türkiye
e-mail: cerenorak@sivas.edu.tr

I. İ. Baysan
Manisa Celal Bayar University, Department of Food Engineering, Manisa, Türkiye

U. Baysan
Dokuz Eylül University, Science and Technology Application and Research Center, İzmir, Türkiye

N. Zencirci et al. (eds.), *Empowering Wheat Cultivation with GIS, Digital Approaches and Artificial Intelligence*, https://doi.org/10.1007/978-3-031-99954-3_11

DBN	deep belief network
FAO	Food and Agriculture Organization
FTIR	Fourier transform infrared spectroscopy
GC	gas chromatography
HPLC	high performance liquid chromatography
i-PLS	interval partial least square
IR	infrared
LC-MS	liquid chromatography-mass spectrometry
MIR	mid-infrared
MSC	multiplicative scatter correction
NIR	near-infrared
NIRS	near-infrared spectroscopy
NMR	nuclear magnetic resonance
PCA	principal component analysis
PLSR	partial least-squares regression
PMSC	piece-wise multiplicative scatter correction
RBFNN	radial basis function neural network
SDS-PAGE	sodium dodecyl sulfate polyacrylamide gel electrophoresis
THz	terahertz
VCPA	variable combination population analysis

11.1 Introduction

According to the Food and Agriculture Organization (FAO) 2022 data, annual global cereal production amounts to approximately 2.5 billion tons, with wheat comprising a substantial portion, around 778 million tons, and recognized as a strategic food commodity (Acevedo et al., 2018). Production levels fluctuate from year to year due to factors including increasing regions, climate shifts, technical advancements, and the like. Thanks to technical advancements, wheat production—which was roughly 750 million tons in 2019—is predicted to rise to an estimated 790 million tons in 2024, despite the effects of unfavorable factors like climate change. Accordingly, the five nations that produce the most wheat globally are France, China, India, Russia, and the United States (Acevedo et al., 2018; Urueña, 2022). Among the grains with the highest yield, wheat has the largest cultivated area due to its growing characteristics. Wheat is unique in this sense since it can readily adapt to a variety of climatic and environmental conditions. Its remarkable adaptability to diverse climates and environmental circumstances across the globe, coupled with its crucial position as a main raw ingredient for the manufacturing of numerous bakery goods, particularly bread, accounts for its unique characteristics (FAO, 2018).

Wheat is used in many different contexts, such as animal feed and regular human diets. Due to decreasing growing areas, caused by rapid population expansion and growing challenges associated with guaranteeing food security in diverse locations, efforts to enhance production are accelerating in many countries (Gandhi & Zhou, 2014). As agricultural areas are reduced and eventually depleted, production per unit area must increase, and product quality must be improved (Max & Hannah, 2023). To avoid future food shortages, population growth and total output must continue to expand in parallel. The characteristics that make wheat so vital include its ease of cultivation, high nutritional value, affordability compared to animal-derived foods, long-term storability, ease of transport, and widespread availability. The predominant consumption of wheat in the form of bread has spurred research focusing particularly on the components influencing the baking quality of wheat flour (Shewry & Hey, 2015).

Throughout history to the present day, humanity has benefited from wheat and its products across various sectors. Wheat grains are used in the production of flour, bulgur, pasta, and starch, while wheat stems find applications in paper, cardboard industries, and animal feed. Consequently, any decrease in wheat production, whether globally or within specific countries, directly impacts consumers by potentially increasing the prices of bread and food items made from wheat flour. Countries reliant on wheat-based diets, and possessing sufficient wheat stocks, deem wheat production strategically significant (Finnie & William, 2016). Wheat cultivation, adapted to vast agricultural areas, has played a significant role in global nutrition for years. Providing over half of the daily caloric intake and approximately half of the required protein, wheat feeds one-third of the world's population (Finnie & William, 2016). Despite agricultural production showing minimal changes compared to the rapidly increasing population, the finite global agricultural land area continues to diminish over the years. To prevent future food shortages, countries must implement and adhere to long-term agricultural plans.

In our daily lives, our bodies encounter numerous harmful agents (such as smoking, alcohol, environmental pollution, and stress), triggering reactions that can lead to irreversible degenerative conditions like cancer. Wheat and its products not only hold significance in human nutrition but also contribute to mitigating and preventing various life-threatening diseases due to beneficial substances that support human health and protection (Ryan et al., 2011). Recent studies emphasize the role of phenolic compounds and antioxidants in reducing the risk of such diseases. Wheat is a natural source of phytochemicals (phenolic compounds, carotenoids, and vitamin E), which exhibit potent antioxidant activities, preventing or delaying the progression of many chronic illnesses in humans and animals (Luthria et al., 2015). Overall, wheat quality in bread-making wheat is defined by the suitability of its physical, chemical, physicochemical, and rheological properties in achieving optimal flour yield (Uthayakumaran & Wrigley, 2017). Particularly, the quantity and quality of protein in wheat are crucial indicators of wheat quality, influenced significantly by wheat variety and growing environment. Factors such as precipitation patterns, nitrogen availability in soil, temperature, and light intensity affect protein content. Sedimentation value, predominantly influenced by genetics, is another criterion determining protein quality (Grausgruber et al., 2000).

Wheat (Triticum spp.) is a cereal plant of great importance worldwide and belongs to the Poaceae family (Shewry & Hey, 2015). In the classification of wheat varieties, factors such as genetic structure, shape characteristics, and agricultural yield play a decisive role (Feldman & Levy, 2009). According to their genetic structure, wheats are divided into three main categories: 42-chromosome hexaploids, 28-chromosome tetraploids, and 14-chromosome diploids (Dubcovsky & Dvorak, 2007). Triticum aestivum, also known as bread wheat, has a high ability to adapt to various conditions thanks to its hexaploid structure and constitutes approximately 95% of global wheat production (Curtis & Halford, 2014). Triticum durum, on the other hand, is a tetraploid hard wheat type characterized by high protein content and a strong gluten network (Sissons, 2008). In recent years, old wheat species rich in nutritional values have become the focus of attention again. These species include Triticum monococcum (einkorn), Triticum dicoccum (emmer), and Triticum spelta (spelt). These old varieties stand out with their higher mineral and antioxidant content compared to today's wheat varieties (Longin & Würschum, 2016). The main goals in wheat breeding include increased yield, resistance to diseases, resistance to environmental stress factors, and high product quality (Hidalgo & Brandolini, 2014). Modern breeding techniques include advanced technologies such as selection with molecular markers and genomic selection, in addition to traditional breeding methods (Reynolds et al., 2009). When evaluated in terms of quality, bread wheat has a protein content of 12–15% and a flexible gluten structure suitable for bread making (Bassi et al., 2016). On the other hand, hard wheat has a higher protein content of 14–18% and a solid gluten structure ideal for pasta production. Conversely, historically significant wheat cultivars provide exceptional nutrient profiles with elevated levels of antioxidant, mineral, and vitamin content (Reynolds et al., 2009). Older wheat varieties are recommended because of their greater nutritional profiles in encouraging healthy dietary choices despite their lower yield potential, even if newer wheat cultivars stand out for their high yield potential and extensive adaptability (Bhatta et al., 2018). In agricultural and dietary contexts, it is critical to assess wheat varieties based on their many applications and quality standards.

11.1.1 Wheat Analysis

Wheat analysis provides a methodological approach that allows the evaluation of the physical and chemical quality characteristics of Triticum species and their advantages and disadvantages for use together (Pomeranz, 1989; Wrigley et al., 2016). With wheat analysis, it is possible to determine which process and for which purpose it can be used by determining the structure and quality characteristics of wheat. Its morphological structure can also be determined with physical analyses applied to wheat. Parameters such as test weight, thousand kernel weight, and kernel hardness are utilized to estimate wheat's plumpness, density, and potential flour yield (Ponce-García et al., 2017). These measurements also determine the storage and transportation characteristics of wheat. Chemical analyses elucidate the biochemical composition of wheat. Protein quantity and quality, particularly the characterization of gluten proteins, are critical in determining the bread-making quality of wheat (Shewry & Halford, 2002). Moisture content influences storage stability and microbial deterioration risk, while ash content provides information about mineral content and is used to estimate flour extraction rates. Rheological analyses assess the dough formation and baking performance of wheat flour. Farinograph, extensograph, and alveograph tests measure properties such as water absorption, development time, stability, extensibility, and elasticity of the dough (Yazar, 2023). These data are utilized in developing appropriate flour formulations for various end products. Enzyme activity analyses, particularly the falling number test measuring alpha-amylase activity, evaluate the germination status and starch degradation of wheat. This is crucial in predicting potential effects on bread volume and texture (Pomeranz, 1989). Microbiological and residue analyses are critical for food safety. Mold, yeast, and bacterial counts, along with pesticide and heavy metal analyses, determine the safety of wheat for consumption and its compliance with legal regulations (Ponce-García et al., 2017). Genetic analyses, using DNA-based techniques, enable the identification of wheat varieties and the detection of genetic modifications. This plays a significant role in breeding programs, variety protection, and monitoring of genetically modified organisms (GMOs) (Wrigley et al., 2016).

Spectroscopic analyses have emerged as powerful nondestructive tools for rapid wheat quality assessment. Near-infrared (NIR) and mid-infrared (MIR) spectroscopy are widely used to determine many quality attributes of wheat, such as moisture content and rheological properties (Osborne, 2006). Applied to wheat, these techniques allow for the simultaneous determination of wheat quality parameters in a short time. As a result of these reasons, these techniques provide a multifaceted approach. It provides a holistic approach to understanding the complex structure of wheat from its cultivation to its final product and to evaluating wheat. As a result of the multifaceted examination of the analysis, it can be used in many application areas such as agricultural production practices and strategies to be followed, the conversion of wheat to flour, bakery practices; the establishment of international standards, and standardization of food safety.

In conclusion, wheat analysis performed in a multidisciplinary manner in the fields of food technology, agricultural practices, agricultural science, and nutrition is of critical importance for sustainable agriculture and food safety goals (Wrigley et al., 2016). By evaluating traditional and advanced techniques together, the evaluation of wheat quality characteristics and their accuracy is increased, supporting rapid decision-making.

Biochemical analyses performed on wheat are important for the technological evaluation of wheat. Protein analysis in food products is one of the basic biochemical analyses. The effect of wheat on bread and other final products can be estimated by analyzing the amount and quality of protein. The total amount of protein is usually measured by the Kjeldahl or Dumas method (Osborne, 2006). In addition, protein composition is evaluated using techniques such as sodium dodecyl sulfate polyacrylamide gel electrophoresis (SDS-PAGE). In carbohydrate analysis, the amylose/

amylopectin ratio and starch content are determined by high-performance liquid chromatography (HPLC) or enzymatic methods. Wheat carbohydrates are examined using HPLC or enzymatic methods, and these analyses reveal the starch content of wheat and the amylose-amylopectin ratio (Shewry & Halford, 2002). In addition, dietary fiber analysis can be done by the same methods. Such analyses help determine the baking quality and energy value of wheat. These data are used to calculate the glycemic index of the final products and to evaluate the texture and structure of the products. The total lipid content and fatty acid profile of wheat are examined by gas chromatography (GC) and liquid chromatography-mass spectrometry (LC-MS) methods (Chiu & Kuo, 2020). These analyses are critical for determining the shelf life of wheat and evaluating its health effects. Protease activity is measured by spectrophotometric methods, while enzyme activity is determined by the alpha-amylase activity decrease test (Pomeranz, 1989). These examinations reveal the quality characteristics and enzyme activity of wheat, which has a direct effect on the final volume, structure, texture, and baking properties of bread. In addition, the vitamin, mineral, and antioxidant activities of wheat are also tested, and their health effects are investigated. These health-related tests also contribute to the development of functional foods (Datta et al., 2019; Dziki et al., 2014; Li et al., 2023; Zehiroglu & Ozturk Sarikaya, 2019). The abovementioned analyses provide a comprehensive assessment of wheat quality, guide the formulation of breeding and agricultural policies, evaluate the quality of final products, and help in the formulation of nutritional guidelines. At the same time, they play an important role in determining the compliance of wheat and wheat products with food safety standards in terms of quality (Wrigley et al., 2016).

11.1.2 Spectroscopic Methods

The methods called spectroscopy are based on the interactions between matter and electromagnetic radiation. These methods usually require minimal sample preparation to analyze the chemical and physical properties of wheat quickly and nondestructively. Spectroscopic methods are becoming increasingly common in the production, processing, and quality control of wheat. Fourier transform infrared (FT-IR), mid-infrared (MIR), and near-infrared (NIR) spectroscopy are the three main types of infrared spectroscopy. NIR spectroscopy is widely used to quickly determine the moisture, protein, fat, and carbohydrate content of wheat in the wavelength range of 750–2500 nm (Osborne, 2006). MIR spectroscopy operates in the wavelength range of 2500–25,000 nm and provides more accurate information about the structures of molecules (Thygesen et al., 2003). High-resolution spectra are obtained using FT-IR, particularly in the MIR region. This method works effectively for analyzing the structural characteristics of wheat proteins and starch (Kizil et al., 2002).

Through the analysis of molecular vibration modes, *Raman spectroscopy* offers information about the chemical composition and visual features of wheat (Schulz et al., 2005).

Wheat's phenolic components, antioxidant capacity, and other fluorescent chemicals are examined using *fluorescence spectroscopy* (Žilić et al., 2010).

UV-Vis spectroscopy measures color in wheat analyses pigments and quantifies proteins within the wavelength range of 190–800 nm (Singh et al., 2017).

With *NMR spectroscopy*, wheat's water content, lipid structure, and starch structure may all be precisely determined. Solid-state NMR is specifically utilized to investigate the molecular mobility of wheat kernel constituents (Hills et al., 2003).

Imaging technologies and spectroscopic measurements are combined in the *hyperspectral imaging technique*. It is employed to investigate the distribution of chemicals within and on the surface of wheat kernels (Mahesh et al., 2008).

The spectral gap between the microwave and infrared areas is filled by *terahertz (THz) spectroscopy*. The crystal structure and moisture content of wheat starch can be investigated using this method (Jiang et al., 2016).

These spectroscopic methods allow for rapid and accurate determination of wheat quality parameters. Moreover, as most of these techniques are nondestructive, they enable repeated analysis of the same sample and real-time process control. Food safety, quality assurance, and wheat breeding have all advanced significantly as a result of the creation and use of spectroscopic techniques. The salient characteristics, uses, and benefits of every spectroscopic technique for wheat analysis are enumerated in Table 11.1. When it comes to wheat analysis, each method has its own advantages and frequently offers complementing data. Based on their unique analytical requirements and the resources at their disposal, researchers and professionals in the business might select any one of these approaches.

11.2 Basis of Infrared Spectroscopy

11.2.1 Near-Infrared Spectroscopy (NIRS)

Among the most popular and reliable methods for assessing, wheat quality is near-infrared (NIR) spectroscopy. This method's foundation is the measurement of the sample's molecular bonds' absorption of near-infrared (NIR) light. The chemical bonds in wheat (C–H in starch molecules, N–H in protein molecules, and O–H in moisture) are linked to particular wavelengths when NIR light interacts with a wheat sample, and they are the primary determinants in assessing the quality of the grain. NIRS creates a spectrum that represents the chemical makeup of the wheat by measuring the quantity of light absorbed by the sample at various wavelengths. For instance, a wavelength of 2100 nm is linked to the protein content. The amount of protein in flour is measured by NIRS, which is a crucial factor in evaluating the quality of flour used in bakeries. NIRS evaluations of moisture content, shelf-life stability, and milling performance are also useful. Furthermore, the structure,

Table 11.1 Characteristics of wheat spectrophotometric techniques

Spectroscopic Method	Wavelength/ Frequency Range	Primary Applications	Advantages	Reference
Near-Infrared (NIR) Spectroscopy	750–2500 nm	Analysis of moisture, protein, fat, and carbohydrate content	Rapid, nondestructive, minimal sample preparation	Osborne (2006)
Mid-Infrared (MIR) Spectroscopy	2500–25,000 nm	Analysis of protein secondary structure	Detailed information about molecular structure	Thygesen et al. (2003)
Fourier Transform Infrared (FT-IR) Spectroscopy	2500–25,000 nm (typically MIR region)	Structural characteristics of starch and proteins	High-resolution, precise measurement	Kizil et al. (2002)
Raman Spectroscopy	Visible-NIR region	Analysis of carotenoids and other pigments	Less affected by water presence, molecular vibration information	Schulz et al. (2005)
Fluorescence Spectroscopy	UV-Visible region	Phenolic compounds, antioxidant capacity	High sensitivity, ability to detect low concentrations	Žilić et al. (2010)
UV-Vis Spectroscopy	190–800 nm	Pigment analysis, protein quantification, color measurements	Simple, widely used, quantitative analysis	Singh et al. (2017)
Nuclear Magnetic Resonance (NMR) Spectroscopy	Radio frequencies	Water content, fat composition, starch structure	Detailed structural information at molecular level	Hills et al. (2003)
Hyperspectral Imaging	Typically, VIS-NIR region	Chemical distribution mapping	Combines spectral and spatial information	Mahesh et al. (2008)
Terahertz (THz) Spectroscopy	0.1–10 THz	Starch crystal structure, moisture content	Information about low-energy bonds	Jiang et al. (2016)

texture, and nutritional value of wheat products are also influenced by the amount of starch and gluten present (Pasquini, 2018; Wang et al., 2019; Zhang et al., 2022).

In analytical processes, it is critical to combine NIRS and chemometric methods. Chemometric techniques are defined as multivariate data analysis tools. Chemometric methods use mathematical and statistical techniques to systematically evaluate and interpret chemical measurement results. Its use in NIR spectral analysis spans three key areas: (*i*) preprocessing of spectral data; (*ii*) developing calibration models for both quantitative and qualitative analysis; and (*iii*) transferring models (Cortés et al., 2019; Yin et al., 2021). To analyze the NIR spectra, chemometric models are created using reference samples with known chemical compositions. These models relate spectral data to specific quality attributes, such as protein, moisture, and starch content. A mathematical model such as partial least squares regression (PLSR) and artificial neural networks (ANN), etc., is used to predict wheat quality from NIR spectra. The accuracy of the model is validated by testing it with additional samples with known properties (Zhang et al., 2022). One of the key benefits of NIRS is its ability to perform rapid analysis without damaging the samples, allowing real-time monitoring of wheat quality during processing. These instruments are also easy to operate and can be seamlessly integrated into production lines for continuous quality control. However, the accuracy of NIRS largely depends on the quality of the calibration model, which needs regular updates to reflect changes in wheat cultivars and growing conditions. NIRS is most effective with homogenous samples, as differences in particle size or composition can impact the precision of the results (Salimi Khorshidi et al., 2018; Wang et al., 2019).

When NIRS is paired with chemometrics, NIR spectral data are used for both qualitative and quantitative analysis of products. The NIR spectroscopy process typically involves (*i*) acquiring spectral data; (*ii*) preprocessing the data; (*iii*) developing a calibration model using a set of samples with known concentrations; (*iv*) validating this model; and (v) applying the model to predict or characterize unknown samples (Cen & He, 2007).

Spectral data can vary depending on the pretreatment method used, and each method offers its benefits, so there is no universal guideline for selecting pretreatment techniques. In practice, determining the best pretreatment method involves trial and error, guided by cross-validation/test set outcomes and experience (Cui & Fearn, 2018). However, various pretreatment processes are frequently applied to biological materials, each with its principles and characteristics. It's important to note that while pretreatment methods can improve the signal-to-noise ratio, they may also cause some loss of information from the raw spectrum. Interestingly, various studies suggest that with the use of optimal datasets and advanced algorithms, the impact of pretreatment on the final model's performance becomes less significant (Mao et al., 2014; Williams, 2020). For instance, the principle of multiplicative scatter correction (MSC) operates by averaging the entire wavelength range and using PLSR to align individual spectra with this overall average, thus mitigating undesirable scatter effects. A related method, piecewise multiplicative scatter correction (PMSC), divides the full spectra into nonoverlapping subranges of equal width and applies MSC to each subrange. It is suitable for nonlinear models based on local linear approaches and easily removes significant differences between spectra (Du et al., 2022).

The goal of spectral wavelength selection also referred to as data dimension reduction is to eliminate redundant information in raw spectral data, as different chemical components typically have specific absorption bands. Spectral wavelength selection can be approached in two primary ways. The first involves choosing a specific range of wavelengths to represent the full spectrum in the raw data. Over time, researchers have increasingly relied on data analysis rather than experience for spectral range selection, transitioning from manual to automated methods, such as interval partial least square (i-PLS). The second approach focuses on selecting individual wavelengths instead of ranges to represent the raw data, using techniques like variable combination population analysis (VCPA). Although this method can reduce redundant information, it has been less frequently applied in wheat NIR spectral wavelength selection (Dong & Sun, 2013; Luo et al., 2015; Yang et al., 2017).

Before developing a model, it's important to address outliers and dataset division. Outliers can occur in both reference values and spectral data. To eliminate outliers from reference values, tools like Dixon's test, Tukey's rule, or Grubbs' studentized mean deviation can be effective (Du et al., 2022). For spectral data, there are two main perspectives on handling outliers. By promoting their identification and elimination, one strategy is to stop spectral outliers from impairing the performance of the calibration model. Principal component analysis (PCA) is the most often used technique for this, usually employing F-residuals or Hotelling's T2 values that surpass the 5% level (Tian et al., 2021). Other techniques include Mahalanobis distance, Monte Carlo cross-validation, Chauvenet's test, and hierarchical clustering combined with PCA (Ye et al., 2018). The opposing viewpoint contends that spectral outliers shouldn't be eliminated since they might contain important information about the raw materials and strengthen the model. Some studies have found that removing outliers does not significantly enhance model performance (Williams, 2020). Ultimately, outlier removal largely depends on expert judgment rather than a universal standard. Regarding dataset division, data are typically split into a calibration set, a validation set, and a prediction set. In the most commonly used cross-validation methods, an independent validation set is not used; only a calibration set and a prediction set are utilized, often in a 3:1 ratio. The key principle in dataset division is to ensure that the measured parameter variations are evenly distributed across each dataset, meaning that the standard deviation values and ranges are similar in both sets (Shi et al., 2019; Zeng et al., 2019).

Calibration model development methods fall into two main categories: (*i*) linear and (*ii*) nonlinear regression methods. Although linear regression methods, particularly PLSR, are widely used, they have limitations and require specific conditions, such as a linear relationship between sample spectra and reference values, which can restrict their applicability. In practice, nonlinearity between sample spectra and reference values is often inevitable due to factors like violations of the Beer-Lambert law, sample inhomogeneity, stray light, errors in reference methods, and the inherent nonlinearity between NIR spectra and the target property (Du et al., 2022). Therefore, assessing the degree of nonlinearity is crucial before applying linear regression methods. Tools such as Run Tests and the Durbin–Watson statistic in augmented partial residual plots are useful for detecting nonlinearity (Zareef et al., 2020). If nonlinearity is minimal, linear models can be effective. However, when nonlinearity is significant, strategies like removing specific wavelengths, adding extra principal components or latent variables, and splitting data should be considered before continuing with linear models (Du et al., 2022). Using nonlinear algorithms in combination with NIRS offers superior accuracy and robustness when dealing with nonlinear data. It typically requires two key preconditions. First, a large and diverse sample set is necessary. Second, certain nonlinear regression methods may need to be adapted to suit wheat NIR spectra (Zareef et al., 2020). For instance, Zhang et al. conducted a study involving 775 wheat samples collected from 1998 to 2005. Examples include the application of an enhanced deep belief network

(DBN) model, a modified convolutional neural network (CNN) model, and an adjusted radial basis function neural network (RBFNN) model, all of which have demonstrated potential in improving model performance (Zhang et al., 2019). After establishing a calibration model, it is optimized using the validation dataset to identify the best-performing model, which is then evaluated for robustness using the prediction dataset. Cross-validation, particularly leave-one-out cross-validation, is the most commonly used validation method. In this approach, one spectrum and its corresponding reference values are excluded from the calibration set and used as the validation set, with this process repeated until every spectrum has been used for validation (Che et al., 2017).

11.2.2 Fourier Transform Infrared (FTIR) Spectroscopy

Fourier transform infrared (FTIR) spectroscopy analyses the biochemical composition of wheat by measuring its absorption of IR light. FTIR spectroscopy works by directing IR light through or reflecting it off a sample. Various chemical bonds in the wheat absorb distinct wavelengths of IR light due to vibrational transitions, such as stretching, bending, and twisting of bonds like C–H, N–H, and O–H. An interferometer modulates the IR light to create an interferogram, a signal that changes with optical path length. This interferogram is then converted into a spectrum using Fourier transform mathematics, which translates the time-domain data into a frequency-domain spectrum of IR wavelengths. Finally, the absorption peaks that are linked to different molecular vibrations in the wheat sample are revealed in the obtained spectrum. In this spectrum, each absorption peak demonstrates a functional group or specific bond (Georget & Belton, 2006; Wang et al., 2012). By examining them, the concentration of various components, such as protein, moisture, and starch can be quantified. For instance, N–H and C=O linkage peaks are frequently seen in proteins. The main structural components of wheat, such as proteins and carbohydrates, show up as well-defined absorption bands in the FTIR spectrum. The spectrum also provides details on molecular structures and functional groups, which make it possible to identify certain wheat biochemical traits. Numerous samples, such as whole grains, flour, and wheat dough, can be subjected to FTIR analysis (Barron & Rouau, 2008; Fanari et al., 2022). Samples can be prepared by milling, pelleting, or thin film creation for FTIR examination. It makes it possible to test the starch content and moisture levels that affect processing and storage, as well as the protein content of the wheat, which is essential for baking performance and flour quality. Wheat's biochemical makeup may be swiftly, accurately, and completely analyzed thanks to FTIR. Generally, it is possible to evaluate more than one component at once without harming the samples. However, sample homogeneity and overlapping peaks may have an impact on the analysis's accuracy. Accurate quantification requires the development and validation of calibration models (Fanari et al., 2022; Infrared Spectroscopy for Food Quality Analysis and Control by Da-Wen Sun (z-Lib.Org), n.d.; Kazarian & Chan, 2013; Levine et al.,

1989). Consequently, it is a versatile tool to understand the biochemical properties of wheat, suggesting valuable and beneficial insights into its quality and characteristics. Amir et al. analyzed four different flour samples via FTIR (Tensor 27, Bruker Optics, Germany) to specify the moisture and starch content as well as crude fat and protein. While the moisture content was determined via FTIR in the mid-IR range, with water peaks detected between 1640 and 3300 cm^{-1}, the FT-IR spectra of starch were assessed across several key regions: below 800–1500 cm^{-1} (fingerprint region), 2800–3000 cm^{-1} (C–H stretch region), and 3000–3600 cm^{-1} (O–H stretch region). The crude fat in flour samples exhibited peaks between 1600 and 1700 cm^{-1}, as well as between 1550 and 1570 cm^{-1} whereas crude protein is expected to show peaks at the amide I band between 1600 and 1700 cm^{-1}, and at the amide II band between 1550 and 1570 cm^{-1} (Amir et al., 2013). Furthermore, similar results were reported in various studies (Che Man & Setiowaty, 1999; Manley et al., 2002).

11.2.3 Attenuated Total Reflectance (ATR) Spectroscopy

Attenuated total reflectance (ATR) spectroscopy is a specialized technique within FTIR spectroscopy used for the biochemical analysis of wheat. ATR spectroscopy directs IR light into a high-refractive-index crystal, where it reflects internally and generates an evanescent wave that penetrates a short distance into the sample in contact with the crystal and interacts with the sample, and the absorbed IR light is reflected into the crystal, revealing details about the sample's molecular composition. The sample is directly in contact with the ATR crystal, enabling the evanescent wave to penetrate its surface. This direct contact facilitates easy sample preparation and accommodates various forms, including powders, liquids, and pastes. Hence, ATR requires minimal sample preparation, often allowing analysis in the sample's natural state, such as solid grains, flour, or dough. It is typically nondestructive, preserving the sample for additional analysis or use (Hell et al., 2016; Martins et al., 2022). For creating a spectrum, the infrared light (IR) that the sample reflects is examined. The peaks in this spectrum show the relative absorption of various infrared wavelengths by the various wheat molecular bonds. These peaks for wheat represent important constituents including lipids (C=O bonds), proteins (N–H bonds), and carbs (C–H bonds). ATR determines the concentrations of these components, protein content, moisture content, and starch content by analyzing the intensity of the absorption peaks (Barron & Rouau, 2008; Kaddour et al., 2008). Wheat's molecular makeup and functional groups are revealed by spectra, which helps to identify the grain's biochemical properties. By analyzing the gluten concentration and other biochemical components, as well as moisture and starch levels that affect processing and storage, ATR offers helpful information about the nutritional and functional characteristics of wheat (Li et al., 2006; Wang et al., 2021). ATR spectroscopy provides rapid, nondestructive analysis with little sample preparation and can handle a variety of sample types. It can provide accurate spectrum data despite its modest penetration depth, which could be troublesome when evaluating very thick or diverse samples. Accurate spectrum interpretation and calibration are essential for this method (Hell et al., 2016; Shi et al., 2019). All things considered, ATR is a useful technique for biochemically analyzing wheat and yields accurate and thorough data regarding its composition and quality. A recent study analyzed the biochemical and phenological traits of multiple wheat species and demonstrated their relationship to drought-resistant genes. The study examined several wheat species, including synthetic hexaploids and landraces, and examined how they responded to drought stress. Important molecular markers for comprehending and enhancing drought resistance in wheat, such as proline concentration, chlorophyll, and antioxidant enzyme activity, were emphasized (Ali et al., 2023). Another study employed cutting-edge methods like mass spectrometry to discover important metabolites involved in stress tolerance. It focused on the metabolomics of wheat under postantecedent heat stress. The research highlighted the significant impact of heat stress on wheat yields and how metabolite expression can serve as biomarkers for breeding heat-tolerant wheat varieties (Thomason et al., 2018).

11.3 Conclusion

In this chapter, the IR sensing methods, including NIR, FTIR, and ATR spectroscopy, which are used in the biochemical analysis of wheat, were focused. It has been proven to be an inevitable tool for the accurate and reliable determination of the biochemical properties of wheat, suggesting insights that are crucial for research, quality control, breeding programs, etc. The use of chemometric techniques alongside them increases their precision, allowing for the extraction of meaningful data from complex spectra. Considering the examples from the literature, proper calibration and validation are fundamental for accuracy and reproducibility. As the demand for more efficient and nondestructive analytical techniques continues to grow, the application of IR spectroscopy in wheat analysis is likely to expand, driving advancements in the field and contributing to the development of more sustainable and productive agricultural practices.

References

Acevedo, M., Zurn, J. D., Molero, G., Singh, P., He, X., Aoun, M., Juliana, P., Bockleman, H., Bonman, M., El-Sohl, M., Amri, A., Coffman, R., & McCandless, L. (2018). The role of wheat in global food security. In U. S. Nagothu (Ed.), *Agricultural development and sustainable intensification: Technology and policy challenges in the face of climate change* (pp. 81–110). Routledge. https://doi.org/10.4324/9780203733301-4

Ali, I., Anwar, S., Ali, A., Ullah, Z., Binjawhar, D. N., Sher, H., Abdel-Hameed, U. K., Khan, M. A., Majeed, K., & Jaremko, M. (2023). Biochemical and phenological characterization of diverse wheats and their association with drought tolerance genes. *BMC Plant Biology, 23*(1), 326. https://doi.org/10.1186/s12870-023-04278-9

Amir, R. M., Anjum, F. M., Khan, M. I., Khan, M. R., Pasha, I., & Nadeem, M. (2013). Application of Fourier transform infrared (FTIR) spectroscopy for the identification of wheat varieties. *Journal of Food Science and Technology, 50*(5), 1018–1023. https://doi.org/10.1007/s13197-011-0424-y

Barron, C., & Rouau, X. (2008). FTIR and Raman signatures of wheat grain peripheral tissues. *Cereal Chemistry, 85*(5), 619–625. https://doi.org/10.1094/CCHEM-85-5-0619

Bassi, F. M., Bentley, A. R., Charmet, G., Ortiz, R., & Crossa, J. (2016). Breeding schemes for the implementation of genomic selection in wheat (Triticum spp.). *Plant Science, 242*, 23–36. https://doi.org/10.1016/J.PLANTSCI.2015.08.021

Bhatta, M., Morgounov, A., Belamkar, V., & Baenziger, P. S. (2018). Genome-Wide Association Study reveals novel genomic regions for grain yield and yield-related traits in drought-stressed synthetic hexaploid wheat. *International Journal of Molecular Sciences, 19*(10). https://doi.org/10.3390/IJMS19103011

Cen, H., & He, Y. (2007). Theory and application of near infrared reflectance spectroscopy in determination of food quality. *Trends in Food Science & Technology, 18*(2), 72–83. https://doi.org/10.1016/j.tifs.2006.09.003

Che Man, Y. B., & Setiowaty, G. (1999). Application of Fourier transform infrared spectroscopy to determine free fatty acid contents in palm olein. *Food Chemistry, 66*(1), 109–114. https://doi.org/10.1016/S0308-8146(98)00254-4

Che, W., Sun, L., Zhang, Q., Zhang, D., Ye, D., Tan, W., Wang, L., & Dai, C. (2017). Application of Visible/Near-Infrared Spectroscopy in the prediction of Azodicarbonamide in wheat flour. *Journal of Food Science, 82*(10), 2516–2525. https://doi.org/10.1111/1750-3841.13859

Chiu, H. H., & Kuo, C. H. (2020). Gas chromatography-mass spectrometry-based analytical strategies for fatty acid analysis in biological samples. *Journal of Food and Drug Analysis, 28*(1), 60–73. https://doi.org/10.1016/J.JFDA.2019.10.003

Cortés, V., Blasco, J., Aleixos, N., Cubero, S., & Talens, P. (2019). Monitoring strategies for quality control of agricultural products using visible and near-infrared spectroscopy: A review. *Trends in Food Science & Technology, 85*, 138–148. https://doi.org/10.1016/j.tifs.2019.01.015

Cui, C., & Fearn, T. (2018). Modern practical convolutional neural networks for multivariate regression: Applications to NIR calibration. *Chemometrics and Intelligent Laboratory Systems, 182*, 9–20. https://doi.org/10.1016/j.chemolab.2018.07.008

Curtis, T., & Halford, N. G. (2014). Food security: The challenge of increasing wheat yield and the importance of not compromising food safety. *The Annals of Applied Biology, 164*(3), 354–372. https://doi.org/10.1111/AAB.12108

Datta, S., Sinha, B. K., Bhattacharjee, S., & Seal, T. (2019). Nutritional composition, mineral content, antioxidant activity and quantitative estimation of water soluble vitamins and phenolics by RP-HPLC in some lesser used wild edible plants. *Heliyon, 5*(3), e01431. https://doi.org/10.1016/J.HELIYON.2019.E01431

Da-Wen Sun (z-lib.org). (n.d.). *Infrared spectroscopy for food quality analysis and control*. Academic Press.

Dong, X., & Sun, X. (2013). A case study of characteristic bands selection in near-infrared spectroscopy: Nondestructive detection of ash and moisture in wheat flour. *Journal of Food Measurement and Characterization, 7*(3), 141–148. https://doi.org/10.1007/s11694-013-9149-0

Du, Z., Tian, W., Tilley, M., Wang, D., Zhang, G., & Li, Y. (2022). Quantitative assessment of wheat quality using near-infrared spectroscopy: A comprehensive review. *Comprehensive Reviews in Food Science and Food Safety, 21*(3), 2956–3009. https://doi.org/10.1111/1541-4337.12958

Dubcovsky, J., & Dvorak, J. (2007). Genome plasticity a key factor in the success of polyploid wheat under domestication. *Science (New York, N.Y.), 316*(5833), 1862–1866. https://doi.org/10.1126/SCIENCE.1143986

Dziki, D., Rózyło, R., Gawlik-Dziki, U., & Świeca, M. (2014). Current trends in the enhancement of antioxidant activity of wheat bread by the addition of plant materials rich in phenolic compounds. *Trends in Food Science & Technology, 40*(1), 48–61. https://doi.org/10.1016/J.TIFS.2014.07.010

Fanari, F., Carboni, G., Desogus, F., Grosso, M., & Wilhelm, M. (2022). A chemometric approach to assess the rheological properties of durum wheat dough by indirect FTIR measurements. *Food and Bioprocess Technology, 15*(5), 1040–1054. https://doi.org/10.1007/s11947-022-02799-z

FAO. (2018). *Crop prospects and food situation. Global information and early warning system on food and agriculture*. Food and Agriculture Organization of the United Nations. https://doi.org/10.4060/cb5603en. https://reliefweb.int/sites/reliefweb.int/files/resources/i8764en.pdf

Feldman, M., & Levy, A. A. (2009). Genome evolution in allopolyploid wheat—A revolutionary reprogramming followed by gradual changes. *Journal of Genetics and Genomics, 36*(9), 511–518. https://doi.org/10.1016/S1673-8527(08)60142-3

Finnie, S., & William, A. A. (2016). *Wheat flour* (pp. 1–16). AACC International Press. https://doi.org/10.1016/B978-1-891127-90-8.50001-2

Gandhi, V. P., & Zhou, Z. (2014). Food demand and the food security challenge with rapid economic growth in the emerging economies of India and China. *Food Research International, 63*, 108–124. https://doi.org/10.1016/J.FOODRES.2014.03.015

Georget, D. M. R., & Belton, P. S. (2006). Effects of temperature and water content on the secondary structure of wheat gluten studied by FTIR spectroscopy. *Biomacromolecules, 7*(2), 469–475. https://doi.org/10.1021/bm050667j

Grausgruber, H., Oberforster, M., Werteker, M., Ruckenbauer, P., & Vollmann, J. (2000). Stability of quality traits in Austrian-grown winter wheats. *Field Crops Research, 66*(3), 257–267. https://doi.org/10.1016/S0378-4290(00)00079-4

Hell, J., Prückler, M., Danner, L., Henniges, U., Apprich, S., Rosenau, T., Kneifel, W., & Böhmdorfer, S. (2016). A comparison between near-infrared (NIR) and mid-infrared (ATR-FTIR) spectroscopy for the multivariate determination of compositional properties in wheat bran samples. *Food Control, 60*, 365–369. https://doi.org/10.1016/j.foodcont.2015.08.003

Hidalgo, A., & Brandolini, A. (2014). Nutritional properties of einkorn wheat (Triticum monococcum L.). *Journal of the Science of Food and Agriculture, 94*(4), 601–612. https://doi.org/10.1002/JSFA.6382

Hills, B., Grant, A., & Belton, P. (2003). NMR characterization of cereal and cereal products. In *Characterization of cereals and flours* (pp. 409–436). CRC Press. https://doi.org/10.1201/9780203911785-13

Jiang, Y., Ge, H., Lian, F., Zhang, Y., & Xia, S. (2016). Early detection of germinated wheat grains using terahertz image and chemometrics. *Nature Communications, 6*, 21299. https://doi.org/10.1038/srep21299

Kaddour, A. A., Mondet, M., & Cuq, B. (2008). Description of chemical changes implied during bread dough mixing by FT-ATR Mid-Infrared Spectroscopy. *Cereal Chemistry, 85*(5), 673–678. https://doi.org/10.1094/CCHEM-85-5-0673

Kazarian, S. G., & Chan, K. L. A. (2013). ATR-FTIR spectroscopic imaging: Recent advances and applications to biological systems. *Analyst, 138*(7), 1940–1951. https://doi.org/10.1039/C3AN36865C

Kizil, R., Irudayaraj, J., & Seetharaman, K. (2002). Characterization of irradiated starches by using FT-Raman and FTIR spectroscopy. *Journal of Agricultural and Food Chemistry, 50*(14), 3912–3918. https://doi.org/10.1021/JF011652P

Levine, S. P., Li-Shi, Y., Strang, C. R., & Hong-Kui, X. (1989). Advantages and disadvantages in the use of Fourier Transform Infrared (FTIR) and Filter Infrared (FIR) spectrometers for monitoring airborne gases and vapors of industrial hygiene concern. *Applied Industrial Hygiene, 4*(7), 180–187. https://doi.org/10.1080/08828032.1989.10390419

Li, W., Dobraszczyk, B. J., Dias, A., & Gil, A. M. (2006). Polymer conformation structure of wheat proteins and gluten subfractions revealed by ATR-FTIR. *Cereal Chemistry, 83*(4), 407–410. https://doi.org/10.1094/CC-83-0407

Li, C., Stump, M., Wu, W., & Li, Y. (2023). Exploring the chemical composition, antioxidant potential, and bread quality effects of the nutritional powerhouse: Wheat bran—A mini-review. *Journal of Agriculture and Food Research, 14*, 100898. https://doi.org/10.1016/J.JAFR.2023.100898

Longin, C. F. H., & Würschum, T. (2016). Back to the future—Tapping into ancient grains for food diversity. *Trends in Plant Science, 21*(9), 731–737. https://doi.org/10.1016/J.TPLANTS.2016.05.005

Luo, Q., Yun, Y., Fan, W., Huang, J., Zhang, L., Deng, B., & Lu, H. (2015). Application of near infrared spectroscopy for the rapid determination of epimedin A, B, C and icariin in Epimedium. *RSC Advances, 5*(7), 5046–5052. https://doi.org/10.1039/C4RA11421C

Luthria, D. L., Lu, Y., & John, K. M. M. (2015). Bioactive phytochemicals in wheat: Extraction, analysis, processing, and functional properties. *Journal of Functional Foods, 18*, 910–925. https://doi.org/10.1016/J.JFF.2015.01.001

Mahesh, S., Manickavasagan, A., Jayas, D. S., Paliwal, J., & White, N. D. G. (2008). Feasibility of near-infrared hyperspectral imaging to differentiate Canadian wheat classes. *Biosystems Engineering, 101*(1), 50–57. https://doi.org/10.1016/j.biosystemseng.2008.05.017

Manley, M., Van Zyl, L., & Osborne, B. G. (2002). Using Fourier Transform near infrared spectroscopy in determining kernel hardness, protein and moisture content of whole wheat flour. *Journal of Near Infrared Spectroscopy, 10*(1), 71–76. https://doi.org/10.1255/jnirs.323

Mao, X., Sun, L., Hui, G., & Xu, L. (2014). Modeling research on wheat protein content measurement using near-infrared reflectance spectroscopy and optimized radial basis function neural network. *Journal of Food and Drug Analysis, 22*(2), 230–235. https://doi.org/10.1016/j.jfda.2014.01.023

Martins, M. S., Nascimento, M. H., Barbosa, L. L., Campos, L. C. G., Singh, M. N., Martin, F. L., Romão, W., Filgueiras, P. R., & Barauna, V. G. (2022). Detection and quantification using ATR-FTIR spectroscopy of whey protein concentrate adulteration with wheat flour. *LWT, 172*, 114161. https://doi.org/10.1016/j.lwt.2022.114161

Max, R., & Hannah, R. (2023). *How has world population growth changed over time? - Our World in Data*. Published Online at OurWorldInData.Org. https://ourworldindata.org/population-growth-over-time?trk=public_post_comment-text

Osborne, B. G. (2006). Applications of near infrared spectroscopy in quality screening of early-generation material in cereal breeding programmes. *Journal of Near Infrared Spectroscopy, 14*(2), 93–101. https://doi.org/10.1255/jnirs.595

Pasquini, C. (2018). Near infrared spectroscopy: A mature analytical technique with new perspectives—A review. *Analytica Chimica Acta, 1026*, 8–36. https://doi.org/10.1016/j.aca.2018.04.004

Pomeranz, Y. (1989). Wheat is unique: Structure, composition, processing, end- use properties, and products. In *Saint* (p. 89573). American Association of Cereal Chemists.

Ponce-García, N., Ramírez-Wong, B., Escalante-Aburto, A., Torres-Chávez, P. I., & Serna-Saldivar, S. O. (2017). Grading factors of wheat kernels based on their physical properties. In *Wheat improvement, management and utilization*. https://doi.org/10.5772/67246

Reynolds, M., Foulkes, M. J., Slafer, G. A., Berry, P., Parry, M. A. J., Snape, J. W., & Angus, W. J. (2009). Raising yield potential in wheat. *Journal of Experimental Botany, 60*(7), 1899–1918. https://doi.org/10.1093/JXB/ERP016

Ryan, L., Thondre, P. S., & Henry, C. J. K. (2011). Oat-based breakfast cereals are a rich source of polyphenols and high in antioxidant potential. *Journal of Food Composition and Analysis, 24*(7), 929–934. https://doi.org/10.1016/J.JFCA.2011.02.002

Salimi Khorshidi, A., Storsley, J., Malunga, L. N., Thandapilly, S. J., & Ames, N. (2018). Advancing the science of wheat quality evaluation using nuclear magnetic resonance (NMR)

and ultrasound-based techniques. *Cereal Chemistry, 95*(3), 347–364. https://doi.org/10.1002/cche.10040
Schulz, H., Baranska, M., & Baranski, R. (2005). Potential of NIR-FT-Raman spectroscopy in natural carotenoid analysis. *Biopolymers, 77*(4), 212–221. https://doi.org/10.1002/bip.20215
Shewry, P. R., & Halford, N. G. (2002). Cereal seed storage proteins: Structures, properties and role in grain utilization. *Journal of Experimental Botany, 53*(370), 947–958. https://doi.org/10.1093/jexbot/53.370.947
Shewry, P. R., & Hey, S. J. (2015). The contribution of wheat to human diet and health. *Food and Energy Security, 4*(3), 178. https://doi.org/10.1002/FES3.64
Shi, H., Lei, Y., Louzada Prates, L., & Yu, P. (2019). Evaluation of near-infrared (NIR) and Fourier transform mid-infrared (ATR-FT/MIR) spectroscopy techniques combined with chemometrics for the determination of crude protein and intestinal protein digestibility of wheat. *Food Chemistry, 272*, 507–513. https://doi.org/10.1016/j.foodchem.2018.08.075
Singh, V. K., Devi, A., Pathania, S., Kumar, V., Tripathi, D. K., Sharma, S., Chauhan, D. K., Singh, V. K., & Zorba, V. (2017). Spectroscopic investigation of wheat grains (Triticum aestivum) infected by wheat seed gall nematodes (Anguina tritici). *Biocatalysis and Agricultural Biotechnology, 9*, 58–66. https://doi.org/10.1016/J.BCAB.2016.11.005
Sissons, M. (2008). Role of durum wheat composition on the quality of pasta and bread. *Food, 2*(2).
Thomason, K., Babar, M. A., Erickson, J. E., Mulvaney, M., Beecher, C., & MacDonald, G. (2018). Comparative physiological and metabolomics analysis of wheat (Triticum aestivum L.) following post-anthesis heat stress. *PLoS One, 13*(6), e0197919. https://doi.org/10.1371/journal.pone.0197919
Thygesen, L. G., Løkke, M. M., Micklander, E., & Engelsen, S. B. (2003). Vibrational microspectroscopy of food. Raman vs. FT-IR. *Trends in Food Science and Technology, 14*(1–2), 50–57. https://doi.org/10.1016/S0924-2244(02)00243-1
Tian, W., Chen, G., Zhang, G., Wang, D., Tilley, M., & Li, Y. (2021). Rapid determination of total phenolic content of whole wheat flour using near-infrared spectroscopy and chemometrics. *Food Chemistry, 344*, 128633. https://doi.org/10.1016/j.foodchem.2020.128633
Urueña, R. (2022). International grains council. In *Handbook of transnational economic governance regimes*. https://doi.org/10.1163/ej.9789004163300.i-1081
Uthayakumaran, S., & Wrigley, C. (2017). Wheat: Grain-quality characteristics and management of quality requirements. In *Cereal grains: Assessing and managing quality* (2nd ed., pp. 91–134). https://doi.org/10.1016/B978-0-08-100719-8.00005-X
Wang, J., Zhu, J., Huang, R., & Yang, Y. (2012). Investigation of cell wall composition related to stem lodging resistance in wheat (Triticum aestivum L.) by FTIR spectroscopy. *Plant Signaling & Behavior, 7*(7), 856–863. https://doi.org/10.4161/psb.20468
Wang, H., Lv, D., Dong, N., Wang, S., & Liu, J. (2019). Application of near-infrared spectroscopy for screening the potato flour content in Chinese steamed bread. *Food Science and Biotechnology, 28*(4), 955–963. https://doi.org/10.1007/s10068-018-00552-x
Wang, R., Wei, X., Wang, H., Zhao, L., Zeng, C., Wang, B., Zhang, W., Liu, L., & Xu, Y. (2021). Development of attenuated Total Reflectance Mid-Infrared (ATR-MIR) and Near-Infrared (NIR) spectroscopy for the determination of resistant starch content in wheat grains. *Journal of Analytical Methods in Chemistry, 2021*(1), 5599388. https://doi.org/10.1155/2021/5599388
Williams, P. C. (2020). Application of chemometrics to prediction of some wheat quality factors by near-infrared spectroscopy. *Cereal Chemistry, 97*(5), 958–966. https://doi.org/10.1002/cche.10318
Wrigley, C., Corke, H., Seetharaman, K., & Faubion, J. (2016). *Encyclopedia of food grains*. https://doi.org/10.1016/c2011-1-04352-7
Yang, M., Chen, Q., Kutsanedzie, F. Y. H., Yang, X., Guo, Z., & Ouyang, Q. (2017). Portable spectroscopy system determination of acid value in peanut oil based on variables selection algorithms. *Measurement, 103*, 179–185. https://doi.org/10.1016/j.measurement.2017.02.037
Yazar, G. (2023). Wheat flour quality assessment by fundamental non-linear rheological methods: A critical review. *Food, 12*(18). https://doi.org/10.3390/FOODS12183353

Ye, D., Sun, L., Zou, B., Zhang, Q., Tan, W., & Che, W. (2018). Non-destructive prediction of protein content in wheat using NIRS. *Spectrochimica Acta Part A: Molecular and Biomolecular Spectroscopy, 189*, 463–472. https://doi.org/10.1016/j.saa.2017.08.055

Yin, J., Hameed, S., Xie, L., & Ying, Y. (2021). Non-destructive detection of foreign contaminants in toast bread with near infrared spectroscopy and computer vision techniques. *Journal of Food Measurement and Characterization, 15*(1), 189–198. https://doi.org/10.1007/s11694-020-00627-6

Zareef, M., Chen, Q., Hassan, M. M., Arslan, M., Hashim, M. M., Ahmad, W., Kutsanedzie, F. Y. H., & Agyekum, A. A. (2020). An overview on the applications of typical non-linear algorithms coupled with NIR spectroscopy in food analysis. *Food Engineering Reviews, 12*(2), 173–190. https://doi.org/10.1007/s12393-020-09210-7

Zehiroglu, C., & Ozturk Sarikaya, S. B. (2019). The importance of antioxidants and place in today's scientific and technological studies. *Journal of Food Science and Technology, 56*(11), 4757. https://doi.org/10.1007/S13197-019-03952-X

Zeng, X.-Y., Zhao, W.-Q., Hu, X.-Z., Li, X.-P., Qiao, Y.-Y., Ma, Z., & Zhang, Q.-A. (2019). Determination of polyphenols in oats by near-infrared spectroscopy (NIRS) and two-dimensional correlation spectroscopy. *Analytical Letters, 52*(6), 962–971. https://doi.org/10.1080/00032719.2018.1508295

Zhang, X., Lin, T., Xu, J., Luo, X., & Ying, Y. (2019). DeepSpectra: An end-to-end deep learning approach for quantitative spectral analysis. *Analytica Chimica Acta, 1058*, 48–57. https://doi.org/10.1016/j.aca.2019.01.002

Zhang, S., Liu, S., Shen, L., Chen, S., He, L., & Liu, A. (2022). Application of near-infrared spectroscopy for the nondestructive analysis of wheat flour: A review. *Current Research in Food Science, 5*, 1305–1312. https://doi.org/10.1016/j.crfs.2022.08.006

Žilić, S., Dodig, D., Šukalović, V. H. T., Maksimović, M., Saratlić, G., & Škrbić, B. (2010). Bread and durum wheat compared for antioxidants contents, and lipoxygenase and peroxidase activities. *International Journal of Food Science and Technology, 45*(7), 1360–1367. https://doi.org/10.1111/j.1365-2621.2010.02251.x

Chapter 12
Modern Approaches to Wheat Diseases

Ali Kadiroğlu and Hakan Hekimhan

Abstract Wheat, as a staple crop in global food systems, is highly susceptible to a broad spectrum of phytopathological threats, resulting in substantial yield reductions annually. To mitigate these losses, contemporary agronomic strategies have increasingly integrated advanced digital technologies with conventional management practices. Central to this transformation are geographic information system (GIS), precision agriculture methodologies, and state-of-the-art artificial intelligence (AI) frameworks. In particular, deep learning architectures including generative adversarial networks (GANs), convolutional neural networks (CNNs), deep convolutional neural networks (DCNNs) and residual neural networks (ResNets) have demonstrated significant efficacy in the automated detection, classification and prognosis of wheat diseases. GIS-enabled spatial data visualization facilitates the dynamic mapping and temporal analysis of disease dispersion patterns, supporting targeted intervention planning. Concurrently, remote sensing technologies such as multispectral and hyperspectral imaging deployed via unmanned aerial vehicles (UAVs) and IoT-enabled sensor networks provide high-resolution, real-time phenotypic monitoring at the micro-field level. Image processing and machine-learning algorithms, trained on annotated crop datasets, enable the early identification of symptomatic features with high precision, thus enhancing the responsiveness and efficiency of disease management protocols. This synergistic integration of digital diagnostics, predictive modelling and geospatial analytics not only optimizes crop health surveillance and intervention timing but also contributes to the broader goals of sustainable agriculture and global food security. The digital transformation of

A. Kadiroğlu
Aegean Agriculture Research Institute, Regional Cereal Rust Diseases Research Center, İzmir, Türkiye

Graduate School of Natural and Applied Sciences, Ankara University, Ankara, Türkiye

H. Hekimhan (✉)
Aegean Agriculture Research Institute, Regional Cereal Rust Diseases Research Center, İzmir, Türkiye

N. Zencirci et al. (eds.), *Empowering Wheat Cultivation with GIS, Digital Approaches and Artificial Intelligence*, https://doi.org/10.1007/978-3-031-99954-3_12

wheat cultivation therefore represents a paradigm shift toward intelligent, data-driven, and adaptive disease management systems.

Keywords Wheat · Phytopathological threats · Geographic information system (GIS) · Artificial intelligence (AI) · Deep learning architectures · Generative adversarial networks (GANs) · Convolutional neural networks (CNNs) · Deep convolutional neural networks (DCNNs) · Residual neural networks (ResNets)

Abbreviations

AI	Artificial intelligence
CNNs	Convolutional neural networks
DCNNs	Deep convolutional neural networks
DL	Deep learning
DS	Disease severity
FAO	Food and Agriculture Organization
GANs	Generative adversarial networks
GIS	Geographic information system
LAB	L, a, b colour values
NDTIs	Normalized difference texture indices
NDVI	Normalized difference vegetation index
PLS-LDA	Partial least-squares linear discriminant analysis
PLSR	Partial least-squares regression
R^2	Coefficient of determination
ResNets	Residual neural networks
RGB	Red green blue
SVI	Spectral vegetation index
SVM	Support vector machine
SVR	Support vector regression
UAV	Unmanned aerial vehicle
VIs	Vegetation indices

12.1 Importance of Wheat Production

Wheat is among the most extensively cultivated food crops worldwide, supplying vital sustenance and nutrition to billions of people globally (Erenstein et al., 2022). As reported by the Food and Agriculture Organization (FAO) of the United Nations, wheat constitutes over one-third of the world's cereal production and is a crucial component of the human food supply (Dadrasi et al., 2023). Wheat stands as a crucial staple crop globally, wielding substantial economic significance. Its paramount role in ensuring global food security and economic stability cannot be overstated. Serving as a dietary cornerstone for a vast majority of the global population, especially in Europe, North America and Asia, wheat cultivation and trade emerge as pivotal factors propelling the agricultural sectors and overall economies of numerous nations. The production and commerce of wheat wield a profound impact on both national and global economic landscapes. Beyond being a key revenue source for farmers, it acts as a catalyst for job creation in rural areas. The international trade of wheat, encompassing both exports and imports, contributes significantly to global trade dynamics, fostering economic growth and providing foreign exchange benefits for exporting nations. As per the Food and Agriculture Organization (FAO, 2025), 2020 witnessed a record-breaking wheat output of 773 million metric tons, with China, India, Russia and the United States taking the lead in this agricultural milestone. The FAO's projection for global wheat production in 2025 has remained largely steady since the initial outlook in March, with the output forecast at 795 million tonnes—matching the previous year's level after recent upward adjustments to the 2024 estimate. Wheat is one of the oldest cultivated crops; archaeological evidence shows that it was grown over 10,000 years ago (Zohary et al., 2012). It is also the most consumed and produced crop worldwide. It is a source of nutrients such as carbohydrates, fibre, and proteins for billions of people. Its cultivation sustains millions of livelihoods ranging from small-scale farmers to large businesses in developed economies. Wheat trade is substantial in the international market, which helps to stabilize food supplies.

and prices. The disease that has significantly affected the yield of wheat over the years is the yellow rust disease; the loss caused by it costs billions of dollars each year. It can cause a reduction not only in the quantity but also the quality of the wheat grain; the loss due to this disease ranges from 10% to 70% (Anonymous, 2025).

12.2 Diseases That Limit Wheat Production Worldwide

Various diseases have exerted a substantial toll on wheat production, leading to significant global reductions in yield. Pathogens, including fungi, bacteria, and viruses, pose a persistent threat to wheat harvests, contributing to diminished production and economic setbacks for farmers and the agricultural sector. In a study by Savary et al. (2012) that delved into the impact of wheat diseases on global productivity, it was revealed that diseases like wheat rusts, powdery mildew, and *Fusarium* head blight can precipitate yield losses ranging from 10% to 50% across different regions. Plants can be adversely affected by many different biotic or abiotic factors. Pathogens, animals, and weeds, which are biotic factors that negatively affect plants, cause 20–40% of crop production to be lost (Oerke, 2006; Savary et al., 2012; Teng, 1987). Approximately 137 pathogens and pests cause damage to wheat, corn, rice, potato and soybean. The loss of wheat caused by these pathogens and pests is between 10% and 28% (Savary et al., 2019).

Rust diseases, blotches, *Fusarium* head blight/scab, spot blotch and blasts are the primary diseases seen in wheat (Figueroa et al., 2018). *Septoria tritici* blotch

(*Zymoseptoria tritici*), one of the wheat blotch diseases, costs wheat growers across Europe 280–1200 million € per year, including direct losses and control (Fones & Gurr, 2015; Kiplagat, 2020) As a result of a survey conducted in Australia, the effect of tan spot (*Pyrenophora tritici-repentis*) disease on yield loss was determined to be 200 million C$ (Murray & Brennan, 2009). It has been determined that *Fusarium* head blight (*Fusarium graminearum*) disease causes an epidemic in many continents every 40–50 years and results in significant damage. This disease caused a yield loss of 3 billion US$ in the United States between 1990 and 2008 (Shattock, 2010). Wheat blast disease caused by the pathogen *Magnaporthe oryzae* pathotype *Triticum* has been reported to cause yield loss between 40% and 100% in wheat (Igarashi, 1990). Stem rust, stripe rust, and leaf rust diseases are caused by *Puccinia graminis* f. sp. *tritici*, *P. striiformis* f. sp. *tritici* (Pst) and *P. triticina* fungi, respectively. Global annual yield loss caused by wheat rust diseases ranges from 4.3 billion US$ to 5.0 billion US$ (Figueroa et al., 2018).

Yellow rust disease is caused by a pathogen named *Puccinia striiformis*; the fungus thrives in cool and moist places and rapidly spreads in areas with frequent dew, with mild temperatures. Initially, this disease was more prevalent in cooler areas, but with time, it has spread to the warmer regions as well, and now it is affecting the production of all the wheat-growing countries all around the world This disease is causing a significant reduction in the production of wheat in underdeveloped countries as they lack the resources to effectively control its spread. Yield losses are proportional to the severity of the disease and the time taken to detect its presence. The sooner the disease is detected and measures are taken to control its damage, the less will be the loss caused by it. Therefore, intelligent systems are introduced that can detect the disease at the earliest stages to minimize the loss. These systems use the significance of controlling the rust disease. These losses not only affect the quantity of wheat produced but also impinge on the quality and marketability of the harvested grain (Khanfri et al., 2018).

12.3 Wheat Field Monitoring and Disease Detection Methods

The manual identification of diseases in wheat fields is a laborious and time-intensive task. As technology progresses, machine-learning approaches have emerged as valuable tools for automating disease diagnosis in agriculture. Deep learning models, particularly convolutional neural networks (CNNs), have shown outstanding performance in image-based classification and detection (Khan et al., 2021; Islam et al., 2022). Leveraging algorithms and models, these technologies analyse extensive databases of plant images, enabling swift and reliable disease identification (Golzarian & Frick, 2020; Ramadan et al., 2023b). CNNs, for instance, exhibit promise in detecting wheat diseases by being trained on labelled datasets containing images of both healthy and diseased wheat plants, allowing them to recognize patterns distinguishing between the two (Dixit & Nema, 2018). Despite the potential benefits, a primary challenge in implementing machine-learning algorithms for wheat disease detection is the scarcity of labelled datasets. The collection and annotation of large-scale datasets for training such models can be both time-consuming and costly, especially considering the diverse kinds and variants of wheat diseases. However, addressing this limitation, the application of data augmentation techniques, including GANs, presents an effective solution. GANs, a type of deep learning model, generate synthetic data samples by learning the underlying patterns and distributions of the training dataset. By employing GANs for data augmentation in wheat disease detection, researchers can create a more extensive and diverse dataset, mitigating the challenge of limited data (Ramadan et al., 2023a).

Farmers are interested in the diagnosis of the disease and in parameters like the time when the disease started to infect the wheat in the field, the rate of spread, and the severity of the disease (Huang et al., 2012). To conduct such measurements, the observation of the entire field is required, and modern technologies can produce the needed evaluations. Images of farm canopies collected through various methods can be used for the early detection of wheat diseases or screening for severity measurements. Drone technology, satellite imaging, field-based or laboratory-based systems have been utilized for image processing (Huang et al., 2012; Vergara-Diaz et al., 2015; Zhang et al., 2019). Throughout the following subsection, studies investigating field monitoring methods for disease diagnosis are described and their relevance in the design of the methods used in the dissertation is discussed.

Multispectral and hyperspectral data collection methods were frequently used in studies where digital photographs of the wheat canopy were evaluated in the field. For example, Ashourloo et al. (2016) used the pictures of wheat canopies from 150 cm height in order to detect wheat leaf rust disease in the field and compared the reliability of three different detection methods. A total of 175 samples were collected in this study conducted under greenhouse conditions. The spectrums of disease-infected- and non-infected leaves in the photographs were determined. The study aims to learn how the symptoms of the disease will affect the detection skills of the three different methods. The first method, partial least-squares regression, is the statistical technique of choice to analyse strongly dependent and noisy variables. The second method, support vector regression (SVR), is a technique that analyses the plant's biological structure. The last technique compared, Gaussian process regression, is a machine-learning technique that analyses nonlinear regressions and is kernel-based. Among these methods, Gaussian process regression gave the best precision (0.98 validation accuracy) with its ability to be less affected by spectral noise and to monitor the nonlinear correlation.

Hansen (1991) investigated the multispectral radiometry data collection technique to detect yellow rust disease in the field objectively and quickly. However, it was stated that the severity of the disease has a critical role in the correct diagnosis because the method was not successful when the disease was not seen clearly on the leaf. Despite that, experiments on hyperspectral images use for detection methods continued.

Huang et al. (2012), in their study, where they conducted an experiment to mapping YR disease in a wheat field, a sensor called the Pushbroom Hyperspectral Imager was used to evaluate multi-time airborne hyperspectral images. Also, The Normalized Difference Vegetation Index (NDVI) method was included in the study to distinguish between leaf and YR symptoms' colours and plant growth deficiency. The study was carried out in a field of 167 ha, and YR disease was artificially infected with the plants. While the obtained method showed an 87% success rate on the test images, some background objects and tree shadows around the field were factors limiting the usability of the study.

Dehkordi et al. (2020) used a method that can detect and classify yellow rust and leaf rust in a wheat field using (UAV) technology. The samples used in this study were naturally infected with YR in the field, and ten selected plants were used in data analysis. In the study, the visual cryptography technique was used to focus only on the diseased area. The technique was made by adjusting the intensity of red, green, blue (RGB) components, where yellow rust and leaf rust were most prominent as disease discrimination parameters. The severity of yellow rust disease and the observed correlation in the study was determined as 0.92.

According to Zhang et al. (2012), YR assessments made with the hyperspectral visual technique can be inaccurate because the technique confuses signs of nutrition stresses with symptoms of yellow rust. In this study on how to distinguish between food stress and YR, the sensitivity of 38 parameters used to differentiate YR was investigated. Results showed that the Physiological Reflectance Index is the only parameter that can distinguish YR symptoms without being sensitive to nutrient stress during the five growth stages of wheat.

Hence, timely and accurate diagnosis and management of wheat diseases are crucial for ensuring food security and sustainable agricultural development (John et al., 2023). Traditional methods of wheat disease diagnosis rely on field observation and empirical judgment of agronomists, which is not only time-consuming and laborious but also involves a certain degree of subjectivity and risk of misjudgement. In modern large-scale agricultural production, there is an urgent need for an efficient and reliable automated disease diagnostic tool to improve diagnostic efficiency and accuracy, to realize early warning and precise prevention and control. Traditional methods of wheat disease diagnosis mainly include visual inspection and laboratory testing. The visual inspection method relies on the experience and knowledge of agricultural experts, and although it can be carried out in the field in real time, it is inefficient and highly dependent on experts (Guerrero et al., 2013). In addition, visual inspection is difficult to cover and detect quickly when dealing with large planting environments and is prone to missing early disease symptoms. Traditional laboratory testing methods, while accurate, are hindered by their cumbersome, time-consuming, and costly nature, making them impractical for large-scale monitoring and real-time diagnosis. In response to these challenges, automated and intelligent methods for diagnosing wheat diseases have emerged as key areas of research focus. Recent advancements in information technology and agricultural techniques have fostered the adoption of image-processing technology for identifying and classifying crop diseases, marking it as a burgeoning diagnostic tool. However, traditional image-processing methods rely on artificially designed features, which are difficult to adapt to the complex and changing field environment and diverse disease symptoms, and the classification effect is limited. Deep learning, a leading technology in artificial intelligence, excels particularly in tasks involving image recognition and classification (Khasim et al., 2024).

Convolutional neural network, as an important model of deep learning, mimics the processing of the human brain's visual system through a hierarchical architecture and can automatically extract multi-level features from image data, thus realizing efficient image classification and recognition. Deep learning methods have already achieved remarkable results in the fields of medical image analysis, automatic driving, security monitoring, etc., and have brought new hope for agricultural image processing. In the field of agriculture, image analysis techniques based on deep learning are gradually applied to crop pest detection and classification. By constructing large-scale image datasets and training deep learning models, automated and intelligent diagnosis of crop diseases can be realized. Deep learning methods exhibit superior accuracy and robustness compared to traditional methods, adeptly adapting to complex and dynamic field environments. There have been some research attempts to apply deep learning to wheat disease image classification. Studies (Lin et al., 2019) have shown that deep learning-based methods for wheat disease classification have achieved better results to some extent. For example, some studies have used convolutional neural networks to classify wheat leaf diseases and achieved high classification accuracy. However, these methods based on convolutional neural networks often neglect the extraction of global features from wheat leaf disease images. Global features play an important role in wheat leaf disease images because they not only contain information about the overall distribution and morphology of the disease, but also capture important information about the environmental background, lighting conditions, and overall leaf morphology.

12.4 Neural Network Architectures for Wheat Field Monitoring

Deep convolutional neural network (DCNN) and residual neural network (ResNet) architectures, which are deep learning methods, were frequently used in the monitoring of wheat fields (Schirrmann et al., 2021; Zhang et al., 2019). For example, in a study by Zhang et al. (2019), yellow rust disease was detected in the field using a method created by combining hyperspectral images visualized using UAV technology with ResNet and DCNN architecture. Spatial information and hyperspectral information were used for 3D data creation. In addition, residual neural network block and inception block layers are used to add depth to the DCNN architecture. The reliability accuracy value of the trained model was determined as 0.85. In this study, it has been shown that deep learning architecture methods can be successful in detecting and quantifying diseases.

Fig. 12.1 Test example of a detection model prepared with ResNet. Purple squares indicate diseased areas on the leaf. (Schirrmann et al. (2021))

In the Schirrmann et al. (2021) study, a total of 3000 canopy photos were taken from a wheat field only 2 m high for a visual classification model based entirely on the use of ResNet architecture. The reason for keeping the photo distance low is to provide a clear view of YR symptoms in the leaves, and the reason for taking many pictures is to provide more data for model training. ResNet is formed by adding the residual blocks that feed the residual values to the following layers of the model, thus providing easier training compared to other neural network architectures. For the model to work better, the number of data has been increased by changing the rotations of the images with the data augmentation method. The created model showed 77% correct results on the test images. The model created showed the areas with YR in square selections as seen in Fig. 12.1 and the measured area covered.

It can be concluded that a neural network-based model is a suitable method for the detection and assessment of YR. Moreover, studies have indicated that better results can be obtained by a closer examination; using multiple images, better results can be achieved.

12.5 Overview of Disease Detection and Classification Methods with Single Leaf Images

Certain automatic image processing methods are applied to a single-leaf photograph for a clearer understanding of the symptoms characteristic of wheat diseases obtained through the reduction of background and noise. Because of the focus they have on the leaf, these methods are similar to the methods adopted throughout this dissertation; they contributed to its development.

Barbedo et al. (2016) developed a pairwise-based system that can classify 82 different symptoms in 12 different plant species, including wheat. The data of this study were photographed under various conditions in different light and angles. The principle of the pairwise-based system is to decompose a classification problem into multiple two-class problems. This system is preferred when there is an uncontrollable problem or multiple diseases to detect. The classification.

method, which tries to recognize the disease by changing the images' colour spaces, succeeded in an overall 58% prediction.

In the study conducted by Li et al. (2010), a wheat disease automatic detection method was used, a distinction was made according to the morphological features of the diseases in the images. As a pre-treatment, the images were converted to greyscale colours. In the study, 14 different parameters were tried to differentiate YR disease. A success rate of 86% was achieved in detecting YR disease of 5 out of 14 parameters (sphere structure, round shape, equivalent radius, Hu1, and Hu2) that indicate the morphological characteristic features of diseases.

Niu et al. (2014) used the LAB colour space and K-means clustering method to identify and distinguish three wheat diseases, including YR disease. The K-means algorithm provides automatic clustering of similar objects in an image. With the LAB space colour adjustments, the colours of the disease symptoms became more easily clusterable in the images. The results indicated that the K-means method was more effective than the greyscale method, and K-means had 100% success in detecting YR diseased plants.

Liu et al. (2013) used a support vector machine (SVM) and a complex algorithm to convert nonlinear data into linear data for disease diagnosis in wheat. In this study, a total of 210 image samples were used for six different disease types. In the study, RGB colours were used as disease detection parameters, and according to the results, a 95–96% accuracy rate was measured in diagnosing these six different wheat diseases.

Ashourloo et al. (2014) prepared a wheat rust disease classification method using spectral vegetation indices (SVI) and a dataset prepared via spectroradiometer. The samples used in this study were examined under greenhouse conditions after being artificially infected with YR. Data images were selected between 400 and 1000 nm. A total of 22 indices, including chlorophyll absorbance, and red, yellow, and blue colour edges, were evaluated. It has been observed that the SVI method gives more accurate results on leaves with higher disease severity.

Wang et al. (2016) developed a method called black paper-based measurement in the study of detection and severity of YR disease and showed 99% accuracy on YR test images at the end of the study. The samples used in the study were prepared by artificially YR infecting wheat grown in the field. For the training and testing of this method, leaf samples were first collected from the field, then pasted on black paper, and the dataset was prepared. The data set was created between 325 and 1075 nm spectra diameter. While SVM and distinguished partial least-squares methods were used to detect disease, SVR and quantitative partial least-squares methods were used to determine disease severity.

Spectrometer ranges at various nanometres are used in YR detection models. Aharoni et al. (2021) conducted a study to determine the effect of spectral wave range change on YR detection. The disease was artificially inoculated, and a total of 955 samples were used. In the research, measurements were made at different spectroradiometer waves between 350 and 2500 nm. According to the results of the study, only the five bands (475, 560, 668, 717, and 840 nm) multispectral sensors gave the best results for YR detection.

Vergara-Diaz et al. (2015) established a trial to identify wheat varieties susceptible to YR disease. First, they used RGB indices to determine disease severity on leaves and measured the values at which rust disease was most prominent. Secondly, he prepared an NDVI equation. He used 660-nm red band and 760-nm infrared band values in the equation. As a result of the measurements made with NDVI and RGB indices, it was determined that the RGB index selection was more successful in measuring the severity of the disease.

To summarize, for the detection of YR disease on a leaf basis, data were previously created with LAB and RGB colour spaces. In the studies; K-means, SVM and black paper-based methods successfully detected YR. The mentioned studies inspired the dissertation's research methods.

Khan et al. (2021) identified that traditional methods of disease detection, which rely on visible symptoms, often lag behind in their effectiveness. In their study, they proposed an early detection model for powdery mildew in wheat using machine learning and hyperspectral imaging. The study first extracted normalized difference texture indices (NDTIs) and vegetation indices (VIs) to enhance the contrast between healthy wheat and wheat infected with powdery mildew. A partial least-squares linear discriminant analysis (PLS-LDA) was then applied to detect powdery mildew using the combined optimal features (VIs and NDTIs). Additionally, a regression model based on partial least-squares regression (PLSR) was developed to estimate disease severity (DS). The results indicated that the discriminant model using the combined VIs and NDTIs significantly improved the early detection of infected leaves, with an overall accuracy of over 82.35% and a Kappa coefficient of 0.56. This model was able to detect symptoms that were not easily recognizable using traditional methods. Furthermore, the disease severity estimation model demonstrated good performance, with coefficients of determination (R^2) of over 0.748 and 0.722 for calibration and validation, respectively. The study concluded that this methodology for disease detection and severity quantification shows great potential for early disease identification in crops.

12.6 Overview of Neural Network Architectures in Leaf-Based Examinations in Wheat

Majumdar et al. (2015) conducted a neural network model algorithm with multilayer perceptrons and made a model for detecting YR disease. In preparing the training dataset, an algorithm called Fuzzy c-means, which collects similar elements in the

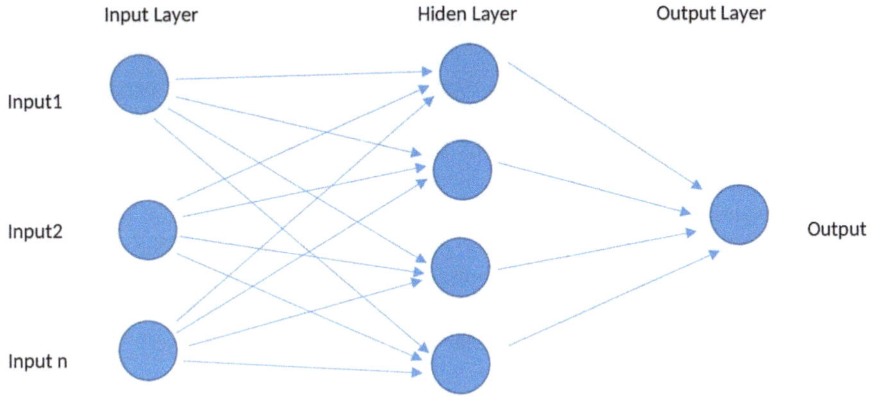

Fig. 12.2 Basic architecture diagram of a backpropagation. (Siddiqui et al. (2011))

image in a single cluster and evaluates the infected area as a whole, was used. With a backpropagation neural network trained with approximately 300 images (including YR-infected leaves), 85% disease detection success was achieved. The backpropagation network is the most popular among the neural network architectures for constructing the YR diagnostic model (Wang et al., 2012). The backpropagation network has a chain derivation structure that improves training by learning from mistakes while learning and training (Jing & Bai, 2019; Yao et al., 2019). Common uses of the backpropagation architecture include the input, hidden and output layers. As seen in the Fig. 12.2, the neurons in the input layer are distributed to all hidden layers, and all hidden layer neurons are connected to a common output layer.

Wang et al. (2012) established a multi-disease diagnostic model that includes the wheat YR disease. In the model, YR symptoms in 50 images segmented by the K-means algorithm were detected using the LAB colour space and trained with a backpropagation network. After that, three hidden layers of transfer points were added to the structure. YR detection feature, in which the number of neurons in the hidden layer is reduced to 6, has brought 100% success in prediction accuracy.

Different types of neural networks allow neurons and data to interact in various ways. When the same data is used in different architectures, the YR detection success of the models will change. In this subsection, a preliminary preparation was made for the YR detection model by investigating which architectures and algorithms are used in neural networks. Jing and Bai (2019) compared the reliability of the backpropagation network to measure the severity of YR disease and the partial least-squares method in an experiment. The samples used in this study were made under field conditions with artificial YR infection. First, the diseased area was determined by combined hyperspectral index and chlorophyll fluorescences. A standard three-layer structure was established for the backpropagation network with 35 diseased, and four healthy samples were used as the training dataset. The backpropagation model gave a more successful result than the partial least-squares method, with an accuracy rate of 0.95 in the measurement of disease severity.

Yao et al. (2019) used the principal component analysis technique for YR detection for data size reduction on imaging leaves collected between 450 and 600 nm. After the samples used in this study were grown under controlled conditions, YR infection was allowed to occur naturally. The detection method continued with the successive projections' algorithm method, which can increase the speed of the model by eliminating overlapping information in the waves during training. Later, the Bayesian regularization algorithm is used to avoid the over-fitting problems in training with 20,000 epochs. At the end of the study, the model trained in the backpropagation network with principal component analysis showed a validation accuracy of 0.9.

To simplify, the backpropagation network has been frequently used in previous studies to detect YR on a leaf basis and was used with data determined by algorithms that better distinguish the diseased region. The detection of a success rate almost equal to 100% indicates that a backpropagation network is an overall sufficient method.

12.7 The Zymomacro_Scanner_v2.1.1 Disease Assessment Macro

In this subsection, the previous uses of a *Z. tritici* assessment macro for measuring disease severity on a leaf basis were examined. Leaf-based disease assessment is necessary to identify plants resistant to disease and particularly to compare their susceptibility. Automatic image analysis methods are being searched to measure the disease spreading rate and severity of the disease in the leaf. Researchers created a macro to measure the severity of the *Z. tritici* infection on the leaf. This macro can measure the percentage of leaf area covered by necrosis, the percentage of leaf area covered by pycnidia, and the percentage of leaf area covered by lesions selected with RGB colour space (Stewart & McDonald, 2014).

Stewart et al. (2016) introduced a new macro named ZymoMacro_Scanner_v2.1.1. In the new macro, the method for imaging the sample was replaced by using a scanner rather than viewing it with a camera. Then, QR codes indicating the name and location of the leaf sample and Code reader were prepared. In addition to the previous study's findings, the total leaf area and disease area measurements were added to the macro. Upgrades to the ZymoMacro_Scanner_v2.1.1 macro have been used successfully for quantitative resistance measurements of the pathogen *Z. tritici*. The final version of the macro was used to rank the resistance of more than 300 winter wheat cultivars with *Septoria tritici* botch disease, and a high correlation was found between the visual scoring measurements and the macro measurements in the study (Karisto et al., 2018). Karisto et al. (2018) stated that the parameters of this macro could be improved, adjusted for disease assessments of different cultivars, and more accurate symptom detection with machine learning.

To summarize, previous studies demonstrated that the ZymoMacro_Scanner_v2.1.1 macro has been able to evaluate in detail the leaf symptoms of the *Z. tritici* pathogen. In addition, the adjustability of the macro has led to the possibility of its adoption for other diseases such as YR.

The main challenges that are being faced by the researchers are as follows:

- Controlled environment dataset: Many publicly available datasets consist of images that are taken in a controlled environment. Images are taken by placing a uniform background behind the leaf to minimize background noise whereas, in natural settings, there will be a significant amount of background interference to the image. Systems that are developed on controlled datasets give good results but are impractical to use in real life.
- Small dataset: Dataset collection is a very costly and time-consuming task. There is limited data available that captures images in natural environments for the identification of wheat rust disease.
- Dataset annotation: Before the detection of the disease, it is important to extract leaves from the image so that different segmentation techniques can be used. Segmentation masks have to be generated to train segmentation models, which is also a time-consuming task.

12.8 Limitation and Future Work

These researchs do have some limitations. First, it focuses on the identification of disease in a single leaf, thereby limiting the scope of the detection process. In real-life scenarios, images captured in natural environments often contain multiple leaves, and these leaves could potentially be used for more comprehensive disease detection. By expanding the scope of these researchs to target multiple leaves within an image, the algorithm could become more versatile and robust. Another limitation to consider is the computational cost of the leaf rotation algorithm (George et al., 2025). One important aspect is to adapt the system to accommodate multiple leaves within a single image. This feature would allow to use all the information in one image, ultimately leading to more comprehensive and informative final reports. It opens the door to a broader understanding of the overall health and disease prevalence among multiple leaves, offering a holistic perspective on the plant's condition. Moreover, there is considerable room for the advancement of both segmentation and classification models. By incorporating more advanced and sophisticated models, it can substantially elevate the precision and efficacy of pipeline. These models can be designed to capture more fine details within the leaf images, enabling even more accurate disease diagnosis and classification (Hassan et al., 2024).

Deep learning (DL), as one of the most recent breakthroughs in computer vision, has emerged as a promising approach in agricultural plant protection by minimizing the subjectivity associated with manually identifying disease symptoms. Studies in this field provide a comprehensive examination of the techniques and tools used for automatic disease identification, the latest DL models and emerging trends in DL-based image analysis. Among the key findings, imaging techniques and sensors—such as RGB, multispectral and hyperspectral cameras—are highlighted for their high efficacy in the early detection of plant diseases. Researchers have also conducted comparative evaluations of various deep learning architectures, including CNNs), vision transformers, GANs, vision-language models and foundation models. Another critical aspect of these studies is the connection they establish between academic advancements in deep learning and their applicability in real-world agricultural settings. By analysing the extent to which these models are suitable for production environments, they offer valuable insights into the practical implementation of DL technologies. As such, these works serve as a comprehensive and informative resource for researchers, academicians and practitioners working in the field of precision agriculture, providing a thorough understanding of the current state and future directions of deep learning in plant disease detection (Ferentinos, 2018; Upadhyay et al., 2025).

12.9 Conclusion

The digitalization of wheat disease management represents a pivotal shift in modern agriculture, enabling more precise, efficient and sustainable responses to plant health threats. The integration of GIS, UAVs, sensor networks and deep learning-based artificial intelligence (AI) has substantially enhanced the capacity for early detection, spatial monitoring and data-driven intervention strategies. These technologies allow for real-time assessment of crop health and more accurate identification of disease outbreaks, thus reducing yield losses and the reliance on blanket chemical applications.

Deep learning architectures, particularly CNNs and GANs, have demonstrated superior performance in symptom detection and disease prediction, often exceeding traditional diagnostic methods in both speed and accuracy. The ability of GANs to generate synthetic training data further addresses the limitations posed by scarce labelled datasets in agricultural domains.

Despite these advancements, widespread adoption remains limited by several challenges, including high initial costs, lack of technical training and limited digital literacy among farmers. Many continue to rely on visual assessments, which delay intervention and contribute to increased crop vulnerability. Addressing these barriers requires comprehensive strategies that combine education, financial support mechanisms and the development of low-cost, user-friendly technological solutions tailored to local agricultural contexts.

Looking ahead, the continued evolution of AI, machine learning and remote sensing will further strengthen disease management frameworks, particularly as technologies become more accessible and affordable. Promoting farmer engagement and capacity building will be essential to bridge the gap between innovation and practice. Ultimately, the adoption of digital tools in wheat cultivation holds immense potential to enhance productivity, ensure food security, and foster climate-resilient agricultural systems in an era of increasing global uncertainty.

References

Aharoni, R., Klymiuk, V., Sarusi, B., Young, S., Fahima, T., Fishbain, B., & Kendler, S. (2021). Spectral light-reflection data dimensionality reduction for timely detection of yellow rust. *Precision Agriculture, 22*(1), 267–286.

Anonymous. (2025). Agriculture and Horticulture Development Board, Yellow rust symptoms and management in wheat. https://ahdb.org.uk/yellowrust [01 Mar 2025].

Ashourloo, D., Mobasheri, M. R., & Huete, A. (2014). Evaluating the effect of different wheat rust disease symptoms on vegetation indices using hyperspectral measurements. *Remote Sensing, 6*(6), 5107–5123.

Ashourloo, D., Aghighi, H., Matkan, A. A., Mobasheri, M. R., & Rad, A. M. (2016). An investigation into machine learning regression techniques for the leaf rust disease detection using hyperspectral measurement. *IEEE Journal of Selected Topics in Applied Earth Observations and Remote Sensing, 9*(9), 4344–4351.

Barbedo, J. G. A., Koenigkan, L. V., & Santos, T. T. (2016). Identifying multiple plant diseases using digital image processing. *Biosystems Engineering, 147*, 104–116.

Dadrasi, A., Chaichi, M., Nehbandani, A., Sheikhi, A., Salmani, F., & Nemati, A. (2023). Addressing food insecurity: an exploration of wheat production expansion. *PLoS One, 18*, e0290684. https://doi.org/10.1371/journal.pone.0290684

Dehkordi, R. H., ELJarroudi, M., Kouadio, L., Meersmans, J., & Beyer, M. (2020). Monitoring wheat leaf rust and stripe rust in winter wheat using high-resolution UAV-based red-green-blue imagery. *Remote Sensing, 12*(22), 1–21.

Dixit, A., & Nema, S. (2018). Wheat leaf disease detection using machine learning method-a review. *International Journal of Computer Science and Mobile Computing, 7*(5), 124–129.

Erenstein, O., Jaleta, M., Mottaleb, K. A., Sonder, K., Donovan, J., & Braun, H.-J. (2022). Global trends in wheat production, consumption and trade. In *Wheat improvement: food security in a changing climate* (pp. 47–66). Springer International Publishing.

FAO, Food and Agriculture Organisation (2025) Cereal Supply and Demand Brief. http://www.fao.org/worldfoodsituation/csdb [06 Mar 2025].

Ferentinos, K. P. (2018). Deep learning models for plant disease detection and diagnosis. *Computers and Electronics in Agriculture, 145*, 311–318. https://doi.org/10.1016/j.compag.2018.01.009

Figueroa, M., Hammond-Kosack, K. E., & Solomon, P. S. (2018). A review of wheat diseases—a field perspective. *Molecular Plant Pathology, 19*(6), 1523–1536.

Fones, H., & Gurr, S. (2015). The impact of *Septoria tritici* blotch disease on wheat: An EU perspective. *Fungal Genetics and Biology, 79*, 3–7.

George, R., Thuseethan, S., Ragel, R. G., Mahendrakumaran, K., Nimishan, S., Wimalasooriya, C., & Alazab, M. (2025). *Past, present and future of deep plant leaf disease recognition: A survey* (Vol. 234., July 2025, 110128, 24 p). Computers and Electronics in Agriculture. https://doi.org/10.1016/j.compag.2025.110128

Golzarian, M. R., & Frick, R. A. (2020). Advances in machine vision applications for plant disease detection. *Annual Review of Phytopathology, 58*, 231–253.

Guerrero, J. M., Guijarro, M., Montalvo, M., Romeo, J., Emmi, L., Ribeiro, A., et al. (2013). Automatic expert system based on images for accuracy crop row detection in maize fields. *Expert Systems with Applications, 40*, 656–664. https://doi.org/10.1016/j.eswa.2012.07.073

Hansen, J. G. (1991). Use of multispectral radiometry in wheat yellow rust experiments. *EPPO Bulletin, 21*(3), 651–658.

Hassan, A., Mumtaz, R., Mahmood, Z., Fayyaz, M., & Naeem, M. K. (2024). Wheat leaf localization and segmentation for yellow rust disease detection in complex natural backgrounds. *Alexandria Engineering Journal, 107*(2024), 786–798. https://doi.org/10.1016/j.aej.2024.09.018

Huang, L. S., Zhao, J. L., Zhang, D. Y., Yuan, L., Dong, Y. Y., & Zhang, J. C. (2012). Identifying and mapping stripe rust in winter wheat using multi-temporal airborne hyperspectral images. *International Journal of Agriculture and Biology, 14*(5), 697–704.

Igarashi, S. (1990) Update on wheat blast (*Pyricularia oryzae*) in Brazil. In A proceeding of the international conference-wheat for the nontraditional warm areas (pp. 480-483).

Islam, S., Sultana, Md, S., Farid, F. A. N., Md., Islam Rashid, M. Bari, B.S, Hashim, N., & Husen, M. N. (2022). Multi-modal hybrid deep learning approach to detect tomato leaf disease using attention based dilated convolution feature extractor with logistic regression classification. *Sensors, 22*(16), 6079.

Jing, X., & Bai, Z. (2019). Remote sensing detection of wheat stripe rust by synergised solar-induced chlorophyll fluorescence and differential spectral index. Proceedings –2019 6th International Conference on Information Science and Control Engineering. *ICISCE, 2019*, 1109–1117.

John, M. A., Bankole, I., Ajayi-Moses, O., Ijila, T., Jeje, O., & Lalit, P. (2023). Relevance of advanced plant disease detection techniques in disease and Pest Management for Ensuring Food Security and Their Implication: A review. *American Journal of Plant Sciences, 14*, 1260–1295. https://doi.org/10.4236/ajps.2023.1411086

Karisto, P., Hund, A., Yu, K., Anderegg, J., Walter, A., Mascher, F., McDonald, B. A., & Mikaberidze, A. (2018). Ranking quantitative resistance to *Septoria tritici* blotch in elite wheat cultivars using automated image analysis. *Phytopathology, 108*(5), 568–581.

Khan, I. H., Liu, H., Li, W., Cao, A., Wang, X., Liu, H., Cheng, T., Tian, Y., Zhu, Y., Cao, W., & Yao, X. (2021). Early detection of powdery mildew disease and accurate quantification of its severity using hyperspectral images in wheat. *Remote Sensing, 13*(18), 3612. https://doi.org/10.3390/rs13183612

Khanfri, S., Boulif, M., & Lahlali, R. (2018). Yellow rust (*Puccinia striiformis*): A serious threat to wheat production worldwide. *Notulae Scientia Biologicae, 10*(3), 410–423.

Khasim, S., Ghosh, H., Rahat, I. S., Shaik, K., & Yesubabu, M. (2024). Deciphering microorganisms through intelligent image recognition: Machine learning and deep learning approaches, challenges, and advancements. *EAI Endorsed Trans. Internet Things, 10*, 1–8. https://doi.org/10.4108/eetiot.4484

Kiplagat, D. K. (2020). *Determination of resistance revels of the 200 of 1091 recombinant inbred lines of the UK magic winter wheat to Septoria tritici blotch and yellow rust diseases based on disease intensity measurements*. The University of Reading.

Li, J., Gao, L., & Shen, Z. (2010). *Extraction and analysis of digital images feature of three kinds of wheat diseases* (Vol. 6, pp. 2543–2548). Proceedings –2010 3rd International Congress on Image and Signal Processing, CISP 2010.

Lin, Z., Mu, S., Huang, F., Mateen, K. A., Wang, M., Gao, W., et al. (2019). A unified matrix-based convolutional neural network for fine- grained image classification of wheat leaf diseases. *IEEE Access, 7*, 11570–11590. https://doi.org/10.1109/Access.6287639

Liu, L., Zhong, Zhang, W., Shu, S. B., & Jin, X. (2013). *Image recognition of wheat disease based on RBF support vector machine, (1201)* (pp. 307–310).

Majumdar, D., Kole, D. K., Chakraborty, A., & Majumder, D. D. (2015). *An integrated digital image analysis system for detection, recognition and diagnosis of disease in wheat leaves* (pp. 400–405). ACM International Conference Proceeding Series, 10–13 August 2015.

Miao, Y., Meng, W., & Zhou, X. (2025). SerpensGate-YOLOv8: An enhanced YOLOv8 model for accurate plant disease detection. *Front Plant Science, 15*, 1514832. https://doi.org/10.3389/fpls.2024.1514832

Murray, G. M., & Brennan, J. P. (2009). Estimating disease losses to the Australian wheat industry. *Australasian Plant Pathology, 38*(6), 558–570.

Niu, X., Wang, M., Chen, X., Guo, S., Zhang, H., & He, D. (2014). *Image segmentation algorithm for disease detection of wheat leaves* (pp. 270–273). International Conference on Advanced Mechatronic Systems, ICAMechS.

Oerke, E. C. (2006). Crop losses to pests. *Journal of Agricultural Science, 144*(1), 31–43.

Ramadan, S. T. Y., Sakib, T. Haque, Md.M.U., Sharmin, N, & Rahman, Md. M. (2023a). Wheat leaf disease synthetic image generation from limited dataset using gan. In International Conference on Human-Centric Smart Computing, pages 501–511. Springer, 2023.

Ramadan, S.T. Y., Sakib, T., Rahat, Md. A., & Mosharrof, S. (2023b). Cyclegan-based data augmentation with cnn and vision transformers (vit) models for improved maize leaf disease classification. In 2023 IEEE 64th International Scientific Conference on Information Technology and Management Science of Riga Technical University (ITMS), pages 1–6, 2023.

Savary, S., Ficke, A., Aubertot, J. N., & Hollier, C. (2012). Crop losses due to diseases and their implications for global food production losses and food security. *Food Security, 4*(4), 519–537.

Savary, S., Willocquet, L., Pethybridge, S. J., Esker, P., McRoberts, N., & Nelson, A. (2019). The global burden of pathogens and pests on major food crops. *Nature Ecology and Evolution, 3*(3), 430–439.

Schirrmann, M., Landwehr, N., Giebel, A., Garz, A., & Dammer, K. H. (2021). Early detection of stripe rust in winter wheat using deep residual neural networks. *Frontiers in Plant Science, 12*(March), 1–14.

Shattock, R. (2010). Diseases of herbaceous perennials. *Plant Pathology, 59*(3), 594–594.

Siddiqui, M. N., Saini, V., Ahuja, R., Wyld, D. C., Wozniak, M., Chaki, N., Meghanathan, N., & Nagamalai, D. (2011). *Advances in Network Security and Applications, 196*(February 2015), 649–657.

Stewart, E. L., & McDonald, B. A. (2014). Measuring quantitative virulence in the wheat pathogen *Zymoseptoria tritici* using high-throughput automated image analysis. *Phytopathology, 104*(9), 985–992.

Stewart, E. L., Hagerty, C. H., Mikaberidze, A., Mundt, C. C., Zhong, Z., & McDonald, B. A. (2016). An improved method for measuring quantitative resistance to the wheat pathogen *Zymoseptoria tritici* using high-throughput automated image analysis. *Phytopathology, 106*(7), 782–788.

Teng, P. S. (Ed.). (1987). *Crop loss assessment and pest management*. APS Press.

Upadhyay, A., Chandel, N. S., Singh, K. P., et al. (2025). Deep learning and computer vision in plant disease detection: A comprehensive review of techniques, models, and trends in precision agriculture. *Artificial Intelligence Review, 58*, 92. https://doi.org/10.1007/s10462-024-11100-x

Vergara-Diaz, O., Kefauver, S. C., Elazab, A., Nieto-Taladriz, M. T., & Araus, J. L. (2015). Grain yield losses in yellow-rusted durum wheat estimated using digital and conventional parameters under field conditions. *Crop Journal, 3*(3), 200–210.

Wang, H., Li, G., Ma, Z., & Li, X. (2012). *Image recognition of plant diseases based on backpropagation networks* (pp. 894–900). 2012 5th International Congress on Image and Signal Processing, CISP. 2012, (October 2012).

Wang, H., Qin, F., Ruan, L., Wang, R., Liu, Q., Ma, Z., Li, X., Cheng, P., & Wang, H. (2016). Identification and severity determination of wheat stripe rust and wheat leaf rust based on hyperspectral data acquired using a black-paper-based measuring method. *PLoS One, 11*(4), 1–25.

Yao, Z., Lei, Y., & He, D. (2019). Early visual detection of wheat stripe rust using visible/near-infrared hyperspectral imaging. *Sensors (Switzerland), 19*(4).

Zhang, J., Pu, R., Huang, W., Yuan, L., Luo, J., & Wang, J. (2012). Using in-situ hyperspectral data for detecting and discriminating yellow rust disease from nutrient stresses. *Field Crops Research, 134*, 165–174.

Zhang, X., Han, Liangxiu, Dong, Y., Shi, Y., Huang, W., Han, Lianghao, González-Moreno, P., Ma, H., Ye, H., & Sobeih, T. (2019). A deep learning-based approach for automated yellow rust disease detection from high-resolution hyperspectral UAV images. *Remote Sensing, 11*(13), 1–16.

Zohary, D., Hopf, M., & Weiss, E. (2012). *Domestication of plants in the Old World.* Oxford University Press.

Chapter 13
Weed Management Approaches with GIS, Digital Approaches, and Artificial Intelligence in Wheat Cultivation

Ender Şahin Çolak, Kübra Kale, and Doğan Işik

Abstract Weeds in wheat (*Triticum aestivum* L.) farming are one of the important biotic stress factors that cause yield loss. Chemical herbicides, which are widely used in traditional weed management, cause problems such as herbicide resistance, environmental pollution, and high costs. Therefore, the need for sustainable and efficient agricultural practices is increasing. In recent years, smart technologies such as artificial intelligence (AI), machine learning (ML), unmanned aerial vehicles (UAVs), robotic systems, and remote sensing have offered innovative solutions for weed management in wheat. In this section, traditional and modern methods used in weed management in wheat farming are compared, and the effectiveness of artificial intelligence-supported systems, hyperspectral imaging, geographic information systems (GIS), and sensor-based technologies is evaluated. Artificial intelligence algorithms and deep learning models can detect problematic weeds in wheat with high accuracy and contribute to developing targeted control strategies. UAVs optimize herbicide use by performing weed mapping with high-resolution imaging and spectral analysis methods. Robotic systems reduce labor costs and support agricultural sustainability by directly controlling problem weeds in wheat through mechanical or selective chemical applications. As a result, integrating artificial intelligence and other advanced smart technologies offers environmentally friendly and economical solutions for wheat weed management. With the widespread use of these smart technologies, it is anticipated that it will be possible to reduce chemical inputs in wheat, increase wheat yield, and protect the ecosystem balance.

Keywords Wheat · Weeds · Artificial intelligence · UAV · GIS

E. Ş. Çolak · K. Kale · D. Işik (✉)
Erciyes University Faculty of Agriculture Department of Plant Protection, Kayseri, Türkiye
e-mail: dogani@erciyes.edu.tr

N. Zencirci et al. (eds.), *Empowering Wheat Cultivation with GIS, Digital Approaches and Artificial Intelligence*, https://doi.org/10.1007/978-3-031-99954-3_13

Abbreviations

AHP	Analytical hierarchy process
AI	Artificial intelligence
ANN	Artificial neural network
CNN	Convolutional neural network
COVID-19	COronaVIrus Disease of 2019
FAO	Food and Agriculture Organization
GIS	Geographic information system
GPS	Global positioning system
IMP	Image processing
IoT	Internet of things
ML	Machine learning
POST	Post-emergence herbicide application
RGB	Red green blue
RILA	Ratio of infected leaf area
RS	Remote sensing
UAV	Unmanned aerial vehicle
YOLO	You only look once

13.1 Introduction

Considering the current growth rate of the world population, it is expected to reach approximately ten billion in 2050 (Kirkaya, 2020). It is known that there is a decrease in the population in agricultural areas due to migration of people living from rural to urban areas. Therefore, given the projected increase in the global human population and the simultaneous decline in rural populations—trends further highlighted by the impacts of the COVID-19 pandemic—the need to improve both the quality and quantity of agricultural production has become inevitable (Kırkaya, 2020). Plant protection factors (weeds, diseases and pests) have economic importance in agriculture as they cause yield and quality losses in commercially grown crops. Regarding sustainability, technology-based agricultural production is needed because of the increasing population and the difficulties in accessing resources (Sharma et al., 2021). Agricultural scientists are researching ways to increase the productivity of basic food products while minimizing damage to natural resources and focusing on sustainable agricultural practices in this context (Tyagi, 2016).

In the traditional production-based crop cultivation systems, weed management is done with experience and sampling. In this case, loss of yield in the product, resistance to herbicides applied to the product, and irregular or inappropriate doses of chemical applications cause environmental pollution and make it difficult to manage weeds. However, the development of artificial intelligence-based technology has created new strategies in agricultural production and offers long-term solutions (Li & Wang, 2024). In this case, it can be said that the actor of the acceleration in agricultural technology is artificial intelligence (Sharma et al., 2021). Artificial intelligence imitates the experience-based production system used in traditional agriculture with artificial neural networks based on learning and is generally a part of today's informatics (Sharma et al., 2021). The algorithm of agricultural production is created with tools that store long-term data, such as sensors and aerial images (Kirkaya, 2020). This algorithm can process the images obtained with new digital approaches and contribute to agricultural activities.

The main task of producers in agriculture is to provide the food supply, fuel needs, and natural products needed for the continuation of human life (Weiss et al., 2020; Meghraoui et al., 2022). Agriculture is a sector where crops are grown economically globally, so yield values are important (Meghraoui et al., 2022). Product yield and quality estimates of wheat (*Triticum aestivum* L.), which is one of the most widely grown agricultural products and one of the internal economic dynamics of countries, are important in terms of international food trade (Gill et al., 2024).

Wheat cereal yields 798.97 million tons from 220.40 million hectares worldwide (FAO, 2023). Wheat is important in human nutrition due to its nutritional value and plays a role in the nutrition of almost one-fifth of the world's population (McFall & Fowler, 2009; Li et al., 2020; Shahzad et al., 2021). It forms the basis of nutrition systems worldwide and faces threats such as high food prices, climate change, and the depletion of natural resources. Due to the impact of these threats, increasing wheat production efficiency has become one of the priority research topics of the scientific community. The increasing demand for wheat and environmental changes especially necessitate increasing production (Lipper et al., 2014). Since wheat has a long-term development period, it faces factors that reduce yield. Living and nonliving factors cause quality and quantity loss in the product (Gupta, 2004; Ali et al., 2014). Weeds are an important factor that prevents yield in wheat cultivation (Shahzad et al., 2021). It is stated that weeds can cause approximately 30% to 50% yield loss in wheat during the production season. (Oerke, 2006; Al-Baldawi & Al-Naqib, 2011; Safi et al., 2022).

Especially narrow- and broad-leaved weed species invade wheat planting areas early, inhibiting seedling development and reducing product quality (Harker & O'Donovan, 2013). Traditional methods used to control weeds include chemical, mechanical, and cultural methods, and the integrated application of these methods is of great importance for sustainable wheat production (Chauhan et al., 2017). However, the spread of herbicide-resistant weed species reduces the effectiveness of existing management strategies and necessitates the development of alternative management methods (Heap, 2021).

The importance of technology in weed management has increased in recent years. Precise herbicide applications with technologies that provide accurate and reliable automatic weed detection can significantly reduce herbicide use and weed management costs (Fennimore & Cutulle, 2019; Wang et al., 2019a). Various detection methods play an important role in weed detection; these include fluorescence, visible or near-infrared spectroscopy, hyperspectral and multispectral imaging, and machine vision. However, the widespread use of smart sprayers in wheat agriculture has not yet been fully achieved. The biggest obstacle to this situation is the lack of robust detection systems that can provide reliable weed detection (Fennimore & Cutulle, 2019). Especially in the case of wheat, different weed species around the plant make detection difficult. Recently, a method based on deep convolutional neural networks (DCNNs) has been developed for real-time detection of interrow-germinated *Lolium* sp. in wheat (Su et al., 2021; Zhuang et al., 2021). However, detecting interrow weeds in the seed growth stage still poses a significant challenge.

The agricultural sector seeks smart solutions to meet the increasing population and food demand. In this context, weeds are a significant problem in wheat production. While traditional methods involve the use of broad-spectrum herbicides, these methods face problems such as environmental impacts, costs, and resistance development. In recent years, smart technologies such as artificial intelligence (AI) and machine learning (ML) have offered revolutionary solutions for weed management in wheat.

13.2 Use of GIS (Geographic Information System) Technology in Wheat Cultivation

In agricultural production, along with Industry 4.0, (internet of things (IoT), artificial intelligence (AI), remote sensing, and IMP (remote sensing and image processing) techniques, GIS (geographic information systems) have been used in recent years (Kirkaya, 2020). GIS is the systematic determination of data according to the regions obtained. It evaluates data according to location, enables management and analysis, and enables data mapping; GIS offers a detailed perspective on data and locations (Kirkaya, 2020).

The GIS software method is used to develop manual mapping of regional distributions of weeds with herbologists' support and Global Positioning System (GPS) technology (Slaughter et al., 2008b). When machine learning technology is integrated with weed detection, automation in weed management is also possible (Slaughter et al., 2008b). Weeds are usually mapped by reconnaissance or remote sensing methods, and a GIS, a positioning system, carries out the mapping process and, if there is a sprayer, customizing the sprayer control system (Loghavi & Mackvandi, 2008). Here, the accuracy of the positioning system is determined by the accuracy of the GPS, and the weed location data is processed using GIS software (Loghavi & Mackvandi, 2008).

In a study by Loghavi and Mackvandi (2008), weed location determination was done stop-and-go. In the study, data was processed with GIS software, and an infrared displacement sensor was developed to detect weeds at their locations precisely and send signals to the nozzles. The study used a non-GPS (NGPS) herbicide application system. According to the results obtained from the study, there could be a significant decrease in the amount of herbicide use.

In a study conducted in Iran, the suitability of herbicides used to manage weeds in wheat-growing areas was tested. The acidity or water hardness of the land was examined. For this purpose, the GIS system was used, and techniques were based on acidity and water hardness. Within the scope of the study, the usability of groundwater used in the application of herbicides for narrow-leaved, broad-leaved, or both was tested. The results obtained from the study showed that the groundwater was alkaline and hard in character. In this case, it was observed that the application of herbicides limited weed management (Kamkar & Mijani, 2023).

Another study addressed the spatial and temporal detection of weeds. In the study, UAVs were used to map weeds in the study area, and GIS software version 7.6.2 was tested for the classification of weeds. After the study, maps that could be considered a prescription for regional weed intervention were created. The most suitable flight period was also determined temporally in the study (Nikolić et al., 2021).

In another study, the production potential of winter wheat was investigated using a GIS-supported growth simulation in wheat-growing areas in the northern regions of China. The research results contributed to discovering the factors that limit productivity (Wu DingRong et al., 2006).

In a study conducted in India, GIS and remote sensing systems were used to determine weed densities. Spectral measurements and ground measurement analysis were performed to distinguish between weedy and weed-free crops (Kaur et al., 2009).

In a study conducted in Çanakkale Province, a GIS-based system implemented peach cultivation's analytical hierarchy process (AHP). The suitability for cultivation in the land was studied, and parameters such as EC, pH, soil organic matter content, nitrogen, phosphorus values, land slope, and elevation were considered as factors. As a result of the evaluations, it was determined that 367.50 ha of the land was very suitable for cultivation, 7085.25 ha was moderately suitable, 6341.25 ha was marginally suitable, and 378.50 ha was not suitable for cultivation. Using the GIS technique in this study saved time and was an economically beneficial application (Everest & Gur, 2022).

In a study conducted on corn cultivation, the mapping of weeds, among the factors affecting yield, was discussed using UAV and GIS. In the study, a prescription map was created to close the sprayer nozzles in areas with no weeds. As a result of the study, it was concluded that area-specific weed management was important in weed management (Kosugi et al., 2023).

Hyperspectral imaging, remote sensing, and GIS are effectively used in wheat field weed management. These technologies increase agricultural productivity and support environmental sustainability through the real-time detection of weeds, mapping their density, and optimizing herbicide applications.

By analyzing their spectral signatures, hyperspectral sensors allow weeds to be distinguished from wheat. Lee et al. (2021) detected weeds in wheat fields with 95% accuracy using hyperspectral imaging technology, allowing immediate intervention. Similarly, Johnson et al. (2019) optimized targeted herbicide applications and reduced chemical use by 30% by determining the density of weeds with multispectral imaging.

GIS and remote sensing technologies play an important role in weed management. Garcia et al. (2018) mapped the distribution of weeds in wheat fields, making herbicide applications more targeted and thus reducing herbicide use by 25%, increasing agricultural productivity. Zhang and Kovacs (2012) stated that GIS makes agricultural pesticide and herbicide applications more efficient and environmentally friendly.

The integration of these technologies forms the basis of precision agriculture practices, enabling intervention only in areas with high weed density. Thus, unnecessary chemical use is prevented, and environmental impacts are minimized. Tsouros et al. (2019) emphasized that remote sensing and GIS-based systems provide sustainable and effective solutions for weed management.

In conclusion, integrating hyperspectral and multispectral imaging, remote sensing, and GIS technologies enables the development of more sensitive and efficient management strategies for weed management in wheat fields. These approaches support environmental sustainability by reducing herbicide use and enabling more informed decisions in agricultural production.

13.3 Current Digital Approaches and Artificial Intelligence Applications in Wheat Cultivation

Current digital systems are adopted to reduce the labor force requirement in agricultural production stages, reduce losses that may occur in the product, and use natural resources efficiently (Anh & Hieu, 2024). These systems are considered a kind of mechanization and can also be perceived as the full automation of agricultural devices (Anh & Hieu, 2024). Combating weeds in wheat farming is of critical importance in terms of increasing productivity and maintaining ecosystem balance. Traditional weed management methods are generally based on the use of chemical herbicides. However, these methods can lead to environmental pollution, herbicide resistance, and increased agricultural costs (Westwood et al., 2018). In recent years, AI and ML technologies have offered revolutionary, innovative approaches to combating weeds within the scope of precision agriculture practices. For example, Smith et al. (2020) used deep learning methods to detect weeds in wheat fields and detected weeds at an early stage with 90% accuracy using algorithms such as convolutional neural networks (CNN).

Image processing techniques and deep learning algorithms can precisely recognize weeds by analyzing information from data sources such as satellite images, drone cameras, and hyperspectral sensors (Mahlein et al., 2019). These methods, especially deep learning models such as YOLO (you only look once) and CNN, make it possible to detect weeds in wheat fields with up to 95% accuracy (Liu et al., 2022).

In recent years, the role of digital technologies in agricultural production has been increasing, and artificial intelligence and machine learning-based approaches offer innovative solutions for weed management (Liakos et al., 2018). These technologies allow for the early detection of weeds using UAVs, image processing techniques, and sensor data and enable the development of targeted control strategies (Kamilaris & Prenafeta-Boldú, 2018; Cakmakci & Cakmakci, 2023). Deep learning algorithms allow for automatically classifying weeds in wheat fields and optimizing herbicide use (Dyrmann et al., 2016). Thanks to AI-supported systems, chemical inputs can be reduced, and sustainable agricultural practices can be encouraged (Shakoor et al., 2021). These developments offer great potential to increase efficiency in agricultural production while minimizing environmental impacts.

IoT technologies have enabled the development of a recommendation system for monitoring, predicting, and controlling diseases, pests, and weeds in wheat (Zhang et al., 2014). Nevavuori et al. (2019) stated that convolutional neural networks (CNNs) are effective in wheat yield prediction with data obtained from UAVs. With the UAVs' spectral and digital images, winter wheat development processes could be monitored (Zhang et al., 2019).

IoT technologies can filter and use data from different sources by enabling data storage, management, and analysis, and they can be widely applied in smart agriculture. Precision and smart agriculture technologies based on the variable rate application principle can reduce input costs and increase production efficiency and quality (Wolfert et al., 2017; Boursianis et al., 2022). Wheat yield can also be predicted with artificial intelligence tools such as artificial neural networks and recurrent neural networks (Bali & Singla, 2021).

The capacity of machine learning models to detect weeds at the seedling growth stage is quite important. Because POST herbicides can generally control weeds more effectively in the early stages than in the later stages (Yu & McCullough, 2016), in a study conducted by Zhuang et al. (2022), object detection neural networks such as CenterNet, Faster R-CNN, TridentNet, VFNet, and YOLOv3 could not reliably detect broadleaf weed seedlings growing in wheat. The precision and recall values in both validation and test datasets remained low. Low precision values indicate that object detection neural networks are more likely to misidentify wheat as a weed, which may lead to incorrect application of herbicides. Low recall values increase the probability of missing weeds, which leads to poor herbicide coverage and poor weed management.

Studies are being carried out on prediction and modeling systems based on available data in agricultural processes. In one study, machine learning was organized with available product data, weather conditions, and remote sensing data (Chlingaryan et al., 2018; Anh & Hieu, 2024). In another study, a proposal was made for precision agriculture by integrating sensor data obtained from the field, environmental factors, and genetic information. According to the proposal, the effect of environmental factors on product yield was found to be higher than for genetic factors (Khaki & Wang, 2019; Anh & Hieu, 2024).

There are innovations in terms of digitalization in agriculture today. In old applications, estimates were made based on plants in randomly selected squares in the field, but today there have been developments in remote sensing techniques (Meghraoui et al., 2022). In a study conducted in Morocco, data was remotely detected using UAVs with high image resolution to estimate yield in wheat fields. After detection, yield estimation was provided with 3D CNN (3D convolutional neural network). The study emphasizes the importance of images obtained before harvest and the need to obtain images effectively throughout the wheat development period (Meghraoui et al., 2022).

Artificial intelligence is a set of applications that emerged with adopting a new understanding of the information technology sector. It is increasing its presence in the agricultural sector daily. When AI is integrated with agriculture, it has a supporting function in food supply problems and future problems, as well as sustainability and the safety of existing foods (Anh & Hieu, 2024).

Artificial intelligence aims to maximize the efficiency of agricultural processes with the correct analysis of existing data (Anh & Hieu, 2024). This technology provides technological facilities and opportunities where small-scale agricultural enterprises in the far corners of the world can easily be reached (Anh & Hieu, 2024). In the production phase, with AI, controlled application is provided in interventions that can affect human health, such as fertilizer and pesticide management. In essence, integrating AI into the agricultural sector is not an increase in efficiency but rather the provision of a more beautiful and livable environment for future generations (Anh & Hieu, 2024).

Identification and localization of weeds have led to the design of sprayers with smart sprayer nozzles with ML and AI (Charles et al., 2020; Su, 2020; Vijayakumar et al., 2023; Sow et al., 2024). With this system, a sufficient amount of herbicide can be applied directly to the weeds defined as sensitive (Olsen et al., 2019a; Sow et al., 2024). The obstacle to the widespread use of this system is economic. That is, the cost of commercialization is relatively high (Zhang, 2020; Sow et al., 2024).

Generally, there are data collection systems, data processing mechanisms, and decision-making algorithm components in integrating AI and robotic systems. For example, the development of robotic cooperation between Bosch and Bayer companies in weed management can be mentioned. In this system, all stages, from data collection to the decision-making algorithm, are available and work accurately (Meena et al., 2024). Convolutional neural network (CNN) and artificial neural network (ANN) systems are used in AI technology (Sow et al., 2024). CNNs are designed according to the data learning system (Asad & Bais, 2020; Sow et al., 2024). It has three stages: convolution, pooling, and fully connected layers. While feature determination is performed in convolution and pooling, classification is performed in fully connected layers (Bajwa et al., 2015; Sow et al., 2024). Artificial neural networks designed according to the human brain structure are also said to be processing algorithms or hardware systems (Shah et al., 2020; Sow et al., 2024). Since the human brain brings together complex neurons and passes on nerve messages, artificial neural networks adopt this principle, have a layered structure, and enable the learning of complex relationships (Abdipour et al., 2019; Sow et al., 2024).

Mondal and Kole (2016) studied the detection of rust disease on wheat leaves using the Pearson correlation coefficient. The images taken from the leaves were evaluated using the ratio of infected leaf area (RILA) on 100 diseased and disease-free plant leaves. Using the Pearson correlation coefficient, one of the most managed features in the disease situation on plant leaves was selected. As a result of the study, the detection accuracy rate was high.

Cotton is a cultivated plant that is difficult to grow and faces yield-inhibiting factors during the cultivation phase. AI -based algorithms and machine learning techniques were used in the study conducted on weeds, which are one of the factors that most affect yield in cotton. The applied models offer the opportunity for real-time precision weed management. The margin of error is very low in systems supported by advanced current sensors (Meena et al., 2024). Weed growth and damage status are estimated using robotic systems where weather conditions, soil conditions, and historical data are received and evaluated. In this way, timely and effective control is possible (Charles & Taylor, 2021; Meena et al., 2024).

A study was conducted in Australia to prevent damage caused by weeds in cotton-growing areas. Weeds in the area were identified using drone-assisted robotic systems equipped with artificial intelligence, enabling the implementation of precision weed management. As a result of the study, it is stated that there was a decrease of approximately 30% in herbicide use and an increase of approximately 20% in the product (Al-Shammari et al., 2020; Meena et al., 2024).

In a study, an experiment was conducted on the real-time detection of visible weeds in wheat fields. Within the scope of the study, approximately 6000 images were collected from the field to store data throughout the season. The study used two frameworks named TensorFlow and PyTorch to implement learning algorithms. The PyTorch learning algorithm was found to be more successful. The factors affecting success here are the detection accuracy and the time to obtain inference (Haq et al., 2023).

UAVs are increasingly used to map and manage wheat fields' weeds. Johnson et al. (2019) determined the density and distribution of weeds using multispectral imaging technology, targeted herbicide applications with this information, reduced their use by 30%, and minimized environmental impacts. With high-resolution imaging systems and spectral sensors, UAVs detect weeds early, map their spread in the field, and distinguish them from wheat using hyperspectral and multispectral cameras. In this way, more precise and targeted herbicide applications can be made (Zhang & Kovacs, 2012; Lottes et al., 2017; Cakmakci & Cakmakci, 2023).

Studies have shown that UAV-supported precision spraying systems significantly reduce chemical use and minimize environmental impacts, promoting sustainable agriculture (Garcia-Ruiz et al., 2015). In addition, data obtained from UAVs can be analyzed with AI and ML algorithms to classify weed species with 90% accuracy and develop targeted spraying strategies (Peña et al., 2013). Thus, drug waste is prevented, and costs are reduced. Zhang and Kovacs (2012) stated that UAVs have great potential in agricultural applications and are especially useful in pesticide and herbicide applications.

UAVs offer an environmentally friendly agricultural approach by reducing the unnecessary use of chemical pesticides and create economic benefits by saving labor and time. Torres-Sánchez et al. (2015) emphasize that UAV-based systems optimize resource use by increasing agricultural productivity.

13.3.1 Artificial Intelligence in Weed Management in Wheat Cultivation

The use of robotic systems in weed management in wheat farming is becoming increasingly widespread with precision agriculture applications. While traditional mechanical and chemical methods require high labor and chemical inputs, robotic systems offer environmentally friendly and efficient alternatives (Slaughter et al., 2008a). Autonomous agricultural robots can detect weeds with image processing and AI algorithms and apply targeted mechanical cleaning or chemical spraying (Fennimore & Cutulle, 2016). Weeds can be diagnosed, and targeted management strategies can be applied thanks to LIDAR, RGB, and hyperspectral imaging systems (Olsen et al., 2019b). Studies show that robotic systems can reduce chemical use and increase efficiency by being programmed correctly (Shamshiri et al., 2018). In the future, weed management will become more sustainable and economical with the more widespread use of autonomous robots.

Robotic systems can automatically detect and classify weeds by integrating AI, ML, and computer vision technologies. These systems determine the location and type of weeds using data collected through cameras and sensors. Slaughter et al. (2008a) stated that robotic systems identify weeds with high accuracy, and thus targeted management strategies can be developed. Bawden et al. (2017) emphasized that AI -based robotic systems can classify weeds with 95% accuracy and have great potential in agricultural applications.

Robotic systems can mechanically remove weeds or apply chemical herbicides only in targeted areas. For example, robots can use mechanical arms to uproot weeds or spray herbicides only in areas with weeds. This minimizes environmental impacts by reducing herbicide use. Fennimore and Cutulle (2019) reported that robotic systems can reduce herbicide use by up to 90% through precision spraying.

Autonomous robots offer a promising solution for mechanically removing weeds in wheat fields. (Brown et al. 2021) developed autonomous robots that detect and mechanically remove weeds in wheat fields. These robots offer an environmentally friendly solution by detecting weeds with computer vision and AI algorithms.

Currently, herbicides such as synthetic auxins (e.g., 2,4-D, dicamba), sulfonylureas (e.g., florasulam), and photosystem II inhibitors (e.g., metribuzin) are used to control broadleaf weeds in wheat by POST application. Autonomous robots, through the use of advanced and reliable machine vision technologies, have the potential to significantly reduce the application of POST herbicides. However, using image classification neural networks, smart sprayers can target herbicides only to small areas containing weeds, saving more herbicides than networks trained with larger images. This demonstrates a more efficient and sustainable approach to herbicide use (Zargar et al., 2020; Zhuang et al., 2021).

13.3.2 Big Data Analytics in Weed Management in Wheat Cultivation

Big data analytics is an important tool to optimize weed management in wheat farming and support agricultural sustainability. Data analytics integrated with remote sensing, GIS, and ML techniques provide detailed information on the detection, density, and distribution of weeds, allowing the development of effective control strategies (Shamshiri et al., 2018).

Technological advancements have enabled the use of artificial intelligence (AI) and machine learning in weed management. For instance, Peña et al. (2013) showed that the system they developed could classify weed species in wheat fields with 90% accuracy. In this way, unnecessary chemical use was reduced, and environmental impacts were minimized by applying targeted spraying strategies.

Advanced data analysis methods, especially convolutional neural networks (CNNs), successfully detect and map weeds with images obtained from UAVs. Nevavuori et al. (2019) found that ESA models identify weeds more accurately than vegetation indices. Similarly, Bawden et al. (2017) emphasized that ML -based data analytics systems play an important role in agricultural decision support mechanisms by classifying weeds with 95% accuracy.

One of the greatest advantages of data analytics is understanding the dynamics of weed populations by analyzing time-dependent changes. Time series analyses integrated with GIS determine weeds' annual and seasonal distributions, helping farmers develop more proactive control strategies (Garcia et al., 2018).

As a result, big data analytics, supported by technologies such as AI and ML, reduces costs in weed management and increases environmental sustainability. In the future, with the widespread use of data analytics-based systems, it is expected that efficiency in agricultural production will increase, and precision management practices will be used more effectively (Tsouros et al., 2019).

13.4 Advantages of Current Technology-Based Applications

- With AI, it is possible to monitor and detect weed development in advance (Poornappriya & Gopinath, 2022).
- Digital image processing techniques make it possible to make accurate diagnoses (Li & Wang, 2024).

- Digital image processing techniques accurately diagnose (Poornappriya & Gopinath, 2022).
- These technologies offer promising applications for the sustainable use of natural resources (Kirkaya, 2020).
- With the possibility of remote sensing of the situation in the field, there is no damage to the crop in field controls (Wang et al., 2019a, b; Meghraoui et al., 2022).
- Another advantage of using UAVs in current technological agricultural activities is the ability to capture images from different altitudes (Mateen & Zhu, 2019).
- Artificial intelligence applications are a technology that can optimize herbicide use and be economical to the producer (Vasileiou et al., 2024).
- With AI, the product is monitored throughout the growing period, and opportunities to intervene in any situation occur (Meghwanshi, 2024).
- Artificial intelligence can combine the past and current situation and make predictions of product supply and demand balance (Meghwanshi, 2024).
- With artificial intelligence, there is a decrease in labor requirements (Meghwanshi, 2024).

13.5 Disadvantages of Current Technology-Based Applications

- The relatively high initial setup costs of AI -based machines and equipment can make access difficult (Meghwanshi, 2024).
- Artificial intelligence applications can sometimes be slow or faulty, and sometimes both, which can cause the result to be incorrect (Hossen et al., 2023).
- Like most technological devices, systems that work using AI also work dependent on the internet. For this reason, security problems can occur regarding the accessibility of data (Meghwanshi, 2024).
- Since AI -based systems require coordination and cooperation, disruptions can occur (Meghwanshi, 2024).
- Depending on whether production is done on a large or small scale, in a city center or in a rural area, there may be inequality in access to technology (Meghwanshi, 2024).

13.6 Conclusion

Artificial intelligence applications, especially machine learning and deep learning algorithms, enable faster and more accurate detection of weeds by performing predictive analysis in agriculture. This way, yield losses can be prevented with early intervention and reduced pesticide use. In addition, it is possible to obtain healthier and higher quality products thanks to these technologies.

This section discusses the importance of wheat cultivation and the difficulties encountered during the cultivation phase. Weeds are one of the most important difficulties encountered during the cultivation phase. If weeds are not effectively managed, they can result in almost no yield. It is also understood within the scope of the section that there are current applications other than traditional weed management and that they are becoming widespread daily.

It is understood that the way to overcome the difficulties and possible problems experienced in combating weeds with traditional methods is through ML, using GIS systems, and integrating AI. In AI, neural network systems that are adopted worldwide are also used, and the success rate is increased. These current weed control methods, also called AI -based systems, are suitable for producers if the initial installation costs are left aside. Namely, after appropriate data is collected and weed mapping is done, success in combating weeds is high with ML and AI. The need for manpower will also decrease with artificial intelligence technology and robots that can be developed in the advanced stage.

There are also negative aspects to AI -based systems. These negative situations mostly occur during the application, in addition to machine and tool equipment failures that may be encountered in the data stage. Machines and UAVs that collect visual data from the field during the day or even during the growing season of the product can be affected by irregularities in the field. Plants and soil in the field are not always in the same condition, and sometimes they can be mixed up with each other due to rainfall and different factors. This reduces the success rate of image detection. It can disrupt the algorithm designed in artificial intelligence in the stage after image detection. As a result, digital approaches and AI offer powerful tools to increase efficiency in wheat cultivation, support sustainable agricultural practices, and reduce production costs. It is expected that with the further development of these technologies in the future, a more efficient, environmentally friendly, and economical production process will be provided in the agricultural sector.

Agricultural production continues to be a system where traditional methods are predominantly applied today. However, the need for AI -based systems in wheat cultivation is a clear reality, especially in agricultural production. Wheat production is a laborious and necessary production. It is well-established that wheat cultivation requires continuous monitoring from sowing to harvest, with human laborbeing essential at each stage of the production process. With the COVID-19 pandemic experienced worldwide, it is seen that some practices in agricultural cultivation will change, and the perception of the labor force has also changed. It is anticipated that, in the future, agricultural monitoring and practices will be carried out by computer-based robotic systems equipped with artificial intelligence, capable of performing autonomous observations andoperations in fields and grain storage facilities. Since the need for labor and the need for working personnel will decrease, access to affordable products will be provided for the consumer in accordance with the supply and demand balance. In order to keep the labor force at a minimum level and to obtain optimal economic efficiency, it will be appropriate for both the producer and the consumer to adapt to current technology rather than the traditional production model.

References

Abdipour, M., Younessi-Hmazekhanlu, M., & Ramazani, S. H. R. (2019). Artificial neural networks and multiple linear regression as potential methods for modeling seed yield of safflower (*Carthamus tinctorius* L.). *Industrial Crops and Products, 127*, 185–194.

Al-Baldawi, M. H. K., & Al-Naqib, M. A. S. (2011). *The weed and methods of control, the practical part* (p. 278). Scientific Publisher, University of Baghdad.

Ali, H., Tahir, M., & Nadeem, M. A. (2014). Determining critical period of weed competition in wheat under different tillage systems. *life, 12*(2), 74–79.

Al-Shammari, D., Fuentes, I., Whelan, M., & B., Filippi, P., & FA Bishop, T. (2020). Mapping of cotton fields within-season using phenology-based metrics derived from a time series of landsat imagery. *Remote Sensing, 12*(18), 3038.

Anh, N. T. M., & Hieu, T. Q. (2024). Exploring the convergence of artificial intelligence and biotechnology: Autonomous robotic solutions for efficient crop harvesting and data-driven agricultural management. *Quarterly Journal of Computational Technologies for Healthcare, 9*(1), 20–33.

Asad, M. H., & Bais, A. (2020). Weed detection in canola fields using maximum likelihood classification and deep convolutional neural network. *Information Processing in Agriculture, 7*(4), 535–545.

Bajwa, A. A., Mahajan, G., & Chauhan, B. S. (2015). Nonconventional weed management strategies for modern agriculture. *Weed Science, 63*(4), 723–747.

Bali, A., & Singla, A. (2021). Wheat yield prediction using artificial neural networks and recurrent neural networks. *Agricultural Engineering International: CIGR Journal, 23*(1), 100–108.

Bawden, D., McDonald, G., & Steel, P. (2017). Machine learning for the classification of weeds. *Agricultural Systems, 68*(2), 112–126.

Boursianis, A. D., Malindretos, G., & Rigas, F. (2022). Smart farming technologies and their impact on crop production: A review of recent developments. *Journal of Agricultural Engineering, 53*(4), 298–310.

Brown, J., Patel, R., & Hawkins, R. (2021). Autonomous robots for weed control in wheat fields. *Precision Agriculture, 22*(4), 438–452.

Cakmakci, M. F., & Cakmakci, R. (2023). Remote sensing, artificial intelligence and smart agriculture technology trends of the future. *European Journal of Science and Technology, 52*, 234–246.

Charles, G. W., & Taylor, I. N. (2021). Extending the critical period for weed control model to better include weed succession using common sunflower as a mimic weed in high-yielding cotton. *Weed Technology, 35*(6), 1029–1037.

Charles, G. W., Sindel, B. M., Cowie, A. L., & Knox, O. G. (2020). Determining the critical period for grass control in high-yielding cotton using Japanese millet as a mimic weed. *Weed Technology, 34*(2), 292–300.

Chauhan, B. S., Mahajan, G., & Gill, G. (2017). Weed management strategies for wheat: Challenges and future directions. *Field Crops Research, 211*, 90–100.

Chlingaryan, A., Sukkarieh, S., & Whelan, B. (2018). Machine learning approaches for crop yield prediction and nitrogen status estimation in precision agriculture: A review. *Computers and Electronics in Agriculture, 151*, 61–69.

Dyrmann, M., Jørgensen, R. N., & Mathiesen, B. (2016). Deep learning for weed detection and management. *Computers and Electronics in Agriculture, 127*, 223–234.

Everest, T., & Gur, E. (2022). A GIS-based land evaluation model for peach cultivation by using AHP: A case study in NW Turkey. *Environmental Monitoring and Assessment, 194*(4), 241.

FAO. (2023). FAOSTAT. Food and Agriculture Organization of the United Nations, https://www.fao.org/faostat/en/#data/QCL/visualize.

Fennimore, S. A., & Cutulle, M. (2016). Robotic and machine vision systems for precision weed management. *Precision Agriculture, 20*(5), 1235–1246.

Fennimore, S. A., & Cutulle, M. (2019). Precision herbicide application for weed management in wheat. *Weed Technology, 33*(1), 1–12.

Garcia, A., Krull, L., & Otero, L. (2018). Geographic information systems (GIS) for weed mapping in wheat fields. *Environmental Management, 60*(3), 487–495.

Garcia-Ruiz, R., González-de-Santiago, E., & Jiménez-Ruiz, D. (2015). UAV applications in agriculture: Precision spraying and weed control. *Journal of Field Robotics, 32*(1), 15–25.

Gill, K. K., Bhatt, K., Kaur, B., & Sandhu, S. S. (2024). Comparative analysis of wheat yield prediction through artificial intelligence, simulation modelling and statistical analysis in Central Punjab. *Journal of Agrometeorology, 26*(3), 377–379.

Gupta, O. P. (2004). *Modern weed management* (Vol. 5, 2nd ed., pp. 18–23). Agrobios Jodhpur, India.

Haq, S. I. U., Tahir, M. N., & Lan, Y. (2023). Weed detection in wheat crops using image analysis and artificial intelligence (AI). *Applied Sciences, 13*(15), 8840.

Harker, K. N., & O'Donovan, J. T. (2013). Integrated weed management in wheat. *Weed Science, 61*(4), 555–563.

Heap, I. (2021). Herbicide-resistant weeds: Global issues and trends. *Pesticide Management Science, 77*(2), 1311–1320.

Hossen, M. I., Fahad, N., Sarkar, M. R., & Rabbi, M. R. (2023). Artificial intelligence in agriculture: A systematic literature review. *Turkish Journal of Computer and Mathematics Education (TURCOMAT), 14*, 137–146.

Johnson, C. A., Flores, M., & Hamilton, T. (2019). Precision weed management with UAVs in wheat farming. *Computers and Electronics in Agriculture, 156*, 278–285.

Kamilaris, A., & Prenafeta-Boldú, F. (2018). Smart agriculture systems: AI and machine learning in farming. *Agricultural Systems, 166*, 215–230.

Kamkar, B., & Mijani, S. (2023). Herbicide selection through GIS-based groundwater quality maps. *Journal of Cleaner Production, 409*, 137131.

Kaur, R., Mahey, R. K., Mukherjee, J., & Sood, A. (2009). Derivation of grassy weed intensity maps in wheat using spatial data with GIS in the central districts of Punjab. *Indian Journal of Weed Science, 41*(3&4), 136–141.

Khaki, S., & Wang, L. (2019). Crop yield prediction using deep neural networks. *Frontiers in Plant Science, 10*, 621.

Kırkaya, A. (2020). Smart farming-precision agriculture technologies and practices. *Journal of Scientific Perspectives, 4*(2), 123–136.

Kosugi, Y., Kramar, D., Ostlie, M., & Flores, P. (2023). Integration UAS-imagery and a commercial-size sprayer to implement site-specific weed control in corn. In *2023 ASABE annual international meeting* (p. 1). American Society of Agricultural and Biological Engineers.

Lee, H., Hwang, D., & Lee, S. (2021). Hyperspectral imaging for weed detection in wheat fields. *Sensors, 21*(1), 34–48.

Li, C., & Wang, M. (2024). Pest and disease management in agricultural production with artificial intelligence: Innovative applications and development trends. *Advances in Resources Research, 4*(3), 381–401.

Li, J., Xuan, J., & Cai, R. (2020). Wheat, a popular cereal crop. *Field Crop, 3*.

Liakos, K. G., Aigbavboa, C. O., & Koenig, D. (2018). AI and machine learning applications in precision farming. *Sustainability, 10*(10), 15–27.

Lipper, L., Thornton, P., & Campbell, B. (2014). Climate change impacts on agriculture: Implications for food security. *FAO Agriculture Studies, 38*, 47–61.

Liu, Y., Zhang, D., & Liang, S. (2022). Weed detection in wheat using deep learning models. *Computers and Electronics in Agriculture, 185*, 106197.

Loghavi, M., & Mackvandi, B. B. (2008). Development of a target oriented weed control system. *Computers and Electronics in Agriculture, 63*(2), 112–118.

Lottes, P., Hörferlin, M., Sander, S., & Stachniss, C. (2017). Effective vision-based classification for separating sugar beets and weeds for precision farming. *Journal of Field Robotics, 34*(6), 1160–1178.

Mahlein, A. K., Kaufmann, H., & Schultz, D. (2019). Precision farming for weed detection using multispectral imaging. *Agricultural Systems, 165*, 19–30.

Mateen, A., & Zhu, Q. (2019). Weed detection in wheat crop using UAV for precision agriculture. *Pakistan Journal of Agricultural Sciences, 56*(3), 809–817.

McFall, K. L., & Fowler, M. E. (2009). Overview of wheat classification and trade. In *Wheat science and trade* (pp. 437–454).

Meena, S. B., Karthickraja, A., & Saravanane, P. (2024). Harnessing AI and robotics for smart weed management in cotton. In *Recent Trends in Agriculture and Allied Sciences* (pp. 87–94). JPS Scientific Publications.

Meghraoui, K., Sebari, I., Bensiali, S., & El Kadi, K. A. (2022). On behalf of an intelligent approach based on 3D CNN and multimodal remote sensing data for precise crop yield estimation: Case study of wheat in Morocco. *Advanced Engineering Science, 2*, 118–126.

Meghwanshi, S. (2024). Artificial intelligence in agriculture: A review. *International Research Journal of Modernization in Engineering Technology and Science, 6*, 4358–4363.

Mondal, D., & Kole, D. K. (2016). A time efficient leaf rust disease detection technique of wheat leaf images using Pearson correlation coefficient and rough fuzzy C-means. In *Information systems design and intelligent applications: Proceedings of third international conference India, volume 1* (pp. 609–618). Springer India.

Nevavuori, P., Li, L., & Kallio, K. (2019). Application of convolutional neural networks for wheat yield prediction based on UAV-derived data. *Remote Sensing, 11*(15), 1740.

Nikolić, N., Rizzo, D., Marraccini, E., Gotor, A. A., Mattivi, P., Saulet, P., et al. (2021). Site-and time-specific early weed control is able to reduce herbicide use in maize-a case study. *Italian Journal of Agronomy, 16*(4), 1780.

Oerke, E. C. (2006). Crop losses to pests. *Journal of Agricultural Science, 144*(4), 31–43.

Olsen, A., Konovalov, D. A., Philippa, B., Ridd, P., Wood, J. C., Johns, J., et al. (2019a). DeepWeeds: A multiclass weed species image dataset for deep learning. *Scientific Reports, 9*(1), 2058.

Olsen, S. J., Lilley, J., & Jorgensen, B. (2019b). Use of robotic systems for weeding in wheat production. *Agricultural Robotics, 9*(4), 301–315.

Peña, J., Pérez, J., & Castillo, J. (2013). UAV-assisted precision spraying for weed control. *Journal of Precision Agriculture, 17*(5), 183–195.

Poornappriya, T. S., & Gopinath, R. (2022). Rice plant disease identification using artificial intelligence approaches. *International Journal of Electrical Engineering and Technology, 11*(10), 392–402.

Safi, S. M. A., Al-Khaldy, R. A. A., & Hammood, W. F. (2022). Evaluation of the efficiency of some herbicides in controlling weeds in wheat (*Triticum aestivum* L.). *Research on crops, 23*(1), 21–25.

Shah, K., Patel, H., Sanghvi, D., & Shah, M. (2020). A comparative analysis of logistic regression, random forest and KNN models for the text classification. *Augmented Human Research, 5*(1), 12.

Shahzad, M., Jabran, K., Hussain, M., Raza, M. A. S., Wijaya, L., El-Sheikh, M. A., & Alyemeni, M. N. (2021). The impact of different weed management strategies on weed flora of wheat-based cropping systems. *PLoS One, 16*(2), e0247137.

Shakoor, N., Bennett, J., & Peterson, T. (2021). AI-based solutions in weed control: A sustainable approach. *Weed Science, 69*(3), 202–215.

Shamshiri, R. R., Kalantari, F., & Sadeghi, M. (2018). Remote sensing technologies for weed detection in agricultural fields. *Agricultural Systems, 167*, 1–10.

Sharma, T., Singh, J., Singh, A., & Chauhan, G. (2021). Artificial intelligence in water management. *RASSA Journal of Science for Society, 3*(3), 186–189.

Slaughter, D. C., Giles, D. K., & Downey, D. (2008a). Agricultural robots for weed management. *Biosystems Engineering, 101*(1), 31–41.

Slaughter, D. C., Giles, D. K., & Downey, D. (2008b). Autonomous robotic weed control systems: A review. *Computers and Electronics in Agriculture, 61*(1), 63–78.

Smith, J., Williams, A., & Carter, T. (2020). Early detection of weeds using deep learning in wheat fields. *Field Crops Research, 241*, 107–118.
Sow, S., Ranjan, S., Seleiman, M. F., Alkharabsheh, H. M., Kumar, M., Kumar, N., et al. (2024). Artificial intelligence for maximizing agricultural input use efficiency: Exploring nutrient, water and Weed Management strategies. *Phyton (0031-9457), 93*(7), 1569.
Su, W. H. (2020). Advanced machine learning in point spectroscopy, RGB-and hyperspectral-imaging for automatic discriminations of crops and weeds: A review. *Smart Cities, 3*(3), 767–792.
Su, D., Qiao, Y., Kong, H., & Sukkarieh, S. (2021). Real time detection of inter-row ryegrass in wheat farms using deep learning. *Biosystems Engineering, 204*, 198–211.
Torres-Sánchez, J., Peña, J. M., López-Granados, F., & Peña, J. M. (2015). A review on remote sensing in precision agriculture. *European Journal of Remote Sensing, 48*(1), 124–134.
Tsouros, D. C., Zazias, G. K., & Moshou, D. (2019). The role of data analytics in precision agriculture: Prospects and challenges. *Precision Agriculture, 20*(3), 789–808. https://doi.org/10.1007/s11119-019-09631-1
Tyagi, B. (2016). Sustainable agriculture practices: A global perspective. *Agricultural Sustainability Journal, 22*(4), 75–89.
Vasileiou, M., Kyrgiakos, L. S., Kleisiari, C., Kleftodimos, G., Vlontzos, G., Belhouchette, H., & Pardalos, P. M. (2024). Transforming weed management in sustainable agriculture with artificial intelligence: A systematic literature review towards weed identification and deep learning. *Crop Protection, 176*, 106522.
Vijayakumar, V., Ampatzidis, Y., Schueller, J. K., & Burks, T. (2023). Smart spraying technologies for precision weed management: A review. *Smart Agricultural Technology, 6*, 100337.
Wang, F., Wang, F., Zhang, Y., Hu, J., Huang, J., & Xie, J. (2019a). Rice yield estimation using parcel-level relative spectral variables from UAV-based hyperspectral imagery. *Frontiers in Plant Science, 10*, 453.
Wang, C., Zhang, J., & Liu, W. (2019b). UAV-based precision weed management in wheat farming. *Field Robotics, 36*(1), 56–66.
Weiss, M., Jacob, F., & Duveiller, G. (2020). Remote sensing for agricultural applications: A meta-review. *Remote Sensing of Environment, 236*, 111402.
Westwood, J. D., Sipes, B. S., & Schultz, P. B. (2018). Integrated pest management and precision agriculture: The convergence of two technologies. *Pest Management Science, 74*(5), 1112–1122. https://doi.org/10.1002/ps.4659
Wolfert, S., Ge, L., Verdouw, C., & Bogaardt, M. J. (2017). Big data in smart farming – A review. *Agricultural Systems, 153*, 69–80.
Wu DingRong, W. D., Yu QiAng, Y. Q., Lu ChangHe, L. C., & Hengsdijk, H. (2006). Quantifying production potentials of winter wheat in the North China Plain. *European Journal of Agronomy, 24*(3), 226–235.
Yu, Q., & McCullough, P. (2016). Herbicide resistance and management strategies in wheat. *Weed Technology, 30*(3), 1–10.
Zargar, M., Bayat, M., & Astarkhanova, T. (2020). Study of postemergence-directed herbicides for redroot pigweed (Amaranthus retroflexus L) control in winter wheat in southern Russia. *Journal of Plant Protection Research, 60*(1).
Zhang, J. (2020). Research on digital image processing and recognition technology of weeds in maize seedling stage based on artificial intelligence. *Journal of Physics: Conference Series, 1648*(4), 042058. IOP Publishing.
Zhang, L., & Kovacs, J. (2012). UAVs in precision agriculture: Weed detection and management. *Journal of Field Robotics, 29*(4), 1–10.
Zhang, X., Wang, J., & Zhang, Q. (2014). Application of IoT technology in the monitoring, prediction and control of diseases, pests and weeds in wheat. *Sensors, 14*(12), 21940–21957.
Zhang, Y., Xu, L., & Li, Z. (2019). Monitoring wheat development stages using vegetation indices derived from UAV imagery. *Remote Sensing, 11*(12), 147.

Zhuang, J., Li, X., Bagavathiannan, M., Jin, X., Yang, J., Meng, W., Li, T., Li, L., Wang, Y., Chen, Y., & Yu, J. (2021). Evaluation of different deep convolutional neural networks for detection of broadleaf weed seedlings in wheat. *Pest Management Science, 77*(2), 776–784. https://doi.org/10.1002/ps.6656

Zhuang, F., Liu, D., & Wei, H. (2022). Weed detection challenges in wheat fields using deep learning models. *Computers and Electronics in Agriculture, 184*, 106175.

Chapter 14
Smart Farming for Sustainable Wheat Intensification

Vinoth Kumar Govintharaj, Periyasamy Eniyavan, Arumugam Pillai M, and Ephrem Habyarimana

Abstract Smart farming technologies are revolutionizing wheat production, enhancing sustainability by optimizing resource use and minimizing environmental impacts. This chapter explores the integration of precision agriculture, Internet of things (IoT), artificial intelligence (AI), remote sensing, and advanced biotechnology in wheat intensification. These technologies enable data-driven decision-making to maximize yield while reducing input waste. Precision agriculture techniques, like GPS-based mapping and variable rate technology, allow precise input applications tailored to crop needs, thus boosting productivity. Remote sensing and UAVs provide timely monitoring of crop health, enabling early detection of diseases and nutrient deficiencies. Smart irrigation systems powered by IoT sensors and real-time data analytics, address water scarcity by conserving water while ensuring optimal soil moisture. Advanced machine learning algorithms analyze imagery and environmental data, supporting predictive models that help anticipate yield outcomes, disease outbreaks, and optimal resource allocation. The adoption of these technologies in wheat farming offers sustainable solutions to climate and resource

V. K. Govintharaj
Department of Genetics and Plant Breeding, V.O.C Agricultural College & Research Institute, Killikulam, TNAU, Coimbatore, Tamil Nadu, India

Sorghum Breeding, International Crops Research Institute for the Semi Arid Tropics, Patancheru, Telangana, India

P. Eniyavan
Sorghum Breeding, International Crops Research Institute for the Semi Arid Tropics, Patancheru, Telangana, India

Department of Genetics and Plant Breeding, CPBG, TNAU, Coimbatore, Tamil Nadu, India

Arumugam Pillai M
Department of Genetics and Plant Breeding, V.O.C Agricultural College & Research Institute, Killikulam, TNAU, Coimbatore, Tamil Nadu, India

E. Habyarimana (✉)
Sorghum Breeding, International Crops Research Institute for the Semi Arid Tropics, Patancheru, Telangana, India
e-mail: Ephrem.Habyarimana@icrisat.org

N. Zencirci et al. (eds.), *Empowering Wheat Cultivation with GIS, Digital Approaches and Artificial Intelligence*, https://doi.org/10.1007/978-3-031-99954-3_14

challenges, ensuring high productivity with minimized ecological impact. Through smart farming, wheat intensification is achieved in a way that aligns with sustainable agricultural practices, supporting food security while protecting natural resources.

Keywords Smart farming technologies · Artificial intelligence · Sustainability · Disease monitoring · Smart irrigation

Abbreviations

AI	Artificial intelligence
AR	Augmented reality
CA	Conservation agriculture
CDI	Controlled deficit irrigation
CNN	Convolutional neural network
CO_2	Carbon dioxide
CRISPR	Clustered regularly interspaced short palindromic repeats
CTS	Conservation tillage systems
CWA	Continuous wavelet analysis
DI	Deficit irrigation
ET	Evapotranspiration
FAO	Food And Agriculture Organization
FAOSTAT	Fao Statistical Database
FHB	Fusarium Head Blight
FMIS	Farm Management Information System
GIS	Geographic information system
GLCM	Gray-level co-occurrence matrix
GPS	Global positioning system
ICT	Information and Communication Technologies
IoT	Internet of things
ML	Machine learning
NDVI	Normalized difference vegetation index
PLSR	Partial least-squares regression
PRI	Photochemical reflectance index
R^2	Coefficient of determination
RF	Random forest
RGB	Red green blue
RS	Remote sensing
RS	Rain sensors
SDIs	Spectral disease indices
SF	Smart farming
SI	Sustainable intensification
SMS	Soil moisture sensors
SOC	Soil organic carbon

SVIs	Spectral vegetation indices
SVM	Support vector machine
TF	Texture features
UAV	Unmanned aerial vehicle
USDA	United States Department Of Agriculture
VIs	Vegetation indices
VR	Virtual reality
VRT	Variable rate technology
YOLO	You only look once

14.1 Introduction

Wheat (*Triticum aestivum* L.) is a staple food crop that belongs to the grass family Poaceae and plays an indispensable role in global food security, providing approximately 20% of the world's calories and protein (Shewry et al., 2015; Acevedo et al., 2018). However, wheat production faces mounting challenges due to population growth, resource limitations, and climate change. Traditional input-intensive farming practices, which rely heavily on chemical fertilizers, pesticides, and irrigation, are increasingly unsustainable. These methods not only strain natural resources but also contribute to environmental degradation, including soil depletion, water scarcity, and greenhouse gas emissions. This is particularly concerning in regions like South Asia, where surface water and groundwater irrigation play a crucial role in supporting agriculture, accounting for nearly 40% of food production (MacDonald et al., 2016; Krupnik et al., 2016). Countries like India and Pakistan are especially vulnerable to climate change, facing rising temperatures and altered monsoon patterns that threaten wheat and rice yields (Auffhammer et al., 2012; Saleem et al., 2022). Furthermore, climate change poses a significant threat to wheat production, with rising temperatures, changing precipitation patterns, and elevated atmospheric CO_2 levels, which could reduce yields by 6% for each 1 °C increase in temperature (Zhao et al., 2017).

Sustainable intensification (SI) of wheat production is essential for addressing the growing demand for food while minimizing environmental impact. SI focuses on increasing productivity on existing agricultural land by using resource-efficient methods. Conservation agriculture (CA) and conservation tillage systems (CTS), which include practices like minimal tillage, mulching, and crop rotation, are critical components of this approach. These techniques improve soil health, increase soil organic carbon (SOC), reduce greenhouse gas emissions, and enhance crop resilience, making them integral to climate-smart agriculture (Abegunde et al., 2019; Michler et al., 2019). Research shows that adopting resource-conserving technologies can significantly lower greenhouse gas emissions, enhance farm profitability, and ensure long-term sustainability, particularly in wheat cultivation (Aravindakshan et al., 2022).

Smart farming (SF) technologies represent a transformative solution for achieving sustainable wheat intensification. These innovations, driven by information and communication technologies (ICT), enable precision agriculture through data-driven sustainable practices (Alfred et al., 2021). SF integrates tools such as sensors, the Internet of things (IoT), and drones to monitor crop health, soil conditions, and resource use with high accuracy. These technologies facilitate efficient irrigation, nutrient management, and pest control, reducing waste and environmental impact while maximizing productivity (Araus & Cairns, 2014). Hyperspectral imaging and remote sensing platforms, such as unmanned aerial vehicles (UAVs) and ground-based systems, are particularly effective in detecting wheat diseases like yellow rust and Fusarium Head Blight, allowing for timely intervention (Ferreira et al., 2024; Jin et al., 2020). By leveraging these technologies, farmers can improve decision-making, optimize input use, and enhance crop yields, contributing to both food security and climate change mitigation.

14.2 Global Wheat Production Trends

Global wheat production is primarily dominated by *Triticum aestivum* L. which accounts for 85–90% of total production, while *Triticum durum* contributes around 5% (Goncharov et al., 2008; Goncharov, 2018; Lyapunova et al., 2020). Globally, wheat is cultivated on 219.15 million hectares, with a production of 808.44 million tonnes (FAOSTAT, 2022). Wheat is cultivated on approximately 60 million hectares of irrigated land, accounting for 27.2% of the global wheat area, while rainfed areas cover around 160.4 million hectares, or 72.8% of the total wheat cultivated area worldwide. Major irrigated wheat-producing countries include India, China, Pakistan, Iran, Turkey, Afghanistan, Egypt, and the United States, while rainfed wheat areas are primarily in Russia, the United States, Kazakhstan, Australia, China, Canada, Ukraine, Turkey, India, and France (Dadrassi et al., 2023). By 2050, the world population is expected to reach 9.7 billion; wheat production will need to increase by 50% to ensure food security (United Nations, 2024). According to FAO estimates, by 2050, global wheat demand will reach around 840 million tonnes, while current production stands at 750 million tonnes (FAO, 2021). Wheat holds a significant place in global grain trade, accounting for one-third of it (Erenstein et al., 2022).

Data from FAOSTAT and USDA have been analyzed to assess global wheat production trends over the past decade. China is currently the largest producer, followed by India, Russia, the United States, Canada, Pakistan, Australia, and Ukraine (Fig. 14.1). The leading countries contribute substantially to global production: China (17%), the European Union (17%), India (14%), Russia (12%), the United States (6%), Canada (4%), Pakistan (4%), Australia (3%), Ukraine (3%), and Turkey (3%) (USDA Wheat, 2023, Fig. 14.2). These trends highlight the importance of sustainable wheat production strategies across major producers to ensure global food security.

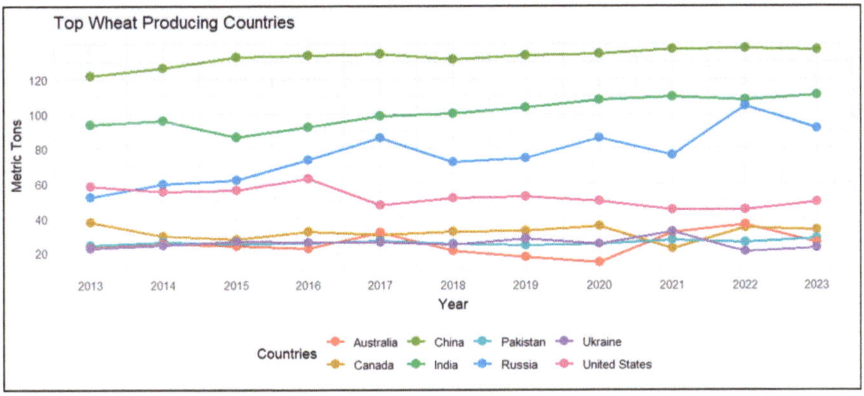

Fig. 14.1 Trends in leading wheat producing countries

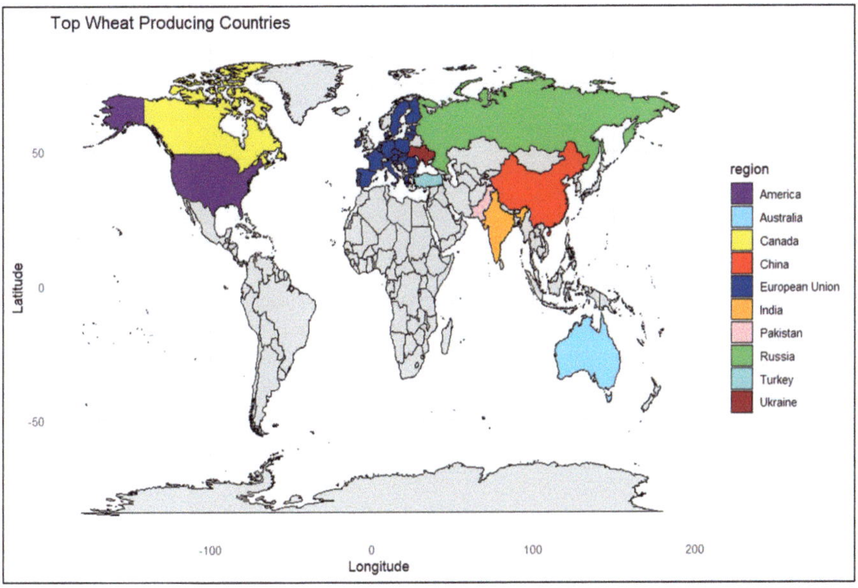

Fig. 14.2 Top wheat producing countries worldwide

14.3 Smart Farming Technologies in Wheat Production

Smart farming technologies are critical to meet the increasing demand for sustainable wheat production. The integration of advanced technologies such as precision agriculture, Internet of things (IoT), artificial intelligence (AI), cloud computing, and biotechnology has enabled wheat farming to achieve higher efficiency, precision, and sustainability. These technologies enable data-driven decisions, allowing farmers to optimize input usage, improve crop health, and maximize yield while

Table 14.1 Smart farming technologies and their applications in wheat production

Technology	Application	Result	Reference
Precision agriculture (GPS, GIS, and remote sensing)	Site-specific management	Enhanced yield and resource optimization through targeted soil and crop management	Saleh et al. (2014); Toscano et al. (2019)
Variable rate technology (VRT)	Targeted input application	20% reduction in labor costs and 53% decrease in pesticide usage with precise input application	Finco et al. (2023)
Automation and robotics	Mechanical weeding and monitoring	7.2% savings in seed use, 6.5% in fertilizer, and 8.8% in plant protection products	Rakintsev et al. (2022)
Drones or UAVs	Disease detection and crop stress monitoring	Early detection of crop stress and pest infestation, reducing yield losses	Gómez-Candón et al. (2014)
Farm management information systems (FMIS)	Data-driven decision-making	Improved yield estimation and optimized resource allocation through real-time data tracking	Paraforos et al. (2016)
IoT sensors	Soil and crop health monitoring	Precise control of irrigation and nutrient application, enhancing soil quality and crop growth	Köksal et al. (2019)
Artificial intelligence (AI)	Yield prediction and disease management	Automated insights to optimize inputs and predict crop health, supporting sustainable practices	Patel et al. (2020)
Blockchain and cloud computing	Supply chain transparency and data management	End-to-end traceability in wheat production, boosting consumer confidence and transparency	Rathore et al. (2022)
AR and VR	Training and digital twin simulation	Improved farmer decision-making through virtual training environments and predictive simulations	Han et al. (2022), Verdouw et al. (2017)
Advanced biotechnology (CRISPR)	Disease resistance breeding	Enhanced resistance to powdery mildew in wheat, reducing need for chemical inputs	Wang et al. (2014)

minimizing environmental impact. Also, these technologies offer a robust management system that addresses challenges like climate variability, resource optimization, pest control, and labor shortages. Table 14.1 provides an overview of various smart farming technologies and their applications in wheat production.

Precision agriculture, a foundational element in smart farming, utilizes global positioning system (GPS), geographical information system (GIS), and remote sensing (RS) to tailor field management according to specific soil and crop needs. With GPS and GIS, fields are mapped and analyzed at high precision, facilitating targeted soil sampling and nutrient management (Saleh et al., 2014). Remote sensing data often captured by drones or satellites can help monitor soil conditions, crop health, and stress indicators like drought or nutrient deficiencies (Sarig et al., 2021; Toscano et al., 2019). Variable rate technology (VRT) further refines the precision approach by enabling the application of inputs like fertilizers and pesticides in variable quantities based on real-time data. Finco et al. (2023) demonstrated that VRT application in durum wheat production in Central Italy reduced pesticide use by 53% while maintaining yield, showcasing how precision agriculture can enhance productivity and sustainability in wheat farming. Agricultural automation, including the use of robotics, has revolutionized traditional wheat farming, reducing labor reliance, inputs, and enabling precise execution of management activities (Rakintsev et al., 2022). Automated tractors equipped with GPS enhance planting and harvesting efficiency. Drones or unmanned aerial vehicles (UAVs) equipped with sensors, multispectral or hyperspectral cameras detect early signs of crop stress, diseases, and pest infestations, enabling timely intervention and efficient crop management. For instance, Gómez-Candón et al. (2014) found that drone monitoring of wheat fields allowed for precise weed control, which improved crop quality and reduced herbicide dependence. Farm management information systems (FMIS) platforms integrate diverse data streams, from IoT sensors to historical field records, offering real-time data that assist in decision making for tasks like irrigation, fertilizer application, disease control, and insect control. These systems formerly supported record-keeping but now play a critical role in modern data management, compliance, and cost control. By organizing data from various farm activities, FMIS helps optimize resource use, track expenses, and improve overall profitability. Paraforos et al. (2016) demonstrated that FMIS application in winter wheat improved operational efficiency and provided farmers with actionable insights for better resource allocation.

Internet of things (IoT) networks interlink various devices and sensors across the farm, facilitating real-time monitoring of conditions like soil moisture, temperature, pH, and crop health. IoT sensors provide data that enables timely irrigation and nutrient management, directly impacting yield and resource conservation. Köksal et al. (2019) illustrated how IoT-based FMIS in wheat farming enabled efficient data collection and guided irrigation, fertilization, and pest management practices, making IoT a vital component in modern wheat farming. Blockchain technology ensures traceability across the wheat supply chain, recording each stage of production for transparency and quality assurance. Rathore et al. (2022) discussed a blockchain-based model to trace the Indian wheat supply chain, highlighting its benefits in reducing food losses and ensuring food quality. Cloud computing further aids data storage, management, and analysis, enabling efficient resource allocation and decision-making. AI enhances decision-making in wheat farming by analyzing complex datasets to identify trends and predict outcomes. Machine learning (ML) algorithms support yield forecasting, pest and disease prediction, and optimal resource allocation. Patel et al. (2020) highlighted how AI, particularly through ML, improves wheat yield predictions, enabling precision farming approaches that minimize resource waste.

Augmented reality (AR) and Virtual reality (VR) provide immersive training tools for farmers and agronomists, simulating scenarios such as machinery operation or disease diagnosis to improve decision-making. This technology, noted by Han et al. (2022), can also enhance consumer engagement, promoting sustainable practices in food choices. Digital twin technology, which creates virtual replicas of wheat fields, supports predictive modeling and helps farmers optimize resource use (Verdouw et al., 2017). Skobelev et al. (2020) implemented a smart digital twin specifically for wheat, designed as an intelligent system that uses agronomic expertise and real-time weather data to project yield and predict each growth stage's timeline.

Advanced biotechnology, including gene-editing techniques like CRISPR-Cas9, enables the development of resilient wheat varieties that resist pests, diseases, and environmental stressors. Biotechnology supports sustainable wheat farming by reducing dependence on chemical inputs. Wang et al. (2014) demonstrated the application of CRISPR technology in wheat to enhance resistance to powdery mildew, a common wheat disease, thus reducing the need for fungicides and supporting sustainable crop protection.

14.4 Smart Farming for Wheat Disease Monitoring

Smart farming integrates advanced technologies like remote sensing, unmanned aerial vehicles (UAVs), and machine learning to monitor and manage wheat diseases with high precision. One major area of focus in this domain is the identification and management of wheat diseases such as yellow rust (*Puccinia striiformis*), Fusarium head blight (FHB), and stripe rust, which are significant threats to wheat production worldwide. Traditional methods for monitoring wheat diseases often depend on visual assessments, which are not only labor-intensive but also subject to human error and inconsistency (Bock et al., 2010). With the advent of UAVs and remote sensing technologies, more accurate and efficient methods for disease detection have emerged. For example, hyperspectral and multispectral imaging, when integrated with UAVs, provide valuable opportunities for precision agriculture by capturing high-resolution images that can identify disease symptoms at various stages, even before they become visible to the naked eye (Khot et al., 2016; Su et al., 2018; Yao et al., 2019). Recent advancements have optimized these systems by integrating machine learning algorithms such as random forests, improving the identification and classification of disease severity (Zheng et al., 2018).

Reflectance measurements play a crucial role in the automated detection of diseases like yellow rust. By analyzing the spectral signatures of wheat leaves, researchers have developed robust models for disease classification. For instance, Moshou et al. (2004) demonstrated the use of neural networks with reflectance data to classify wheat leaves, achieving up to 99% accuracy. Similarly, Huang et al. (2007) utilized hyperspectral imaging to quantify disease severity, showing that spectral indices like the photochemical reflectance index (PRI) are highly effective ($r^2 = 0.91$). These indices are particularly sensitive to changes in chlorophyll and pigment levels, making them reliable indicators of stress caused by wheat diseases (Ashourloo et al., 2014). UAVs equipped with multispectral and hyperspectral sensors enable rapid, large-scale assessments of wheat fields, detecting diseases such as yellow rust and FHB, well before visible symptoms appear. Xiao et al.

(2021) utilized UAVs to detect FHB by combining spectral and texture features, optimizing the gray-level co-occurrence matrix (GLCM) window size to improve the classification of disease severity, achieving an accuracy of 90%. This method underscores the importance of precise feature extraction and selection in building effective agricultural monitoring systems. Another study by Guo et al. (2021) demonstrated the utility of UAV-based hyperspectral imaging for detecting wheat yellow rust at the field scale. By extracting vegetation indices (VIs) and texture features (TFs), they developed partial least-squares regression (PLSR) models to monitor disease during different infection periods. The integration of VIs and TFs provided the highest monitoring accuracy ($R^2 = 0.82$), showcasing the effectiveness of combining multiple features for improved disease detection. Furthermore, their study evaluated the impact of spatial resolution on monitoring accuracy, finding that a resolution of 10 cm was optimal for their VI-TF-based model. Bohnenkamp et al. (2019) compared hyperspectral imaging on both ground and UAV scales to assess yellow rust in wheat, highlighting challenges in adapting laboratory methods for field conditions. Their research emphasized the need for real-time data collection to guide effective disease management, demonstrating that UAV technology can provide critical insights into disease dynamics, contributing to optimized resource utilization.

The integration of machine learning algorithms into disease detection models has further enhanced the capabilities of smart farming. Machine learning models like support vector machines (SVM), random forests (RF), and convolutional neural networks (CNN) have been used to process and analyze spectral data collected by UAVs. For instance, Zhang et al. (2019) applied a deep learning-based approach using high-resolution hyperspectral images to detect yellow rust. Their model, optimized with multiple Inception-ResNet layers, incorporated both spatial and spectral features, achieving an accuracy of 85% across various wheat growth stages. Liu et al. (2020) advanced the monitoring of FHB using UAV hyperspectral imagery by developing an improved backpropagation neural network model, achieving an accuracy of 98%. This study underscores the capability of UAVs for rapid, nondestructive assessments of wheat diseases. Similarly, the application of continuous wavelet analysis (CWA) in disease detection has shown promising results. Ma et al. (2019) utilized CWA on spectral data to identify FHB, achieving an accuracy of 88.7%. This approach highlights the potential of spectroscopic techniques in detecting diseases early, enabling targeted and timely interventions that minimize yield losses and pesticide use. Heidarian Dehkordi et al. (2020) investigated UAV-based RGB imagery to differentiate between healthy and diseased wheat plants, focusing on stripe and leaf rusts. Their findings revealed significant correlations between UAV-derived indices and disease severity, showcasing the potential of RGB imagery as a cost-effective tool for continuous disease monitoring throughout the growing season.

Smart farming technologies extend beyond the detection of individual wheat diseases. UAVs, equipped with RGB imaging capabilities, can also classify farming progress in rice-wheat rotation fields. Song et al. (2022) proposed a method for classifying farming progress using UAV RGB images, achieving significantly higher accuracy compared to traditional machine learning methods. This approach emphasizes the versatility of UAV technology in understanding broader agricultural practices, which is essential for disease prevention and crop management. Su et al. (2021) further explored the dynamics of stripe rust in wheat populations through UAV remote sensing. Their research highlighted the temporal aspects of disease spread and emphasized the utility of UAVs in monitoring disease dynamics over time, thereby informing timely interventions. Recent advancements have also focused on developing lightweight and scalable models for disease detection. Hong et al. (2022) introduced a lightweight model for detecting FHB using UAV-enabled edge computing. Their model, based on YOLOv4 and MobileNet architectures, achieved an impressive accuracy of 93.69%, demonstrating its potential for real-time monitoring and disease control. By optimizing the network parameters and reducing computing complexity, their model is well-suited for deployment on UAVs, allowing farmers to receive timely alerts and take action quickly.

The integration of UAV technology with advanced machine learning techniques supports sustainable farming practices by optimizing pesticide use and minimizing environmental impact (Table 14.2). The combination of spectral imaging and machine learning models enhances detection accuracy, enabling precise interventions. For example, Su et al. (2018) developed a system using UAVs equipped with multispectral cameras to monitor yellow rust, achieving an accuracy of 89%. This system utilized spectral vegetation indices (SVIs) such as NDVI and OSAVI to automate disease detection, proving effective at a large scale. Moreover, recent studies have demonstrated the potential of UAV technology in monitoring other diseases that indirectly affect wheat through crop rotations, such as Southern Corn Rust (Meng et al., 2020). By developing spectral disease indices (SDIs) and utilizing in situ leaf reflectance spectra, researchers have created models that classify disease severity and support sustainable crop management strategies. In conclusion, the integration of UAVs and advanced imaging techniques within smart farming is revolutionizing wheat disease monitoring. By enhancing detection capabilities, these technologies not only help secure wheat yield but also promote sustainable agricultural practices by reducing the need for excessive pesticide application. As smart farming technology continues to evolve, future research will focus on refining these methods and expanding their applications to different geographical regions and wheat varieties, ensuring global food security in the face of climate and environmental challenges.

14.5 Smart Irrigation for Wheat Cultivation

Water is an essential resource in agriculture, with approximately 70% of global water withdrawals used for irrigation (Simionesei et al., 2020). Wheat productivity heavily depends on adequate water availability and proper timing of irrigation. Insufficient water during critical growth stages such as flowering and grain filling can significantly reduce yields. Water scarcity is a prominent issue in wheat production areas

Table 14.2 The references dealing with the monitoring of diseases in wheat

Disease	Sensor	Model input	Model output	Model type	Reference
Yellow rust	Hyperspectral	Spectral signatures	Disease classification	Neural networks	Moshou et al. (2004)
	Multispectral	Vegetation indices (SVLs like NDVI, OSAVI)	Disease severity	Random forest	Su et al. (2018)
	Hyperspectral	Spectral indices, Texture features (TFs)	Disease monitoring	Partial least squares Regression (PLSR)	Guo et al. (2021)
Stripe rust	Hyperspectral	Spectral and spatial features	Disease detection	Deep learning (Inception-ResNet)	Zhang et al. (2019)
	RGB imagery	RGB-derived indices	Disease monitoring	Correlation coefficients	Heidarian Dehkordi et al. (2020)
	Multispectral	Vegetation indices (VIs)	Disease severity	Random forest, support Vector machine (SVM)	Su Baofeng et al. (2021)
Fusarium head blight (FHB)	Hyperspectral	Vegetation indices (VIs) and wavelet features	Disease monitoring	Continuous wavelet analysis (CWA)	Ma et al. (2019)
	Hyperspectral	Vegetation indices (VIs) and texture features (TFs)	Disease monitoring	Backpropagation neural network (BPNN)	Liu et al. (2020)
	Hyperspectral	Spectral and texture features (TFs)	Disease severity	Gray-level co-occurrence matrix (GLCM)	Xiao et al. (2021)

worldwide (Davarpanah & Ahmadi, 2021). Thus, irrigation is the most important input needed to meet the water deficit of wheat worldwide for maximum productivity. Farmers often resort to excessive flood irrigation, which not only exceeds wheat's water needs but also leads to deep percolation, low water productivity, and declining groundwater tables (Memon et al., 2021). Such practices highlight the need for efficient irrigation methods to balance water supply with crop demands and conserve water resources.

Traditional irrigation methods often lead to water wastage, uneven distribution, and soil degradation, ultimately affecting resource sustainability, as they apply uniform water amounts without considering specific crop needs (Abioye et al., 2020). Therefore, proper irrigation management is essential for ensuring global water security and optimizing crop production (Nazari et al., 2018). Adopting water-efficient irrigation techniques is crucial for sustainable wheat cultivation, particularly in arid and semiarid regions, as these methods optimize water use, maintain soil health, and maximize crop yields while conserving water resources. Addressing water management challenges in wheat cultivation through modern irrigation strategies is vital for achieving global food security and promoting sustainable agriculture.

Smart irrigation systems are innovative, sustainable methods designed to maximize water use efficiency in agricultural fields. These systems aim to optimize water usage, ensuring high crop yield while minimizing the negative environmental impacts often associated with traditional irrigation practices (Mason et al., 2019). Conventional methods, such as deficit irrigation (DI) and controlled deficit irrigation (CDI), are frequently adopted by farmers to conserve water resources (Rathore et al., 2021). DI involves applying less water than the crop typically requires, which helps reduce water usage while maintaining wheat yields at acceptable levels (Kovalenko et al., 2021; Razzaghi et al., 2020). Studies have shown that DI, when used during specific wheat growth stages like jointing to shooting and flowering to soft dough, can reduce water usage by up to 39% with minimal impact on yield. Irrigation restricted only to the first joint and anthesis stages yields the highest water savings with a modest 16% reduction in yield (Xue et al., 2006). However, continued use of DI may negatively affect grain yield (Rady et al., 2021). Alternatively, CDI, which applies water stress only during noncritical growth phases, has been shown to minimize yield loss and optimize water usage (Kassaye et al., 2020; Zhang et al., 2006).

The modernization of irrigation systems, including the use of drip irrigation and precision sprinklers, plays a critical role in improving water-use efficiency of crops. These systems deliver water directly to the roots, minimizing wastage and ensuring an even distribution across the field. Such techniques not only optimize soil moisture levels but also maintain high yields with reduced water input. Effective irrigation scheduling, considering soil moisture levels, crop growth stages, and evapotranspiration rates, is critical, particularly in regions with limited water availability. Proper scheduling ensures that crops receive adequate water at critical growth stages, promoting both productivity and sustainability. Smart irrigation strategies increasingly incorporate advanced sensors like soil moisture sensors (SMS), evapotranspiration (ET) controllers, rain sensors (RS), and optical sensors to prevent water wastage during irrigation. The integration of these sensors with the Internet of things (IoT) allows real-time data collection and monitoring, enabling precise and informed irrigation management (Gulati & Thakur, 2018; Mousavi et al., 2021). The IoT-based systems, coupled with humidity and rain sensors, have been successfully applied in wheat fields (Jha et al., 2017). ET-based irrigation controllers with digital electronic controllers, ET sensors, and modules have been used for the irrigation management of wheat and tomatoes. ETc for local microclimate was calculated based on the modified Penman equation. This information allows the system to adjust irrigation times, saving up to 27% of water (Al-Ghobari & Mohammad, 2011). In Algeria, a similar smart irrigation system managed wheat irrigation based on soil moisture data, resulting in a 5% reduction in water usage compared to traditional methods (Debauche et al., 2020). While effective, these sensor-based systems require robust hardware and reliable connectivity, particularly for underground sensors, which may face signal issues due to disconnection (Al-Naji et al., 2021). To overcome such challenges, new technologies like optical sensors, drones, and UAVs with RGB cameras are being employed to manage irrigation, avoiding the need for soil-buried sensors (Ajith et al., 2018). These advancements highlight the potential of smart irrigation systems to revolutionize wheat production by optimizing water use and ensuring sustainable agricultural practices.

14.6 Future Prospects and Conclusion

Smart farming technologies will play an increasingly pivotal role in securing and intensifying wheat production to meet the demands of a growing global population. Innovations in precision agriculture, IoT-enabled sensors, AI-driven analytics, and drone-based monitoring systems have opened avenues for optimizing resource usage, detecting crop health, and implementing timely interventions, all of which support sustainable agricultural practices. As the effects of climate change continue to disrupt traditional farming systems, adapting to these changes by incorporating resilient, data-driven practices will be essential. This shift will not only improve yield stability in diverse environmental conditions but also promote environmentally conscious farming practices that preserve soil health, reduce pesticide dependency, and conserve water resources.

The future of sustainable wheat production depends on continuous investment in research and development for smarter, more resilient farming practices. Breeding programs that focus on developing wheat varieties resistant to climate-related stresses and major pests will be vital for maintaining yields in a changing climate. Furthermore, the adoption of smart irrigation systems, enhanced through AI and real-time monitoring, will allow farmers to manage water resources with greater efficiency. Advances in automation and machine learning are already reducing labor demands and allowing farmers to shift from manual to mechanized systems, ultimately enhancing productivity and reducing operational costs. To ensure widespread adoption of these technologies, supportive policies and collaboration among governments, private sectors, and agricultural communities are essential. Policies aimed at incentivizing smart technology adoption, providing training for farmers, and ensuring access to digital infrastructure can accelerate this transition. Public–private partnerships and investments in digital infrastructure will further drive innovation in wheat farming, making technology accessible to smallholder farmers who might otherwise lack the resources for initial investments. In conclusion, the path to sustainable wheat intensification lies in embracing smart farming technologies, fostering innovation, and building resilience into agricultural systems. By advancing research, policy, and collaborative frameworks, the global community can ensure wheat productivity remains robust and sustainable, addressing food security challenges while conserving environmental resources.

References

Abegunde, V. O., Sibanda, M., & Obi, A. (2019). The dynamics of climate change adaptation in Sub-Saharan Africa: A review of climate-smart agriculture among small-scale farmers. *Climate, 7*(11), 132.

Abioye, E. A., Abidin, M. S. Z., Mahmud, M. S. A., Buyamin, S., Ishak, M. H. I., AbdRahman, M. K. I., et al. (2020). A review on monitoring and advanced control strategies for precision irrigation. *Computers and Electronics in Agriculture, 173*, 105441.

Acevedo, M., Zurn, J. D., Molero, G., Singh, P., He, X., Aoun, M., et al. (2018). The role of wheat in global food security. In *Agricultural development and sustainable intensification* (pp. 81–110). Routledge.

Ajith, G., Bharadwaj, C. N., Nag, T. S., & Gururaj, C. (2018). Uav aided irrigation using object detection through wireless communication technology. In *2018 2nd International Conference on Trends in Electronics and Informatics (ICOEI)* (pp. 714–719). IEEE.

Alfred, R., Obit, J. H., Chin, C. P. Y., Haviluddin, H., & Lim, Y. (2021). Towards paddy rice smart farming: a review on big data, machine learning, and rice production tasks. *IEEE Access, 9*, 50358–50380.

Al-Ghobari, H. M., & Mohammad, F. S. (2011). Intelligent irrigation performance: evaluation and quantifying its ability for conserving water in arid region. *Applied Water, Science, 1*(3–4), 73–83.

Al-Naji, A., Fakhri, A. B., Gharghan, S. K., & Chahl, J. (2021). Soil color analysis based on a RGB camera and an artificial neural network towards smart irrigation: A pilot study. *Heliyon, 7*(1), e06078.

Araus, J. L., & Cairns, J. E. (2014). Field high-throughput phenotyping: the new crop breeding frontier. *Trends in Plant Science, 19*(1), 52–61.

Aravindakshan, S., AlQahtany, A., Arshad, M., Manjunatha, A. V., & Krupnik, T. J. (2022). A meta frontier approach and fractional regression model to analyze the environmental efficiency of alternative tillage practices for wheat in Bangladesh. *Environmental Science and Pollution Research, 29*(27), 41231–41246.

Ashourloo, D., Mobasheri, M. R., & Huete, A. (2014). Evaluating the effect of different wheat rust disease symptoms on vegetation indices using hyperspectral measurements. *Remote Sensing, 6*, 5107–5512.

Auffhammer, M., Ramanathan, V., & Vincent, J. R. (2012). Climate change, the monsoon, and rice yield in India. *Climatic Change, 111*, 411–424.

Bock, C. H., Poole, G. H., Parker, P. E., & Gottwald, T. R. (2010). Plant disease severity estimated visually, by digital photography and image analysis, and by hyperspectral imaging. *Critical Reviews in Plant Sciences, 29*, 59–107.

Bohnenkamp, D., Behmann, J., & Mahlein, A.-K. (2019). *In-field detection of yellow rust in wheat on the ground canopy and UAV scale, 11*, 2495.

Dadrasi, A., Chaichi, M., Nehbandani, A., Sheikhi, A., Salmani, F., & Nemati, A. (2023). Addressing food insecurity: An exploration of wheat production expansion. *PLoS One, 18*(12), e0290684.

Davarpanah, R., & Ahmadi, S. H. (2021). Modeling the effects of irrigation management scenarios on winter wheat yield and water use indicators in response to climate variations and water delivery systems. *Journal of Hydrology, 598*, 126269.

De Gómez-Candón, D., Castro, A. I., & López-Granados, F. (2014). Assessing the accuracy of mosaics from unmanned aerial vehicle (UAV) imagery for precision agriculture purposes in wheat. *Precision Agriculture, 15*, 44–56.

Debauche, O., Mahmoudi, S., Elmoulat, M., Mahmoudi, S. A., Manneback, P., & Lebeau, F. (2020). Edge AI-IoT pivot irrigation, plant diseases, and pests identification. *Procedia Computer Science, 177*, 40–48.

Erenstein, O., Jaleta, M., Mottaleb, K. A., Sonder, K., Donovan, J., & Braun, H. J. (2022). Global trends in wheat production, consumption and trade. In *Wheat improvement: food security in a changing climate* (pp. 47–66). Springer International Publishing.

FAO. Food and Agriculture Organization of the United Nations. 2021. https://www.fao.org/statistics/en/

Ferreira, L. D. C., Carvalho, I. C. B., Jorge, L. A. D. C., Quezado-Duval, A. M., & Rossato, M. (2024). Hyperspectral imaging for the detection of plant pathogens in seeds: recent developments and challenges. *Frontiers in Plant Science, 15*, 1387925.

Finco, A., Bentivoglio, D., Belletti, M., Chiaraluce, G., Fiorentini, M., Ledda, L., & Orsini, R. (2023). Does Precision Technologies Adoption Contribute to the Economic and Agri-Environmental Sustainability of Mediterranean Wheat Production? An Italian case study. *Agronomy, 13*(7), 1818.

Food and Agriculture Organization of the United Nations. (2022). FAOSTAT statistical database. Retrieved from https://www.fao.org/faostat

Goncharov, S. V., & Kurashov, M. Y. (2018). *Prospects for the development of the Russian durum wheat market* (Vol. 2). Vestnik of Voronezh State Agrarian University.

Goncharov, N. P., & Ya, E. (2008). *Kondratenko, Origin, domestication and evolution of wheat* (Vol. 12, pp. 159–179). Information Bulletin of VOGiS.

Gulati, A., & Thakur, S. (2018). Smart irrigation using Internet of Things. In *In: 2018 8th International Conference on Cloud Computing, Data Science & Engineering (Confluence)* (pp. 819–823). IEEE.

Guo, A., Huang, W., Dong, Y., Ye, H., Ma, H., Liu, B., Wu, W., Ren, Y., Ruan, C., & Geng, Y. (2021). Wheat Yellow Rust Detection Using UAV-Based Hyperspectral Technology. *Remote Sensing, 13*, 123.

Han, D. I. D., Abreu e Silva, S. G., Schröder, K., Melissen, F., & Haggis-Burridge, M. (2022). Designing immersive sustainable food experiences in augmented reality: a consumer participatory co-creation approach. *Food, 11*(22), 3646.

Heidarian Dehkordi, R., El Jarroudi, M., Kouadio, L., Meersmans, J., & Beyer, M. (2020). Monitoring wheat leaf rust and stripe rust in winter wheat using high-resolution UAV-based red-green-blue imagery. *Remote Sensing, 12*, 3696.

Hong, Q., Jiang, L., Zhang, Z., Ji, S., Gu, C., Mao, W., Li, W., Liu, T., Li, B., & Tan, C. (2022). A lightweight model for wheat ear fusarium head blight detection based on RGB images. *Remote Sensing, 14*, 3481.

Huang, W., Lamb, D. W., Niu, Z., Zhang, Y., Liu, L., & Wang, J. (2007). Identification of yellow rust in wheat using in-situ spectral reflectance measurements and airborne hyperspectral imaging. *Precision Agric., 8*, 187–197.

Jha, R. K., Kumar, S., Joshi, K., & Pandey, R. (2017). Field monitoring using IoT in agriculture. In *In: 2017 International Conference on Intelligent Computing, Instrumentation and Control Technologies (ICICICT)* (pp. 1417–1420). IEEE.

Jin, X., Zarco-Tejada, P. J., Schmidhalter, U., Reynolds, M. P., Hawkesford, M. J., Varshney, R. K., et al. (2020). High-throughput estimation of crop traits: A review of ground and aerial phenotyping platforms. *IEEE Geoscience and Remote Sensing Magazine, 9*(1), 200–231.

Kassaye, K. T., Yilma, W. A., Fisha, M. H., & Haile, D. H. (2020). Yield and water use efficiency of potato under alternate furrows and deficit irrigation. *International Journal of Agronomy, 2020*, 1–11.

Khot, L. R., Sankaran, S., Carter, A. H., Johnson, D. A., & Cummings, T. F. (2016). UAS imaging-based decision tools for arid winter wheat and irrigated potato production management. *International Journal of Remote Sensing, 37*, 125–137.

Köksal, Ö., & Tekinerdogan, B. (2019). Architecture design approach for IoT-based farm management information systems. *Precision Agriculture, 20*, 926–958.

Kovalenko, Y., Tindjau, R., Madilao, L. L., & Castellarin, S. D. (2021). Regulated deficit irrigation strategies affect the terpene accumulation in Gewürztraminer (Vitis vinifera L.) grapes grown in the Okanagan valley. *Food Chemistry, 341*, 128172.

Krupnik, T. J., Schulthess, U., Ahmed, Z. U., & McDonald, A. J. (2016). Sustainable crop intensification through surface water irrigation in Bangladesh? A geospatial assessment of landscape-scale production potential. *Land Use Policy, 60*, 206–222.

Liu, L., Dong, Y., Huang, W., Du, X., & Ma, H. (2020). Monitoring wheat fusarium head blight using unmanned aerial vehicle hyperspectral imagery. *Remote Sensing, 12*, 3811.

Lyapunova, O. A. O. A., & Andreeva, A. S. (2020). Varieties and lines that replenished the durum wheat gene pool of VIR in 2000-2019. *Works on Applied Botany, Genetics and Selection, 181*(1), 7–16.

Ma, H., Huang, W., Jing, Y., Pignatti, S., Laneve, G., Dong, Y., Ye, H., Liu, L., Guo, A., & Jiang, J. (2019). Identification of fusarium head blight in winter wheat ears using continuous wavelet analysis. *Sensors, 20*, 20.

MacDonald, A. M., Bonsor, H. C., Ahmed, K. M., Burgess, W. G., Basharat, M., Calow, R. C., Dixit, A., Foster, S. S. D., Gopal, K., Lapworth, D. J., Lark, R. M., Moench, M., Mukherjee, A., Rao, M. S., Shamsudduha, M., Smith, L., Taylor, R. G., Van Tucker, J., Steenbergen, F., & Yadav, S. K. (2016). Groundwater quality and depletion in the Indo-Gangetic Basin mapped from in situ observations. *Nature Geoscience, 9*(10), 762–766.

Mason, B., Rufí-Salís, M., Parada, F., Gabarrell, X., & Gruden, C. (2019). Intelligent urban irrigation systems: Saving water and maintaining crop yields. *Agricultural Water Management, 226*, 105812.

Memon, S. A., Sheikh, I. A., Talpur, M. A., & Mangrio, M. A. (2021). Impact of deficit irrigation strategies on winter wheat in semi-arid climate of sindh. *Agricultural Water Management, 243*, 106389.

Meng, R., Lv, Z., Yan, J., Chen, G., Zhao, F., Zeng, L., & Xu, B. (2020). Development of spectral disease indices for southern corn rust detection and severity classification. *Remote Sensing, 2020*(12), 3233.

Michler, J. D., Baylis, K., Arends-Kuenning, M., & Mazvimavi, K. (2019). Conservation agriculture and climate resilience. *Journal of Environmental Economics and Management, 93*, 148–169.

Moshou, D., Bravo, C., West, J., Wahlen, S., McCartney, A., & Ramon, H. (2004). *Automatic detection of 'yellow rust' in wheat using reflectance measurements and neural networks* (pp. 173–188).

Mousavi, S. K., Ghaffari, A., Besharat, S., & Afshari, H. (2021). Improving the security of internet of things using cryptographic algorithms: A case of smart irrigation systems. *Journal of Ambient Intelligence and Humanized Computing, 12*, 2033–2051.

Nazari, B., Liaghat, A., Akbari, M. R., & Keshavarz, M. (2018). Irrigation water management in Iran: Implications for water use efficiency improvement. *Agricultural Water Management, 208*, 7–18.

Paraforos, D. S., Vassiliadis, V., Kortenbruck, D., Stamkopoulos, K., Ziogas, V., Sapounas, A. A., & Griepentrog, H. W. (2016). A farm management information system using future internet technologies. *IFAC-Papers OnLine, 49*(16), 324–329.

Patel, K., Mehta, D., Mistry, C., Gupta, R., Tanwar, S., Kumar, N., & Alazab, M. (2020). Facial sentiment analysis using AI techniques: state-of-the-art, taxonomies, and challenges. *IEEE Access, 8*, 90495–90519.

Rady, M. O., Semida, W. M., Howladar, S. M., & Abd El-Mageed, T. A. (2021). Raised beds modulate physiological responses, yield and water use efficiency of wheat (Triticum aestivum L) under deficit irrigation. *Agricultural Water Management, 245*, 106629.

Rakintsev, D. S. & Seitov, S. K. (2022). Economic advantages of using agricultural robots in precision farming.

Rathore, V. S., Nathawat, N. S., Bhardwaj, S., Yadav, B. M., Kumar, M., Santra, P., et al. (2021). Optimization of deficit irrigation and nitrogen fertilizer management for peanut production in an arid region. *Scientific Reports, 11*(1–14), 5456.

Rathore, S., Gupta, N., Rathore, A. S., & Soni, G. (2022). Blockchain-based smart wheat supply chain model in Indian context. In *Agri-food 4.0* (Vol. 27, pp. 77–96). Emerald Publishing Limited.

Razzaghi, F., Bahadori-Ghasroldashti, M. R., Henriksen, S., Sepaskhah, A. R., & Jacobsen, S. E. (2020). Physiological characteristics and irrigation water productivity of quinoa (Chenopodium quinoa Willd.) in response to deficit irrigation imposed at different growing stages—A field study from Southern Iran. *Journal of Agronomy and Crop Science, 206*, 390 404.

Saleem, F., Arshad, A., Mirchi, A., Khaliq, T., Zeng, X., Rahman, M. M., et al. (2022). Observed changes in crop yield associated with droughts propagation via natural and human-disturbed agro-ecological zones of Pakistan. *Remote Sensing, 14*(9), 2152.

Saleh, A., & Belal, A. (2014, February). Delineation of site-specific management zones by fuzzy clustering of soil and topographic attributes: A case study of East Nile Delta, Egypt. In *IOP Conference Series: Earth and Environmental Science* (Vol. 18, No. 1, p. 012046). IOP Publishing.

Sarig, S., Shlevin, E., Zilberman, A., Richker, I., Dudai, M., Nezer, S., & Ben-Asher, J. (2021). Proximal sensing of nitrogen needs by spring wheat. *Agronomy, 11*(3), 437.

Shewry, P. R., & Hey, S. J. (2015). The contribution of wheat to human diet and health. *Food and Energy Security, 4*(3), 178–202.

Simionesei, L., Ramos, T. B., Palma, J., Oliveira, A. R., & Neves, R. (2020). IrrigaSys: A web-based irrigation decision support system based on open source data and technology. *Computers and Electronics in Agriculture, 178*, 105822.

Skobelev, P., Laryukhin, V., Simonova, E., Goryanin, O., Yalovenko, V., & Yalovenko, O. (2020, September). Multi-agent approach for developing a digital twin of wheat. In *2020 IEEE International Conference on Smart Computing (SMARTCOMP)* (pp. 268–273). IEEE.

Song, X., Wu, F., Lu, X., Yang, T., Ju, C., Sun, C., & Liu, T. (2022). The classification of farming progress in rice–wheat rotation fields based on UAV RGB images and the regional mean model. *Agriculture, 12*, 124.

Su, J., Liu, C., Coombes, M., Hu, X., Wang, C., Xu, X., Li, Q., Guo, L., & Chen, W. H. (2018). Wheat Yellow Rust Monitoring by Learning from Multispectral UAV Aerial Imagery. *Computers and Electronics in Agriculture, 155*, 157–166.

Su, B., Liu, Y., Huang, Y., Wei, R., Cao, X., & Han, D. (2021). Analysis for stripe rust dynamics in wheat population using UAV remote sensing. *Trans. Chin. Soc. Agric. Eng., 37*, 127–135.

Toscano, P., Di Castrignanò, A., Gennaro, S. F., Vonella, A. V., Ventrella, D., & Matese, A. (2019). A precision agriculture approach for durum wheat yield assessment using remote sensing data and yield mapping. *Agronomy, 9*(8), 437.

United Nations, Population. https://www.un.org/en/global-issues/population, 2024.

USDA. (2023). Available on. https://fas.usda.gov/data/production/commodity/0410000

Verdouw, C. N. & Kruize, J. W. (2017, October). Digital twins in farm management: illustrations from the FIWARE accelerators SmartAgriFood and Fractals. In *Proceedings of the 7th Asian-Australasian Conference on Precision Agriculture Digital, Hamilton, New Zealand* (pp. 16–18).

Wang, Y., Cheng, X., Shan, Q., Zhang, Y., Liu, J., Gao, C., & Qiu, J. L. (2014). Simultaneous editing of three homoeoalleles in hexaploid bread wheat confers heritable resistance to powdery mildew. *Nature Biotechnology, 32*(9), 947–951.

Xiao, Y., Dong, Y., Huang, W., Liu, L., & Ma, H. (2021). Wheat fusarium head blight detection using UAV-based spectral and texture features in optimal window size. *Remote Sensing, 13*, 2437.

Xue, Q., Zhu, Z., Musick, J. T., Stewart, B., & Dusek, D. A. (2006). Physiological mechanisms contributing to the increased wateruse efficiency in winter wheat under deficit irrigation. *Journal of Plant Physiology, 163*, 154–164.

Yao, Z., Lei, Y., & He, D. (2019). *Early visual detection of wheat stripe rust using visible/near-infrared hyperspectral imaging, 19*, 952.

Zhang, B., Li, F.-M., Huang, G., Cheng, Z.-Y., & Zhang, Y. (2006). Yield performance of spring wheat improved by regulated deficit irrigation in an arid area. *Agricultural Water Management, 79*, 28.

Zhang, X., Han, L., & Dong, Y. (2019). *A deep learning-based approach for automated yellow rust disease detection from high-resolution hyperspectral UAV images, 11*, 1554.

Zhao, C., Liu, B., Piao, S., Wang, X., Lobell, D. B., Huang, Y., Huang, M., Yao, Y., Bassu, S., Ciais, P., Durand, J. L., Elliott, J., Ewert, F., Janssens, I. A., Li, T., Lin, E., Liu, Q., Martre, P., Müller, C., Peng, S., Peñuelas, J., Ruane, A. C., Wallach, D., Wang, T., Wu, D., Liu, Z., Zhu, Y., Zhu, Z., & Asseng, S. (2017). Temperature increase reduces global yields of major crops in four independent estimates. *Proceedings of the National Academy of Sciences of the United States of America, 114*(35), 9326–9331.

Zheng Q, Huang W, Cui X, Dong Y, Shi Y, Ma H, Liu L (2018). Identification of wheat yellow rust using optimal three-band spectral indices in different growth stages.

Correction to: Empowering Wheat Cultivation with GIS, Digital Approaches and Artificial Intelligence

Nusret Zencirci, Faheem Shehzad Baloch, Jin-Ying Gou, Velimir Mladenov, Sotirios Fragkostefanakis, Marta da Silva Lopes, Ephrem Habyarimana, Hakan Ulukan, and Asuman Kaplan Evlice

Correction to:
Chapters 5 and 6 in: N. Zencirci et al. (eds.), *Empowering Wheat Cultivation with GIS, Digital Approaches and Artificial Intelligence,*
https://doi.org/10.1007/978-3-031-99954-3

The initial version of this book was inadvertently published with one of the chapter authors name (Chapters 5 and 6) being incorrectly spelled as "Muhammad Tatar." This has now been corrected to "Muhammed Tatar."

The updated version of these chapters can be found at
https://doi.org/10.1007/978-3-031-99954-3_5
https://doi.org/10.1007/978-3-031-99954-3_6

N. Zencirci et al. (eds.), *Empowering Wheat Cultivation with GIS, Digital Approaches and Artificial Intelligence*, https://doi.org/10.1007/978-3-031-99954-3_15

Index

N. Zencirci et al. (eds.), *Empowering Wheat Cultivation with GIS, Digital Approaches and Artificial Intelligence*, https://doi.org/10.1007/978-3-031-99954-3

MIX
Papier aus verantwortungsvollen Quellen
Paper from responsible sources
FSC® C105338

If you have any concerns about our products,
you can contact us on
ProductSafety@springernature.com

In case Publisher is established outside the EU,
the EU authorized representative is:
Springer Nature Customer Service Center GmbH
Europaplatz 3, 69115 Heidelberg, Germany

Printed by Libri Plureos GmbH
in Hamburg, Germany